WA 1307501 2

D1757995

The vector- and rodent-borne diseases of Europe and North America: their distribution and public health of burden

There are a significant number of diseases carried by insects such as mosquitoes or sand flies or by ticks, mites and rodents, and these are far more common than is often realized. New diseases are constantly being discovered and are becoming more widely distributed with the increase in travelling, to and from tropical, disease-endemic countries. Here, Norman Gratz (former Director, Division of Vector Biology and Control, World Health Organization), reviews the distribution of the vector and rodent-borne diseases in Europe, the USA and Canada; their incidence and prevalence, their costs and hence their public health burdens are detailed, and their arthropod vectors and rodent reservoir hosts described. Armed with such information, the individual clinician is more likely to have a degree of epidemiological suspicion that will lead to an earlier diagnosis and correct treatment of these infections. Equally, authorities will more readily understand the measures necessary to control this group of infectious agents.

The vector- and rodent-borne diseases of Europe and North America: their distribution and public health burden

NORMAN G. GRATZ

World Health Organization, Geneva

CAMBRIDGE UNIVERSITY PRESS
Cambridge, New York, Melbourne, Madrid, Cape Town, Singapore, São Paulo

Cambridge University Press
The Edinburgh Building, Cambridge CB2 2RU, UK

Published in the United States of America by Cambridge University Press, New York

www.cambridge.org
Information on this title: www.cambridge.org/9780521854474

© Cambridge University Press 2006

This publication is in copyright. Subject to statutory exception
and to the provisions of relevant collective licensing agreements,
no reproduction of any part may take place without
the written permission of Cambridge University Press.

First published 2006

Printed in the United Kingdom at the University Press, Cambridge

A catalogue record for this publication is available from the British Library

ISBN-13 978-0-521-85447-4 hardback
ISBN-10 0-521-85447-4 hardback

Learning Resources
Centre

13075012

Cambridge University Press has no responsibility for the persistence or accuracy of URLs for
external or third-party internet websites referred to in this publication, and does not
guarantee that any content on such websites is, or will remain, accurate or appropriate.

Contents

Acknowledgements

The author gratefully acknowledges the assistance provided by Professor Mike Service who read the manuscript and provided many valuable suggestions and corrections for the text.

My thanks are due to Dr Graham White for his encouragement to me for beginning the writing of the book.

I much appreciate the information provided to me by Dr Joseph M. Conlon.

Cambridge University Press would like to thank Professor Mike Service for all his efforts in editing this manuscript following the death of Norman Gratz.

Tribute to the author, Norman Gratz

Despite battling with illness Dr Norman Gratz valiantly tried to complete his manuscript of this book and send it to Cambridge University Press for publication. He succeeded in doing this. Sadly, however, he died in Geneva soon afterwards, in November 2005, and so never saw his magnum opus published.

Norman joined the World Health Organization in 1958 and remained with the organization for the rest of his working life. On retirement he continued to serve on WHO committees and act as a consultant for WHO, numerous chemical companies, and government and non-government agencies.

He was one of the few medical entomologists who was also involved with the role of rodent reservoir hosts in disease transmission and soon became recognized worldwide as an authority on vector-borne diseases and their control. Norman travelled extensively for WHO in Africa, Asia and Latin America advising on research and control strategies on the vectors of malaria, filariasis, Chagas disease, murine typhus, plague and various tick-borne diseases.

Norman published more than 100 scientific papers and in 1985 received the Medal of Honour from the American Mosquito Control Association. He built up an amazing database of more than 31 000 abstracts on vector- and rodent-borne infections which became the foundation for this book.

On several occasions I told Norman that because of his vast knowledge of vector- and rodent-borne infections he was probably the only single person who could write this book.

Mike W. Service
Emeritus Professor of Medical Entomology
Liverpool School of Tropical Medicine

Preface

Until the early part of the twentieth century many vector-and rodent-borne infections were very serious public health problems in Europe and North America. Thousands of cases of malaria occurred annually throughout these regions and populations suffered greatly from the disease. Malaria transmission persisted in most of southern Europe and the USA until it was eradicated in the 1950s. Among the arboviruses, dengue transmitted by the mosquito *Aedes aegypti*, was the cause of a great epidemic in Athens, Greece in 1928 with over 650 000 cases and more than a thousand deaths. The same species was also the vector of yellow fever which caused many thousands of deaths in the USA during the nineteenth century; the last epidemic of the disease occurred in New Orleans in 1905 with more than 3000 cases and at least 452 deaths being recorded. Great epidemics of louse-borne typhus occurred in many parts of Europe during World War I accounting for great human mortality. The war-associated louse-borne diseases such as epidemic typhus, epidemic relapsing fever and trench fever disappeared after 1945 due to the applications of the newly discovered DDT and related compounds; at the time, optimism ran high that this group of infections was unlikely to again be a problem, and indeed at the time effective control was obtained of most of the group.

Yet by the end of the twentieth century, vector and rodent-borne infections have again become serious public health problems; there has been a recrudescence of several diseases long thought to have been eradicated or under effective control; at the same time, new vector and rodent-borne diseases have been discovered both in Europe and North America, some of which now occur in high incidence. A number of diseases in this group have been introduced into geographical areas in which they have not previously been found such as the introduction of West Nile Virus into New York in 1999; the virus and diseases it causes have subsequently spread throughout the USA and much of Canada. At the time of writing, six years after the introduction of the virus, there is no sign of a significant diminution in its annual incidence in the USA.

Among the tick-borne infections, Lyme disease was first identified in 1975 as the cause of an epidemic of arthritis occurring near Old Lyme, Connecticut. This infection has now become the most common vector-borne disease in both the USA and Europe and in parts of eastern Canada. Tens of thousands of cases now occur yearly in Europe and the USA. In the years 2001 and 2002, nearly 41 000 cases of Lyme disease were reported to the US Centers for Disease Control and Prevention (CDCP) and, in Europe, it has been estimated that as many as 60 000 cases a year occur in Germany alone.

Tick-borne encephalitis virus transmitted by *Ixodes ricinus* in western Europe is endemic in central, eastern and northern Europe and may cause a wide spectrum of clinical forms, ranging from asymptomatic infection to severe meningoencephalitis. In eastern Europe the virus is transmitted by *I. persulcatus* and its incidence has been increasing. Ecological changes have resulted in an important spread of the tick vectors and they are now commonly found in parks in the middle of many cities throughout Europe. It appears that climate change has resulted in a northward movement of the tick vector and the disease in Sweden.

Human ehrlichiosis and anaplasmosis are tick-borne zoonotic infections that have become increasingly recognized in the USA and Europe. The increased desire of humans to pursue outdoor recreational activities during the summer months has also amplified their potential exposure to pathogenic bacteria that spend a portion of their life cycle in invertebrate bloodsucking enzootic hosts. Just like *Borrelia burgdorferi*, the agent of Lyme borreliosis, *Ehrlichia* and *Anaplasma* species cycle within hard-bodied ticks. Rocky Mountain spotted fever is the most severe and most frequently reported rickettsial illness in the USA. The disease is caused by *Rickettsia rickettsii*.

The number of annual cases of babesiosis, which is transmitted by the same tick vector as Lyme disease, is unknown but in areas of the USA where infected ticks are common, up to 20% of people have antibody results suggesting exposure. Although most of those exposed have no evidence of the disease, about 6% of people with babesiosis severe enough to require hospitalization die.

In Europe, Mediterranean Spotted Fever (MSF), also known as Boutonneuse fever, is transmitted by the dog tick, *Rhipicephalus sanguineus*. The disease is endemic to the Mediterranean area, where, for the last few years, the number of cases has increased, possibly due, in part, to climatic factors; in the last few decades an increased incidence of MSF was reported for Spain, France, Italy, Portugal and Israel. Mediterranean Spotted fever was originally characterized as a benign rickettsiosis. However, there have been recent reports of very severe cases in France, Spain, Israel and South Africa, manifested by cutaneous and neurological signs, psychological disturbances, respiratory problems and acute renal failure. The presence of the infectious agent, *Rickettsia conorii* appears to be spreading north.

Omsk haemorrhagic fever, louping-ill disease and Crimean–Congo haemorrhagic fever are other tick-borne diseases in Europe.

In both Europe and North America, ectoparasite infestations of humans such as head lice and scabies are present in increasing numbers; both of these serious pests have developed resistance to many of the insecticides that have provided effective control in the past. Scabies is the cause of frequent nosocomial outbreaks in health-care facilities.

It is now realized that house dust mites and cockroaches are responsible for an extraordinary amount of allergies including serious cases of asthma virtually everywhere and their control poses a most difficult problem.

In the last few decades a substantial number of newly emerged rodent-borne diseases of man have been recognized. Some of these infections may be due to agents that were not recognized in the past but others are characterized by such dramatic clinical courses, often with significant mortality and rapid spread, that they are to be considered truly 'emerging diseases'. The hantaviruses belong to the emerging pathogens having gained more and more attention in the last decades. Rodent-borne haemorrhagic fever with renal syndrome has spread widely in rodent populations in Europe, the USA and parts of Canada, with an increasing number of human cases. New species of hantaviruses with greater virulence are emerging in both Europe and North America. Transmission to humans occurs by direct contact with rodents or their excreta or by inhalation of aerosolized infectious material, e.g. dust created by disturbing rodent nests.

A new rodent-borne disease syndrome has appeared in the USA; named the hantavirus pulmonary syndrome (HPS), it is frequently associated with a case fatality rate of 30–60%; the disease was at first determined to be due solely to sin nombre virus, and was thought restricted to the western USA. However, new species of viruses giving rise to HPS have now been found throughout the USA and much of the Americas. As human populations grow and spread to suburban rodent-infested areas, it is likely that hantavirus diseases will become more common in the future especially as the elimination or effective control of their wild rodent reservoir hosts can not be considered as realistic.

Many public health authorities and medical practitioners are not aware of the reappearance of this group of diseases nor of the appearance of new diseases transmitted by insects, ticks and rodents. They are not necessarily familiar with the exotic infectious agents of this group that, with increased tourism, are increasingly being imported from disease-endemic countries. This has often resulted in the delayed or mistaken diagnosis of members of this group of infections to the detriment of patients. The failure of public health authorities to recognize this group has often resulted in delays before effective measures have been undertaken to control their arthropod vectors or rodent reservoir hosts.

The following book will review the distribution of the vector and rodent-borne diseases in Europe, the USA and Canada; their incidence and prevalence, their costs and hence their public health burden will be detailed and their arthropod vectors and rodent-reservoir hosts described. Armed with such information, the individual clinician is more likely to have a degree of epidemiological suspicion that will lead to an earlier diagnosis and correct treatment of these infections. Equally, authorities will more readily understand the measures necessary to control this group of infectious agents.

1

Introduction

The vector and rodent-borne diseases are a heavy burden on public health in much of the tropical and semitropical regions of the world; among these diseases are malaria, leishmaniasis, sleeping sickness, many arboviruses and a plethora of other infections. Despite their largely temperate climates, Europe, the USA and Canada have not been spared by this group of infections; moreover, some endemic diseases thought to have been under control in these regions are now resurging and new infections are emerging on both continents. Co-infections of HIV virus and leishmaniasis and of Lyme disease and tick-borne encephalitis, pose diagnostic and treatment problems to clinicians. Ecological and climate changes have favoured increases in the densities of insect and tick vectors and rodent reservoir hosts and in the agents they transmit. With increased travel to tropical disease-endemic areas, the number of imported cases of malaria and other vector-borne diseases has sharply risen and some of these have become established, with grave consequences.

The following chapters will review the status of the vector and rodent-borne diseases which are endemic or imported into Europe and the USA and Canada as well as the literature describing their epidemiology, incidence, distribution, vectors and reservoir hosts. Emphasis will be placed on the epidemiology of the infections rather than on clinical aspects or treatment. To plan the prevention and control of this group of infections, knowledge of their epidemiology and distribution is essential for public health officials and health scientists; clinicians must be aware of the infections they may encounter to ensure a rapid diagnosis and timely treatment, especially those introduced from abroad.

An extensive bibliography is provided as much of the literature on this group of infections is scattered through a wide spectrum of journals. Furthermore, as will be seen, the incidence of many of the vector and rodent-borne diseases in both Europe and North America are actually increasing, some quite seriously, while new diseases are emerging and old ones resurging; to substantiate this, an effort has been made to provide references to as many relevant studies as possible as well as an evaluation of the current public health importance of the infections which will be reviewed. The literature has been reviewed to June 2005.

PART I THE VECTOR- AND RODENT-BORNE DISEASES OF EUROPE

2

Vector- and rodent-borne diseases in European history

Plague

The most serious of the vector- and rodent-borne disease epidemics in European history were the pandemics of plague that swept through the continent decimating the population. The first recorded epidemic began in Arabia at the time of Justinian, reaching Egypt in AD 542; it then spread through Palestine and Syria to Europe and throughout the Roman Empire to the British Isles and Ireland. The most infamous of the plague pandemics was the 'Black death' which ravaged the continent from the middle of the fourteenth century until the end of the sixteenth century. In Great Britain, one-half to two-thirds of the population is believed to have been killed and it is generally believed that 25 million people or as much as a quarter of the European population fell victim to this pandemic. The last outbreak of plague in Europe occurred in Marseilles, France in 1720, probably introduced by a plague-infested ship arriving from Syria. Some 50 000 people died in the city; the disease spread over a great part of Provence but disappeared in 1722 (Pollitzer, 1954).

There has been some controversy as to whether or not the epidemics described above were indeed caused by *Yersinia pestis* inasmuch as diagnosis of ancient septicaemia or other forms of plague solely on the basis of historical clinical observations is not possible. Furthermore, the lack of suitable infected material prevented direct demonstration of ancient septicaemia; thus, the history of most infections such as plague has remained hypothetical. Some recent investigations have supported the contention that these ancient epidemics were indeed caused by infections with *Y. pestis*. Drancourt *et al.* (1998) made DNA extracts from the dental pulp of 12 unerupted teeth extracted from skeletons excavated from sixteenth and eighteenth century French graves of persons thought to have died of plague and from seven ancient negative control teeth. Polymerase chain reaction (PCR) probes incorporating ancient DNA extracts and primers specific for the human beta-globin gene demonstrated the absence of inhibitors in these preparations. The incorporation of primers specific for *Y. pestis rpoB* (the RNA polymerase beta-subunit-encoding gene) and the recognized virulence-associated *pla* (the plasminogen activator-encoding gene) repeatedly yielded products that had a nucleotide sequence indistinguishable from that of modern-day isolates of the bacterium. The specific *pla* sequence was obtained from 6 of 12 plague skeleton teeth but none of seven negative controls thus confirming presence of the disease at the end of the sixteenth century in France. Further DNA extractions of dental pulp were also positive in a study in the south of France (Raoult *et al.*, 2000).

In Germany, in an area which was also struck by the purported plague epidemics, Wiechmann & Grupe (2004) carried out a molecular genetic investigation of a double inhumation, presumably a mother and child burial from Aschheim (Upper Bavaria, sixth century), which included analysis of mitochondrial DNA, molecular sexing and polymorphic nuclear DNA. *Y. pestis*-specific DNA was detected confirming the presence of *Y. pestis* in southern Germany during the first plague pandemic recorded.

Malaria

Malaria has existed in the Mediterranean basin since the prehistoric era and the arrival of man. The ancient Romans associated the malarial fevers with proximity to marshes. The modern term 'malaria' originates from Italy of the Middle Ages when the two words 'mala' and 'aria' became the word known today as 'malaria' which gradually came into common use.

In the nineteenth century malaria transmission extended from part of Norway, through southern Sweden, Finland, Russia, Poland, along the countries of the Baltic coast including northern Germany, through Denmark and the Netherlands, south along the coasts of Belgium, France, Spain and Portugal and all of the countries of the Mediterranean and Adriatic, in the Balkans where it was particularly severe and throughout Greece and the Danube peninsula (Strong, 1944). Coastal southern and eastern England had unusually high levels of mortality from malaria from the sixteenth to the nineteenth century (Dobson, 1994).

In the beginning of the 1950s eradication programmes were launched by national governments with the support of the World Health Organization and successfully eradicated the disease from virtually all the continent. By 1969, Hungary, Bulgaria, Romania, Yugoslavia, Spain, Poland, Italy, the Netherlands and Portugal had completely eradicated endemic malaria.

Typhus

Epidemic or louse-borne typhus *Rickettsia prowazekii* transmitted to humans by the human body louse *Pediculus humanus* may cause a mortality of 60% in untreated people. Widespread epidemics have occurred in Europe. Armies have been virtually wiped out by epidemics of typhus. Between 1915 and 1918, typhus was responsible for one-fifth to one-third of all illnesses in the British forces, and for about one-fifth of the German and Austrian armies. An outbreak of louse-borne typhus occurred in Serbia in November 1914; within six months 500 000 people developed typhus fever. Over 200 000, of whom 70 000 were Serbian troops, died from the disease. One half of the 60 000 Austrian prisoners also died from typhus. At its peak new cases ran at 10 000 per day. Mortality ranged from 20% at the start to 60–70% at the end of the epidemic.

In the 1917–1921 epidemic in Russia the epidemic raged amidst the famine and dislocation of the revolution and 20–25 million cases were estimated to have occurred. For a while it looked as if the fate of the revolution was at the mercy of typhus fever. Lenin, in 1919, put it succinctly: 'Either socialism will defeat the louse, or the louse will defeat socialism' (Tschanz, D. W. http://scarab.msu.montana.edu /historybug /WWI/TEF.htm, accessed 21 June 2005).

Murine typhus caused by *Rickettsia typhi*, is endemic in southern Europe. One tale has it that this milder rickettsial infection transmitted from rats to man by fleas, has an historical basis in the thirteenth century legend of the Pied Piper, who led away the rats from the town of Hamelin, Germany; when refused payment for his services, he led away 130 children and disappeared with them in the mountains. It is suggested that the children actually died in an outbreak of disease and were buried in a common grave at the site of the legendary disappearance. The association with rats points to a rodent-borne infection, and the pied (mottled) coat of the piper seems to indicate a disease causing conspicuous macular lesions (Dirckx, 1980).

Arboviruses

At the end of the eighteenth century and into the nineteenth, yellow fever broke out in Spanish ports, having been brought by vessels mainly from infected ports in the New World. Cadiz suffered five epidemics in the eighteenth century, and Malaga one; from 1800–1821 the disease assumed alarming proportions, Cadiz being still affected (Waddell 1990), while Seville, Malaga, Cartagena, Barcelona (Angolotti, 1980), Palma, Gibraltar (Sawchuk & Burke, 1998) and other ports and their surroundings suffered severely. In the epidemic at Barcelona in the summer of 1821, some 12 000 persons died (Chastel, 1999). Yellow fever also invaded the port of Saint Nazaire in France in the 1860s (Coleman, 1984) and at Lisbon in 1857 some 6000 died. An outbreak of yellow fever occurred in Swansea, UK in 1865; both this outbreak and the Saint Nazaire one were caused by infected mosquitoes flying to the mainland from ships in harbour.

Dengue was long endemic in most of the countries of the Mediterranean; in 1928 approximately 650 000 residents of Athens and Piraeus contracted dengue and 1061 died (Halstead & Papaevangelou, 1980). The disease disappeared from Europe with the disappearance of its vector, *Aedes aegypti*.

From the beginning of the twentieth century, sanitary conditions improved in urban and rural areas, farming and irrigation practices changed, and the great epidemics described above have ceased. However, a substantial number of vector and rodent-borne infections persist both in Europe and North America and new infectious agents are emerging, some of which represent no small threat to public health.

3

The arboviruses

There are between 500 and 600 known arthropod-borne viruses, or arboviruses, in the world of which some 100 may give rise to human disease. There are six families of arboviruses; Togaviridae, Flaviviridae, Bunyaviridae, Reoviridae, Rhabdoviridae and Orthomyxoviridae. By 1996, 51 arboviruses had been reported from Europe – they are the subject of a comprehensive review by Hubalek & Halouzka (1996). Many of these viruses are not known to cause human illness; some have only been isolated from arthropods, birds or other animals and their public health significance is unknown. Others, however, may cause significant human illness and mortality. The arboviruses will be considered by the four groups of arthropods that transmit them, i.e. mosquitoes, sandflies, biting midges and ticks. The epidemiology of the arboviruses is rapidly evolving and their distribution is spreading to areas in which they have not been previously endemic and, in some cases, as appear to be occurring with West Nile virus, increased virulence has been seen in some recent outbreaks.

4

The mosquito-borne arboviruses of Europe

West Nile virus

West Nile virus (WNV), a member of the Japanese encephalitis complex, is a neurotropic flavivirus virus that produces damage of varying severity in human, animal and avian hosts. The virus is amplified in birds and transmitted to humans usually by *Culex* mosquitoes. Most cases of WNV are subclinical, with overt clinical illness affecting 1:100 to 1:150 cases. Meningoencephalitis is the most common diagnosis in hospitalized WNV patients, affecting 50–84%. In the elderly the mortality rate may range as high as 10% though it is much lower in the current outbreak in the USA. The epidemiological cycle of WNV is shown in Figure 4.1.

West Nile virus was first isolated from a febrile woman in the West Nile District of Uganda in 1937 (Smithburn *et al.*, 1940); in 1950 it was found that the virus was present in a large percentage of normal individuals in the vicinity of Cairo, Egypt. The majority of the children from whom the sera were collected appeared to be normal; there was no evidence that children with viremia were severely ill. In 1950 more than 70% of the Cairo inhabitants aged 4 years and over had antibodies to WNV (Melnick *et al.* 1950).

In 1951, WNV was recognized in Israel; the disease had probably already been present in that country for several years. There were large outbreaks in 1950–1951 and it is estimated that the number of cases was in the hundreds (Goldblum *et al.*, 1954); none of the cases was fatal and there were apparently many subclinical cases. Israel is an important path for migrating birds to and from Africa and to Europe and the virus may have been introduced in this manner.

The known distribution of WNV is shown in Figure 4.2.

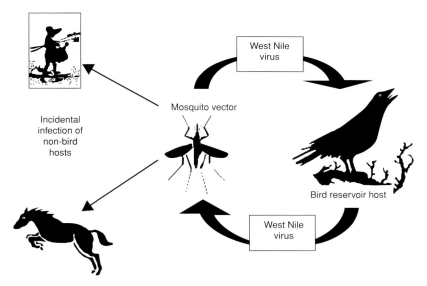

Figure 4.1 Transmission cycle of West Nile virus.

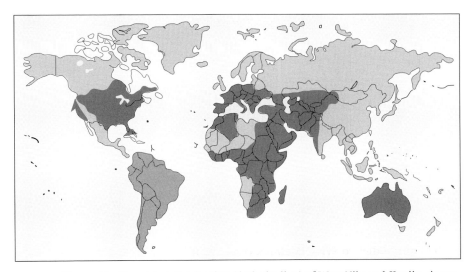

Figure 4.2 The global distribution (dark shading) of West Nile and Kunjin viruses.

Albania

The first report of WNV in Europe was the detection of the virus in 1958, in two Albanians found to have specific WNV antibodies (Bárdos *et al.*, 1959). The virus remains endemic in the country (Eltari *et al.*, 1993).

The subsequent spread of WNV through Europe is reflected in Table 4.1 which records the presence of WNV as indicated by the occurrence of human cases

Table 4.1 *Countries in Europe in which West Nile virus endemic activity has been detected*

Countries	Reported activity or outbreaks		
Albania	1958		
Austria	1964–1977	1988	
Belarus	1972–1973*	1977	
Bulgaria	1960–1970	1978*	
Czech Republic	1978*	1980s*	1990* 1997
France	1962–65,	1975–1980	1999–2003 2004*
Germany	1990s*		
Greece	1970–1978	1980–1981	
Hungary	1970s	1984	
Italy	1966–1969	1981*	1998*
Poland	1998*		
Portugal	1962 1965	1967	1969 1973 1976–1970
Romania	1966–1970	1975	1980 1984 1996–2000
Russia	1962–1976	1977*	1981–1986 1989 1991
			1992* 1999
Slovakia	1970–1973*	1984–1987	1998
Spain	1960s	1979*	1980 1998
Ukraine	1980s	1998	

* Birds, animals or arthropods only.

or isolations or positive serology in humans, animals, birds or arthropods. The following section reviews the literature reporting on the presence and incidence of WNV by country in Europe.

Austria

Serological surveys reported in the 1960s and 70s indicated the presence of WNV in 12 dogs (33.3%), 17 pigs (6.9%), 25 cattle (26.7%) and 21 of 61 hedgehogs (*Erinaceus europaeus*) tested (Sixl *et al.*, 1973). Antibodies to WNV were found in farmers in Eastern Austria (Sixl *et al.*, 1976); 385 horses and 102 free-living birds found dead were all negative and WNV was not considered, at the time, a significant pathogen in Austria (Weissenbock *et al.*, 2003a).

Belarus

In 1985–1994 (see Table 4.1) four strains of WNV were isolated; one from birds, two from *Aedes* mosquitoes and one from a febrile patient. Another isolation was made from *Anopheles* mosquitoes in 1999. Specific WNV antibodies in human blood sera were identified in 1.7% of the Belarusian population. In the Gomel and Brest regions the percentage of seropositive individuals was 5.8 and 15.4, respectively. Antibodies were found in 0.6–5.8% of cattle, in 2.9–6.8% of wild small mammals and in 6.5–16.7% of birds. Sixteen cases were confirmed among patients with a febrile aetiology (Samoilova *et al.*, 2003).

Bulgaria

In 1978 serum samples from 233 cattle on 22 farms in Bulgaria were examined. Fourteen cattle were positive for WNV antibodies (Karadzhov *et al.*, 1982). Little other information is available on WNV in Bulgaria; a survey of mosquitoes reported in 1991 (Kamarinchev *et al.*, 1991) found WNV in several species of *Aedes* and *Culex*.

Czech Republic

The first reports of the presence of WNV were made in 1976 (see Table 4.1) (Prazniakova *et al.*, 1976). Serological surveys in the country have shown that the virus is common in both migratory and non-migratory species of birds including domestic fowl. The first isolation of WNV in the Czech Republic was reported by Hubalek *et al.* (1998); 11 334 mosquitoes from south Moravia were examined in 197 pools. *Aedes vexans*, *Ae. cinereus* and *Culex pipiens* were found positive.

In July 1997 heavy flooding occurred along the Morava river; populations of *Aedes* mosquitoes rapidly increased. During the surveillance, 11 334 mosquitoes were examined and WNV isolates were made from *Aedes vexans* and *Culex pipiens*. West Nile virus antibodies were detected in 13 (2.1%) of 619 persons seeking treatment at hospitals and clinics in the Brelav area and two children had clinical symptoms, the first recorded cases of WNV in central Europe (Hubalek *et al.*, 1999).

Hubalek (2000) characterized WNV in Europe as follows: 'European epidemics . . . reveal some general features. They usually burst out with full strength in the first year, but few cases are observed in the consecutive 1 to 2 (exceptionally 3) years, whereas smaller epidemics or clusters of cases only last for one season. The outbreaks are associated with high populations of mosquitoes (especially *Culex* spp.) caused by flooding and subsequent dry and warm weather, or formation of suitable larval breeding habitats. Urban WNF outbreaks associated with *Culex pipiens* biotype *molestus* are dangerous. Natural (exoanthropic, sylvatic) foci of WNV characterized by the wild bird-ornithophilic mosquito cycle probably occur in many wetlands of climatically warm and some temperate parts of Europe; these foci remain silent but could activate under circumstances supporting an enhanced virus circulation due to appropriate abiotic (weather) and biotic (increased populations of vector mosquitoes and susceptible avian hosts) factors. It is very probable that WNV strains are transported between sub-Saharan Africa and Europe by migratory birds'.

France

West Nile virus was first reported in the Carmargue region of southern France in 1962–1965 and had been apparently introduced in 1962. There was a high mortality rate among horses in the area with some 50 dying during this period, and 13 human cases. West Nile virus was isolated from two mild human cases in 1964. There was an epidemic outbreak in 1962 during which severe cases of the disease were observed; an epidemic occurred again in 1963 at a lower incidence. In 1964 and 1965, only a few mild cases were seen in the Camargue (Panthier *et al.*, 1968). The horse is the principal victim of WNV infection in the French Mediterranean littoral; in 1962–1965 there were 500 clinical cases and 50 equine deaths. Human infection is usually benign. The vector is *Culex modestus*. Infection in the horse ranges from unapparent infection to clinical symptoms. Antibodies to WNV were frequent in man in an area to the west and north of the Camargue and were also often found in horses, wild rabbits and hares but not in birds or small rodents (Joubert, 1975).

In September 2000, WNV again appeared in the south of France when 47 horses in the Herault region, close to the Camargue, developed symptoms of WNV encephalitis. There were 12 equine deaths but no human cases (Zientara, 2000). Equine cases were reported in the Department of Gard (16 horses) and the Department of Bouches du Rhone (3 horses). In total there were 76 clinical cases and 21 deaths. Seroconversions were detected in two sentinel birds in October 2001 in a mallard and in August 2002 in a chicken indicating a low circulation of WNV in the Camargue region. *Cx. pipiens* was considered the main vector.

The most recent case of WNV in France occurred in October 2003, in a man in the department of the Var; his wife also tested positive and there was a report of WNV infection in a horse some 20 km from the human cases. Two suspected equine cases were notified in the Var, in mid-September. The occurrence of 2 human cases (one confirmed and one probable) and 3 equine cases (one confirmed, 2 probable), in the same area in a 5-week period, suggests that infection was contracted in the Var located more than 100 km east from the Camargue. A total of 3 encephalitis and 4 mild illness cases were identified in humans, all living or having stayed in the vicinity of Fréjus city. Four equine cases were identified within 25 km from the human cases (Del Giudice *et al.*, 2005).

Germany

No autochthonous human cases of WNV have been reported from Germany. However, Malkinson & Banet (2002) considered that the findings of anti-WNV antibodies in a population of storks maintained in northern Germany could be evidence for local infection. They pointed out that the unique susceptibility of young domestic geese in Israel in 1997–2000 to WNV and the isolation of similar strains from migrating White storks in Israel and Egypt suggest that the recent isolates are more pathogenic for certain avian species and that migrating birds play a crucial role in geographical spread of the virus. There have been no reports of isolations of WNV from mosquitoes in Germany or other reports of seropositive birds.

Greece

Lundstrom (1999) noted that in serological surveys, antibody prevalence to WNV was more prevalent in Greece than that to tick-borne encephalitis (TBE); Lundstrom also cites the presence of WNV in human serum collected by Pavlatos & Smith (1964) and again in 1972 (Papapanagiotuo *et al.* 1974). In a survey of animals reported from Greece in 1980 (Koptopoulos & Papadopoulos, 1980), antibodies to WNV were found in 8.8% of sheep, 8.7% of goats, 3.9% of cattle, 20.4% of horses, 1.4% of pigs, 24.5% of birds, 29% of humans, in 1 of 2 hares and in 1 of 26 rabbits.

Hungary

The earliest report of the presence of WNV in Hungary was the isolation of the virus from small mammals in the course of surveys for TBE (Molnar *et al.*, 1976). In another survey (Molnar, 1982) reported the presence of antibodies to WNV in ticks and mosquitoes in the country.

Italy

West Nile virus was first reported in Italy in 1966 when three clinical cases were described in children (Gelli Peralta, 1966) and surveys showed antibodies to WNV in small mammals of various regions.

The first equine outbreaks occurred in 1998 in 14 horses in Tuscany. Eight animals recovered without important consequences. West Nile virus was serologically detected in all 14 horses and was isolated from an affected horse (Cantile *et al.*, 2000). The outbreak involved race horses and racehorse breeding stock of high economic value (Autorino *et al.* 2002). No human cases were reported.

In 1998, virus isolated from a horse in Tuscany with encephalitis was sequenced. The strain appeared to share 99.2% nucleotide identity with a

Senegalese strain isolated in 1993 from a mosquito (*Culex neavei*), indicating a link with viruses isolated in France (1965), Algeria (1968) and Morocco (1996), and suggests the hypothesis of an introduction of WNV by migratory birds crossing the Mediterranean Sea.

Moldavia

In 1973–1974, WNV was isolated from ticks in what was then Soviet Moldavia (Chumakov *et al.*, 1974). No human cases of WNV have been reported from the country.

Poland

Sparrows from central Poland were examined by Juricova *et al.* (1998) between 1995 and 1996 and antibodies to WNV were detected in 12.1% of the tree sparrows, *Passer montanus*. As these were non-migratory birds, the virus can be considered as endemic in Poland despite the absence of any reports of human cases.

Portugal

In 1972, Filipe described the first finding of an arbovirus in mosquitoes in Portugal; it was identified as WNV, or a virus antigenically closely related, and was isolated from a female *Anopheles maculipennis* in Beja, in the south of the country in 1969. The site of collection was to the north of the farms where an outbreak of WN encephalitis had occurred in 1962–1965. There have been no reports of human WNV cases or of antibodies in humans.

In 1967, Filipe collected sera from cattle and sheep in south Portugal; antibodies were detected to WNV. Filipe & Pinto (1969) described a survey of cattle and sheep in the south for antibodies to WNV and TBE; of the bovine serum, 16% had antibodies to WNV and also reacted to TBE as an overlap reaction. Several cases of equine encephalitis with 10 or 12 deaths had occurred in horses at a farm near Aljustrel in southern Portugal in 1962–1965; antibodies against WNV were found in animals on nearby farms. Tests in 1970 of 24 horses that had survived the disease in the 1962–65 epizootic revealed antibodies against WNV in 7 of them (Filipe *et al.*, 1973).

Romania

In 1975, Draganescu *et al.* reported antibodies to several flaviviruses in humans and domestic animals in a biotope with a high frequency of migratory birds. There were human antibodies in a proportion of 11.8% to TBE and 25.5% to WNV; in domestic animals, WNV antibodies were detected in 4.9% of the sheep, 4.1% of the cattle and 12% of the goats tested. In 1977, Draganescu *et al.* described an outbreak of a febrile disease affecting 14 of 41 crew members of a ship that passed from Romania through the Suez Canal and the Red Sea on its way to Japan. Serological studies of the crew showed WNV antibodies in 25 of the 35 crew members checked. One patient died. The illness in other patients was mostly benign. The crewmen were probably infected in Romania before departure of the vessel.

High titres of antibodies to WNV were detected in 66 sera from 133 stray dogs in Romania (Rosiu *et al.*, 1980). Antipa *et al.* (1984) took sera from 8 species of migratory birds; 22 serum samples gave positive reactions (titres: 1/20 – 1/80) to WNV and Ntaya viruses. Seroprevalence data suggest that WNV activity in southern Romania dates to the 1960s or earlier (Campbell *et al.*, 2001).

In mid-July 1996, a clustering of meningoencephalitis cases was noted in Bucharest. Several hundred cases were reported in August. By the first week of September, antibodies to WNV had been demonstrated in several patients. The epidemic started at the end of July, peaked in the first week of September, and the last confirmed patient fell ill at the end of September. There were reports of 683 suspected cases; 527 met a clinical case definition of aseptic meningitis or encephalitis. By October, serological testing had been completed on 200 of the 527 suspected cases, yielding 168 laboratory-confirmed cases. Nine confirmed cases were fatal, a case fatality rate of 5.4%, all in people aged 60 years or older. The age-specific incidence in persons above 70 years was more than six times the incidence among children and young adults. It is likely that some of the 31 fatal unconfirmed cases represented true WNV cases. Disease incidence was highest in an agricultural area surrounding Bucharest. The epidemic involved an extensive area of southern and eastern Romania. A serosurvey of 75 blood samples demonstrated IgG antibodies to WNV in 4.1% of blood samples. Extrapolated to the city as a whole, it was estimated that between 90 000 and 100 000 people were infected. *Culex pipiens* is the dominant mosquito in Bucharest and was the most likely vector in 1996 (WHO, 1996). This epidemic was the largest outbreak of WNV reported in Europe and was characterized by a high virulence of the WNV strains involved.

The risk factors for WNV during the epidemic were analysed by Han *et al.* (1999). There was widespread subclinical infection during the outbreak. The risk factors for acquiring infection and for developing clinical meningoencephalitis after infection were mosquitoes in the home, reported by 37 of 38 (97%) asymptomatically seropositive persons compared with 36 of 50 (72%) seronegative controls. Among apartment dwellers, flooded basements were a risk factor (reported by 15 of 24 (63%) seropositive persons vs. 11 of 37 (30%) seronegative controls. Meningoencephalitis was associated with spending more time outdoors.

Surveillance after the 1996 epidemic showed that sporadic human cases continued to occur in Bucharest and the lower Danube delta in 1997 and in Bucharest in 1998, giving evidence of ongoing virus transmission in southeastern Romania.

Ceianu *et al.* (2001) reviewed WNV surveillance in Romania from 1997 to 2000. It was noted that the virus was still circulating in both human and avian cycles – surveillance found 39 clinical human WNV cases during 1997–2000. Retrospective sampling of domestic fowl in the vicinity of patient residences during 1997–2000 demonstrated seroprevalence rates of 7.8–29%. Limited wild bird surveillance showed seroprevalence rates of 5–8%.

Apparently WNV persists locally in poorly understood transmission cycles and the virus remains a serious threat to public health in Romania; the implementation of an effective mosquito control is imperative.

Russia

West Nile virus has been isolated in Russia from species of ticks, mosquitoes and birds in many areas of the country. Seventeen virus strains were isolated from the tick *Ornithodoros capensis* collected in 1970 in the nesting grounds of herring gulls (*Larus argentatus*) in the Baku Archipelago, Azerbaijan. One strain was identical to WNV (Gromashevsky *et al.*, 1973). West Nile virus was isolated from *Ixodes ricinus* in what was then Soviet Moldavia in 1973–1974. A single isolate with both Crimean–Congo haemorrhagic fever CCHF and WNV was obtained from *Dermacentor marginatus* though no human cases were reported in the areas where these viruses were isolated (Chumakov *et al.*, 1974).

In a study in what is now Turkmenistan, Berdyev *et al.* (1975) found antibodies to WNV in 8 of 116 human serum samples, and 9 of 37 serum samples from wild animals. Antibodies were also detected in 6 out of 88 camel samples, 2 out of 3 horse samples, 30 out of 338 sheep samples and 189 out of 534 cattle samples.

In the Volgograd and Krasnodar regions, WNV antibodies were found in 50 out of 64 patients examined. The large number of reported cases suggested that an epidemic caused by the virus occurred in these regions in the summer of 1999 involving as many as 1000 cases and dozens of deaths (L'vov *et al.*, 2000a). Two strains of WNV were isolated from the brain of a dead subject and from a patient during the 1999 outbreak. These strains reacted with convalescent sera proving their aetiological role in this outbreak (L'vov *et al.*, 2000b). From 25 July to 1 October 1999, 826 patients were admitted to Volgograd Region hospitals with acute aseptic meningoencephalitis, meningitis or fever consistent with arboviral infection. Of 84 cases of meningoencephalitis, 40 were fatal. The authors believed that the unusual pathogenic characteristics may have been due to the extension of new pathogenic WNV strain(s) or to the peculiarities of the human host response (Platonov *et al.*, 2001).

In 1963–1993, strains of WNV were isolated from ticks, birds and mosquitoes in southern European Russia and western Siberia and WNV antibody was found in 0.4–8% of healthy adult blood donors. Sporadic human clinical cases were observed in the Volga River delta. In spite of these reports, WNV infection was not considered by the health authorities as a potentially emerging infection, and the large WNV outbreak in southern Russia, starting in late July 1999, was not recognized in a timely fashion. First evidence suggesting WNV circulation was obtained by IgM-capture enzyme-linked immunosorbent assay (ELISA) in September and two weeks later, WNV disease was confirmed in all 14 non-survivors from whom brain tissue samples were available. Moreover, 35 of 56 patients who contracted aseptic meningitis in 1998 had a high titre of WNV antibody, indicating that WNV infection may have been introduced into the Volgograd region before 1999. The Volgograd isolate had the greatest homology (99.6%) with the WN-Romania-1996 mosquito strain. Specimens from a patient in the 2000 outbreak of WNV in Israel indicated a closer relationship of this isolate to 1996 Romanian and 1999 Russian strains than to 1998–99 Israeli or 1999 New York isolates (Briese *et al.*, 2002). The possible role of migratory birds in this pattern must be considered.

Isolations of WNV from migratory and non-migratory birds was indeed made by L'vov *et al.* (2002a) in the Volga region. Four strains were isolated from birds and their ticks in the Volga delta. The strains were isolated from the great cormorant (*Phalacrocorax carbo*), the crow (*Corvus corone*) and *Hyalomma marginatum* nymphs.

The emerging situation in Russia is probably the result of natural and social factors and may also be due to the introduction of more virulent strains or by evolution of the virus. Whatever the source, the increased virulence of WNV in the recent outbreaks in Russia is of serious public health concern.

Slovakia

In 1972, 2043 mosquitoes of 10 species were collected while biting man and were tested in 129 pools. A strain of virus identified as WNV was isolated from a pool of *Aedes cantans* (Labuda *et al.*, 1974). A serological survey of birds found a single mallard duck (*Anas platyrhynchos*) positive for WNV (Emek *et al.*, 1975). In an additional survey, tick-borne encephalitis and WN viruses were isolated from the blood, brain and liver of migrating birds (Emek *et al.*, 1977).

Juricova (1988) investigated sera from 62 Passeriformes birds of 14 species, caught during the autumn migration of 1987 in the Krkonose mountains in Slovakia; 6.4% of the birds were positive for WNV.

Spain

West Nile virus activity was detected in Spain between the 1960s and the 1980s in the regions of Catalunya, Andalucia, Valencia and Galicia (Lozano & Filipe, 1998).

A survey in the northern provinces of La Coruna, Orense, Pontevedra, Leon and Asturias obtained sera from 701 persons. Tests confirmed that there had been earlier infection with WNV (Gonzalez & Filipe, 1977). The sera of 386 small mammals (rodents, insectivores, small carnivores and bats (Chiroptera)) trapped in 1978 and 1979 were tested for antibody against 10 arboviruses. Positive reactions were found against flaviviruses, among them 3.1% to WNV.

Lozano & Filipe (1998) studied the prevalence of WNV and other viruses among the human population of the Ebro Delta; 1037 samples of serum were taken in 10 towns and analysed for the presence of WNV antibodies and 12 other arboviruses. Antibody titres revealed a significant percentage of samples with high titres to WNV and other antigens. In three localities located in the Delta, the prevalence of Flaviviridae antibodies was as high as 30%, with residual levels of WNV-related IgM in some serum samples; these results suggest that WNV is moving throughout the human population, periodically giving rise to epidemic outbreaks. Bearing in mind the high percentage of neurological complications in the most recent outbreaks of WNV infections recorded in the Mediterranean basin it was felt that WNV plays a role in the factors contributing to viral meningitis and encephalitis within the population of risk areas within Spain.

UK

While no human cases of WNV have been reported from the UK, the Health Protection Agency has an annual enhanced surveillance programme for possible human cases. The scheme operates during the summer, when there is WNV activity in other countries and involves looking for WNV in blood and cerebrospinal fluid samples, taken from patients with encephalitis or viral meningitis with no known cause.

Buckley *et al.* (2003) reported the presence of virus-specific neutralizing antibodies to WNV, Usutu virus (USUV) and Sindbis virus (SINV) in the sera of resident and migrant birds in the UK, implying that each of these viruses is being introduced to UK birds, possibly by mosquitoes. This was supported by nucleotide sequencing that identified three slightly different sequences of WNV RNA in tissues of magpies (*Pica pica*) and a blackbird (*Turdus merula*). The detection of specific neutralizing antibodies to WNV in birds provides a plausible explanation for the lack of evidence of a decrease in the bird population in the UK compared with North America. Many birds migrate annually from regions in Africa where WNV, USUV and SINV co-circulate and are actively transmitted between birds and mosquitoes. It is, therefore, possible that they are carried by birds to the UK and transmitted via indigenous mosquitoes to non-migratory birds and to other wildlife species. A relatively high proportion of resident birds were positive, implying efficient transmission from the migrant bird population. The authors concluded that there is no evidence that British citizens suffer from febrile illness, fatal encephalitis or polyarthritis arising from the bite of mosquitoes infected with WNV, USUV or SINV. They also observed that on balance, it seems unlikely that these viruses present significant health problems to humans, birds or horses in the UK, since the likely risk of exposure to WNV-, USUV- or SINV-infected mosquitoes for humans living in urban or peri-urban areas of the country at the present time should be reasonably low. Nevertheless, as the impact of climate change takes effect and as more people spend increasing periods of time in the countryside, where mosquitoes are likely to occur in the highest densities, the risk of human exposure to encephalitic infection by WNV will almost certainly increase.

Ukraine

The first isolation of WNV in Ukraine was made from a rook (*Corvus frugilegus*) caught in May 1980, in the Black Sea area (Vinograd *et al.*, 1982).

Active foci of TBE, WNV, California serologic group (CSG) and Batai viruses have been identified as a result of investigations in the forest-steppe zone of Ukraine. Of the viruses identified, WNV was the leading arboviral infection in the forest-steppe zone with 53.1% of the infections (Lozyns'kyi & Vynohrad, 1998).

Serbia

There is little information available on the presence of WNV in Serbia; Vesenjak Hirjan (1991) mentions that antibodies to WNV have been found in 8% of 397 people in the west of Serbia, 1% of 479 people in the east of Serbia, 1% of 826 people in Montenegro and 1% of 629 people in Kosovo. There have been no reports of human clinical cases.

The vectors of West Nile virus in Europe

Table 4.2 lists the countries in which the vectorial status of an arthropod has been determined or in which isolations of virus have been made from an arthropod.

In countries in which there have been epidemic outbreaks of the disease in man or animals, it can be assumed that the vectors are mosquitoes. Birds may

Table 4.2 *Arthropod species confirmed as vectors or from which West Nile virus was isolated*

Country	Species
Azerbaijan	*Ornithodoros capensis*
Belarus	*Aedes* spp., *Anopheles* spp.
Bulgaria	*Ae. cantans.*
Czech Republic	*Ae. cinereus, Ae. vexans, Culex pipiens, Ixodes ricinus*
France	*Culex modestus, Cx. pipiens*
Italy	*Cx. impudicus? Cx. pipiens?*
Moldavia	*Dermacentor marginatus, Ixodes ricinus*
Portugal	*Anopheles maculipennis*
Slovakia	*Ae. cantans, I. ricinus*
Romania	*Cx. pipiens, Cx. pipiens molestus*
Russia	*Ae. vexans, Cx. modestus, Cx. molestus, Cx. univittatus, D. marginatus, Hyalomma marginatum, Ixodes lividus, I. ricinus*
Ukraine	*An. maculipennis*

be infected by mosquitoes or ticks but it seems quite unlikely that ticks are involved as vectors in large outbreaks.

The reservoir hosts of West Nile virus in Europe

West Nile virus has been isolated from many animals in Europe but most, including horses, cattle, camels, sheep, pigs, boars, hares, dogs and cats, are dead-end hosts and not reservoir hosts of the infection. It has been found in humans, birds and other vertebrates in Africa, eastern and western Europe, western Asia and the Middle East, but until 1999 had not previously been documented in the Americas.

It is likely that WNV was originally introduced into Europe by migratory birds coming from Africa. Malkinson & Banet (2002) reviewed the role of wild birds in the epidemiology of WNV and the following observation is taken from their review. Surveys on wild birds conducted during the last two decades in Europe, notably Poland, the Czech Republic and the UK have revealed endemic foci of infection. Some species of seropositive birds were non-migrators while others were hatchlings of migrating species. Persistently infected avian reservoir hosts are potential sources of viruses for mosquitoes that multiply in the temperate European zone in hot, wet summers. In the past, evidence for geographical circulation of WNV was based on antigenic analysis of strains from different countries while more recent epidemiological studies have relied on analysis of nucleotide sequences of the envelope gene. With the reappearance of epidemic WNV in European countries, interest has been focused again on the African origin of the causal agent carried by migrating wild birds. In some epidemics, isolates were made from human cases or mosquitoes and only serologic evidence for infection was available from domestic and wild bird populations. It remains to be determined whether the European endemic foci of WNV are in themselves sources of infection for other birds that migrate across Europe and do not necessarily reach sub-Saharan Africa.

As has been described above, specific neutralizing antibodies identified by nucleotide sequencing show three slightly different sequences of WNV RNA in tissues of magpies and a blackbird in the UK (Buckley *et al.*, 2003).

Juricova (1988) in the then Czech Republic, examined sera from 62 passeriform birds of 14 species, during the autumn migration of 1987 in the Krkonose mountains; 6.4% were positive for antibodies to WNV.

Much remains to be determined about the role of migratory birds in the introduction of WNV into the countries of Europe. Non-migratory birds are frequently found infected, but unlike North America do not appear to suffer from massive mortality.

Conclusions on the public health importance of West Nile virus in Europe

Although the presence of WNV in Europe has been recorded since at least the early 1960s, the disease is now emerging more frequently in outbreaks with greater virulence. L'vov *et al.* (2004) believes that the emerging WNV situation in Russia has, in part, resulted from the introduction of more virulent strains or by evolution of the virus. In the 1990s and up to the present, encephalitis has been a more prominent feature of WNV infection in Europe, the Middle East and the USA, suggesting the emergence of more neurovirulent strains (Johnson & Irani, 2002). Since 1996, epizootics involving hundreds of humans, horses and thousands of wild and domestic bird cases of encephalitis and mortality have been reported in Europe, North Africa, the Middle East, Russia and the USA. The biological and molecular markers of virus virulence should be characterized to assess whether novel strains with increased virulence are responsible for the proliferating outbreaks. In an outbreak in Volgograd, Russia in 1999, 40 of 84 (48%) patients with encephalitis died (Gelfand, 2003). Mortality appears to have increased as well among infected animals and birds. Many recent outbreaks of WNV show increasing virulence to the elderly; overall case fatality rates in recent epidemics of WNV ranged from 4% to 14% with higher rates in elderly patients. In an outbreak in Israel in 2000, the overall mortality rate in patients aged 70 years or older was 29%, and 32 of 33 deaths occurred in patients older than 68 years.

West Nile virus in Europe is of growing public health importance. The virus appears to be spreading to geographic areas in which human infections have not previously been reported. Effective surveillance programmes must be considered an imperative. In the absence of an effective vaccine, control of the transmission of WNV must depend on effective control of the mosquito vectors. The extent to which tick vectors are of epidemiological importance remains uncertain.

Batai virus (Calovo virus)

Batai virus, also identified as Calovo virus, a bunyavirus, was first isolated in Slovakia from *Anopheles maculipennis* in 1960. It has been isolated in Norway, Sweden, Finland and northern and southern Russia, the Ukraine, the Czech and Slovak Republics, Austria, Germany, Hungary, Portugal and Romania. The principal western European vectors appear to be mainly *An. maculipennis* and *An. claviger*; it has been isolated from *Aedes communis* and other mosquito species in Sweden (Francy *et al.*, 1989).

Lundstrom (1994) considered that Batai or Calovo viruses are not associated with human disease in Western Europe and that their potential for human disease was low (Lundstrom, 1999). Calovo virus has a very low prevalence in humans in the Czech Republic due to the marked zoophilia of its vectors (Danielova, 1990). Surveys in the Czech Republic showed a high prevalence of Batai/Calovo virus in mammals ranging as high as 23% among deer, boars and hares (Hubalek *et al.*, 1993). The virus is present at a low level in house sparrows (*Passer domesticus*) in Poland (Juricova *et al.*, 1998) and at a higher level (10.9%) in 608 sheep in Slovakia (Juricova *et al.*, 1986). The virus has also been identified both in otherwise healthy human subjects and domestic animals in Croatia (Vesenjak Hirjan *et al.*, 1989).

Shcherbakova *et al.* (1997) did not find Batai virus in a survey of healthy residents of the Saratov region of Russia, but 61.2% of 80 bovine samples had antibodies. While Batai/Calovo virus is widespread in Europe, it does not appear, at present, to be of public health importance and no clinical disease associated with it has been reported.

Ockelbo virus

Ockelbo virus is a Sindbis-related virus. First described in Sweden in the 1960s, it is probably also the causative agent of 'Pogosta disease' in Finland and Karelian fever in western Russia (Francy *et al.*, 1989). The many strains of the Sindbis group are widely distributed throughout Europe, Asia and Africa and are closely related to Ockelbo and other viruses in northern Europe. The virus is maintained in nature in a mosquito–bird transmission cycle and is transmitted in endemic areas by migratory birds and ornithophilic *Culex* species and *Culiseta morsitans* as vectors (Lundstrom *et al.*, 1992).

The identification of Ockelbo virus as a Sindbis virus was confirmed when an alphavirus isolated from *Culiseta* mosquitoes was associated with Ockelbo disease, an exanthema arthralgia syndrome occurring in Sweden. The isolate was made from mosquitoes collected in an area of central Sweden with a considerable Ockelbo morbidity and was indistinguishable from Sindbis virus. Patients with Ockelbo disease developed neutralizing antibodies to the virus in their convalescent sera, suggesting that it is the aetiologic agent of the disease (Niklasson *et al.*, 1984).

In the 1980s Ockelbo disease caused outbreaks involving hundreds of cases in parts of northern Europe. Russia reported 200 and Finland 300 laboratory-confirmed cases in 1981. An outbreak occurred again in Finland in 1995 when 1400 laboratory confirmed cases were reported. In Sweden, an annual average of 31 laboratory confirmed cases were diagnosed during the period 1981–1988 although it is believed that as many as 600–1200 cases a year occur (Lundstrom *et al.*, 1991). In 1982 an outbreak of Ockelbo disease with rash, arthralgia and moderate fever reactions occurred in Sweden. A virus, designated Edsbyn 5/82, isolated from mosquitoes and closely related to Sindbis virus, was the probable aetiologic agent. Most cases occurred in central Sweden. The frequency of antibody in healthy individuals (blood donors) within the endemic area was 2–3% and in foci with a high incidence, reached 8%. Arthralgia is the dominating feature of Ockelbo disease and may immobilize patients for a week or up to a month or more (Niklasson *et al.*, 1988).

Ockelbo virus is also identical to Karelian fever, an infection reported from western Russia. L'vov *et al.* (1988) examined the causative agents of Ockelbo disease in Sweden, Pogosta disease in Finland and Karelian fever in Russia; the results indicated that Ockelbo and Karelian fever viruses are essentially identical and that Ockelbo disease, Pogosta disease and Karelian fever are synonyms for the same disease.

The virus has been isolated from many species of mosquitoes including *Aedes cantans, Ae. cinereus, Ae. communis, Ae. excrucians, Ae. intrudens, Culex pipiens, Culiseta morsitans* and *Culex torrentium*, species which feed upon Passeriformes bird reservoir hosts and man.

In Sweden, the incidence of Ockelbo disease and the prevalence of Ockelbo virus antibodies were investigated in 100–250 human sera from out patient volunteers in each of 21 towns in 11 counties. The disease was found to occur throughout most of Sweden but with higher incidence and antibody prevalence in the central part of the country. It generally affects middle-aged men and women and is uncommon in people younger than 20 years of age. It occurs each year between the third week of July and the first week of October, with a peak during the second half of August. During the 8 years studied (1981–1988), an average of 31 Ockelbo patients per year were diagnosed. Antibody prevalence rates were highest in the oldest age groups. It is suggested that many cases are asymptomatic and/or unreported (Lundstrom *et al.* 1991).

Sera from 324 birds collected in an Ockelbo endemic area in central Sweden were examined for specific antibodies to Ockelbo virus. Birds examined belonged to the orders Anseriformes, Galliformes and Passeriformes. Ockelbo virus antibodies were detected in 8% of the specimens, including species from each of the three orders tested. Specific antibodies found in caged birds and in 6- to 10-week-old birds suggested local transmission. The highest prevalence (27%, 14/51) was observed in the Passeriformes in which 5 of 9 species tested contained antibodies. The high antibody prevalence in Passeriformes and the very large populations of this group in relation to other avian groups in Sweden gives them a high potential as amplification hosts for Ockelbo virus (Lundstrom *et al.*, 1992).

In northern Finland, a disease given the name of Pogosta disease and characterized by arthritis, rash and fever was described in 1974. As indicated above, this disease was later found to be closely related if not identical to Ockelbo disease. Some 93% of the patients had joint inflammation, 40% with polyarthritis. Rash was seen in 88% of the patients, and 23% had fever. A follow-up study found that 50% of the patients suffered from chronic muscle and joint pain at least 2.5 years after the initial symptoms. Outbreaks of Pogosta disease in Northern Karelia seem to occur every 7 years. A study was made of antibodies in 2250 serum samples in Finland; 400 of the sera were from healthy blood donors and 1850 samples from patients from different parts of Finland suspected to have a viral infection. Antibody prevalence was almost equally distributed throughout the country, but highest in western Finland (17%). The authors concluded that (1) Pogosta disease is more common than previously thought; (2) it is not restricted to eastern Finland but is spread throughout Finland; and (3) it is also common in children. In Finland the yearly incidence of Sindbis virus is 2.7/100 000 (18 in the most endemic area of northern Karelia). The annual average was 136 (varying from 1 to 1282) with epidemics occurring in August–September with a 7-year interval (Brummer Korvenkontio *et al.*, 2002). As with other virus infections of the group, the main symptom in humans is a febrile arthritis-like disease (Toivanen *et al.*, 2002).

While the number of cases of Ockelbo virus in Sweden has recently declined, the magnitude of the outbreaks in the 1980s demonstrates that there is a considerable potential for further outbreaks. As has been noted in Finland, what has been termed Pogosta disease in that country is much more widely spread than earlier believed (Toivanen *et al.* 2002) and this may also be the case for Ockelbo virus.

Inkoo virus

Inkoo virus is a member of the California serogroup of the bunyaviruses, and has been reported from Estonia, Finland, Norway, Russia and Sweden. Inkoo virus is transmitted by *Aedes communis* and *Ae. punctor* in Scandinavia; it has been isolated from *Ae. communis* in Sweden (Francy *et al.*, 1989). In Russia the virus has been isolated from *Ae. hexodontus* and *Ae. punctor* (Mitchell *et al.*, 1993). The virus is quite common in Finland with its prevalence increasing towards the north where its prevalence rises to 69% of the population (Brummer Korvenkontio & Saikku, 1975). Antibody prevalence is also high in parts of Sweden (Niklasson & Vene, 1996) but in neither country is there any evidence of human disease caused by this virus. However, in Russia, Demikhov (1995) noted that patients with antibodies to Inkoo virus had chronic neurological disease; 16.7% of convalescents after the febrile form of the disease and 30.7% of convalescents after the neuroinfection form had for 1–2.5 years (the follow-up period) after the initial disease developed neurological disturbances and neurological symptoms. Demikhov & Chaitsev (1995) described severe illness ascribed to infection with the virus but there was no mortality. While the true incidence of Inkoo virus is not known, Inkoo and, as will be seen below, Tahyna virus are the most common California group viruses in Eurasia and must remain under close surveillance.

Tahyna virus

In 1958 a virus transmitted by a mosquito was isolated in the Slovakian village of Tahyna (Bardos & Danielova, 1959). The virus was unknown in Europe and was found to belong to the California group and eventually found to occur in most European countries. It was considered possible that young hares (*Lepus* species) are the hosts for this virus, and that it multiplies in them. Foals and suckling pigs are also hosts and possibly also increase the virus. In human patients, infection with the Tahyna virus appears with influenza-like symptoms. In some cases, meningoencephalitis and atypical pneumonia were observed but no fatal cases have been reported (Bardos, 1976). There are no significant clinical differences between Tahyna and Inkoo viruses. The symptoms seen in 41 patients were fever in 25 patients, neuroinfection in 13, and in 3 subjects the infection was unapparent. Of the patients with the neuroinfectious form of the disease,

3 presented with aseptic meningitis, 2 with meningoencephalitis and 5 with encephalitis.

Butenko *et al.* (1995) examined 221 healthy persons and 520 patients suffering from acute fevers and neurological infections in four districts of the Ryazan region of Russia. Of the healthy subjects, 40.7% reacted positively with Tahyna and Inkoo viruses, indicating many contacts of the population with these viruses. The aetiological role of viruses of the California encephalitis complex was demonstrated in 9.8% of 520 patients suffering from acute fevers and neurological infections. The disease was frequently preceded by a visit to woodlands where the patients had been bitten by mosquitoes. The incubation period lasted 3–7 days and the morbidity was sporadic. In 1991, antibodies to Tahyna virus were found in 60% of the population of the Sverdlovsk region (Glinskikh *et al.*, 1994).

Kolman *et al.* (1979) examined a human population from the southern Moravian region of Czechoslovakia for antibodies against Lednice, Sindbis, TBE, WNV, Tahyna and Calovo. No antibodies to Sindbis, WN and Calovo viruses were found but 17.8–42.0% of antibodies to Tahyna virus were detected in all age groups. The total infection rate was 26.0%. Serum samples from 475 men from different districts of the West Bohemian region were examined for the presence of antibodies against viruses of TBE, Tahyna and Tribec; antibodies against Tahyna were detected in 4.4% of the sera (Januska *et al.*, 1990).

After the 1997 floods in the Czech Republic, serosurveys were carried out in the Breclav area. In sera of 619 inhabitants, antibodies to Tahyna virus were detected in 333 (53.8%) and to WNV in 13 (2.1%).

The vectors of Tahyna virus are mainly pasture-breeding *Aedes* species; most isolations of the virus have been made from *Ae. vexans*. The anthropophilic nature of this species accounts for the high antibody rates in humans in countries where the infection is endemic and there is wide distribution of the virus (Danielova, 1990). Table 4.3 presents a listing of countries in which Tahyna virus has been isolated or in which antibodies have been detected.

Usutu virus

In the summer of 2001, a massive die-off of birds occurred in Austria; it was first feared that this was due to WNV as a similar massive die-off had occurred in the USA. A virus was subsequently isolated from birds which exhibited 97% identity to Usutu virus (USUV), a mosquito-borne *Flavivirus* of the Japanese encephalitis virus group; USUV had never previously been observed outside of Africa nor had it been associated with fatal disease in animals or humans. It seems likely that the virus was brought to Europe by migrating birds from Africa. It was first isolated from bird-biting mosquitoes (*Coquillettidia aurites*)

Table 4.3 *Reports of Tahyna virus found in surveys in Europe*

Country	Results – isolations or antibodies	Reference
Austria	*Aedes caspius*	Pilaski & Mackenstein, 1989
Austria	*Aedes caspius, Aedes vexans*	Pilaski, 1987
Croatia	humans	Vesenjak Hirjan *et al.*, 1989
Croatia	bears	Madic *et al.*, 1993
Czech Rep.	*Aedes cinereus, Aedes vexans*	Danielova *et al.*, 1976
Czech Rep.	*Aedes sticticus, Culex modestus*	Danielova & Holubova, 1977
Czech Rep.	birds	Hubalek *et al.*, 1989
Czech Rep.	*Aedes* spp., humans	Danilov, 1990
Czech Rep.	game animals-deer, boars	Hubalek *et al.*, 1993
Czech Rep.	birds-cormorants	Juricova *et al.*, 1993
Czech Rep.	birds-ducks	Juricova & Hubalek, 1993
Czech Rep.	*Aedes cinereus, Aedes vexans*, humans	Hubalek *et al.*, 1999
Czech Rep.	birds-sparrows	Juricova *et al.*, 2000
France	*Aedes caspius*, humans	Joubert, 1975
Germany	*Aedes caspius*	Pilaski & Mackenstein, 1989
Germany	domestic animals, humans	Knuth *et al.*, 1990
Hungary	*Aedes caspius*	Molnar, 1982
Italy	small mammals	Le Lay Rogues *et al.*, 1983
Poland	birds-sparrows	Juricova *et al.*, 1998
Portugal	cattle and sheep	Filipe & Pinto, 1969
Russia	*Anopheles hyrcanus*	L'vov, 1973
Russia	humans	Kolobukhina *et al.*, 1989
Russia	humans	Butenko *et al.*, 1990
Russia	humans	Glinskikh *et al.*, 1994
Russia	*Aedes communis, Aedes excrucians*	L'vov *et al.*, 1998
Romania	*Cx. pipiens*	Arcan *et al.*, 1974
Romania	cattle, sheep, goats, humans	Draganescu & Girjabu, 1979
Slovakia	humans, hares (?)	Bardos, 1976
Serbia	*Aedes vexans*	Gligic & Adamovic, 1976
Slovakia	*Aedes vexans*	Danielova *et al.*, 1978
Slovakia	humans	Kolman *et al.*, 1979
Slovakia	sheep	Juricova *et al.*, 1986
Slovakia	*Culiseta annulata* larvae	Bardos *et al.*, 1978
Spain	rodents	Chastel *et al.*, 1980

in South Africa in 1959 (Williams *et al.*, 1964) and isolations have been made from birds, mosquitoes, from a *Praomys* rat, and one from a man with fever and rash. In Africa, the virus circulates between birds and mosquitoes with mammals being inadvertent hosts if bitten by infected mosquitoes. There are no reports of severe disease in man.

Usutu virus now appears to be established in central Europe, and may have considerable effects on avian populations; whether the virus has the potential to cause severe human disease is unknown (Weissenbock *et al.*, 2002). In 2002 the virus continued to kill birds, predominantly blackbirds (*Turdus merula*); it is estimated 30% of the blackbirds around Vienna have died since spring 2001. High numbers of avian deaths were recorded within the city of Vienna and in surrounding districts of the federal state of Lower Austria, while single mortalities were noticed in the federal states of Styria and Burgenland. Laboratory-confirmed cases of USUV infection originated from the federal states of Vienna and Lower Austria only. Usutu virus has over-wintered and established a transmission cycle in Austria and seems to have become a resident pathogen of Austria with a tendency to spread to other geographic areas (Weissenbock *et al.*, 2003b).

By 2004 the virus had spread across the entire east of the country and probably as far as Slovakia and Hungary. While nuthatches, owls, sparrows, swallows, thrushes and tits were also dying, blackbirds (*Turdus merula*) have been by far the worst-hit species, accounting for 95% of the deaths.

Buckley *et al.* (2003) reported the presence of virus-specific neutralizing antibodies to WNV, Usutu virus and Sindbis virus in the sera of resident and migrant birds in the UK, implying that each of these viruses is being introduced to UK birds; however, no decrease has been seen in avian populations in the country. The appearance of an arbovirus not previously seen in Europe emphasizes the need for continual surveillance.

Dengue virus

Dengue was once endemic in the countries of southern Europe where *Aedes aegypti* was present. The occurrence in 1927–1928 of a massive epidemic of dengue with high mortality in Athens, Greece has already been described. Dengue is the most important arboviral human disease globally, but it has disappeared from Europe as an endemic disease. The disappearance of dengue transmission was mainly due to the virtually universal availability of piped water supplies in Europe and the disappearance of containers such as water jars and barrels which had been used for storing water for household use and which served as larval habitats for *Ae. aegypti*. Grist & Burgess (1994) pointed out that while the spread of *Ae. aegypti* in Europe is limited by its cold intolerance, this is not the case with *Ae. albopictus*; the reintroduction of dengue transmission must be considered as a possibility as *Ae. albopictus*, a species which has been invading Europe, is a vector of dengue in some parts of the world (Gratz, 2004). *Aedes albopictus* strains from Albania have readily transmitted dengue in the laboratory (Vazeille Falcoz *et al.*, 1999). Dengue is often introduced into Europe by travellers from endemic countries; a viraemic traveller bitten by *Ae. albopictus* could be the source of renewed transmission in Europe (Ciufolini & Nicoletti, 1997), particularly in areas where the density of *Ae. albopictus* populations is high.

Each year, an estimated 3 million German residents spend time in dengue-endemic countries. Schwarz *et al.* (1996) studied 249 German tourists returning from tropical areas with a febrile infection. Acute infection with dengue was diagnosed in 26 (10.4%): most infections were acquired in Thailand (57.7%). In a study of 670 German aid workers who had spent 2 years in the tropics, 49 (7.3%) were positive for antibodies to dengue, none (1.3%) to chikungunya and 1 (0.1%) to Sindbis virus. The steady increase in case reports of imported dengue in Germany in early 2001 likely reflects a recently improved surveillance system. The further steep rise in German case reports, particularly during late 2001 and the first half of 2002, corresponds to a surge of local dengue reporting from many dengue-endemic areas and probably reflects a true increase in the number of imported cases. In the number of reports of travel-associated infectious diseases, dengue fever is second only to malaria (about 1000 cases per year, with a 15% drop in cases from 2001 to 2002) in Germany. The increase in cases from the fourth quarter of 2001 to the first quarter of 2002 is mainly due to cases imported from Brazil. During the first quarter of 2002, the state of Rio de Janeiro recorded an incidence that was 6.5 times higher than it had been in January through March of 2001. This state alone accounted for almost 50% of the total cases in Brazil during this period, including an urban epidemic in the city of Rio de Janeiro. The city draws large numbers of German tourists, especially during the festival of Carnival, which may well have contributed to the high number of cases acquired in Brazil in February and March 2002. The peak in cases imported to Germany in the second and third quarter of 2002 reflects the dengue season in Thailand and other parts of South East Asia. In Thailand, dengue transmission is associated with the rainy season, which varies regionally but in most areas of the country starts around April. In mid-April 2002, an out-of-season outbreak of dengue was reported at the island resort of Koh Phangan, which may explain the high incidence among German travellers in March and April (Frank *et al.*, 2004).

Lopez Vélez *et al.* (1996) carried out a clinical study on 37 travellers returning to Spain from dengue-endemic areas. Anti-dengue antibodies were found in 24 of 37 patients. In 71.4% of the cases seen, dengue was acquired in Asia.

The European Network on Imported Infectious Disease Surveillance has noted that 45% of dengue cases among returning travellers were acquired by patients in South East Asia, 19% were imported from South and Central America, 16% from the Indian subcontinent, 12% from the Caribbean and 8% from Africa. This distribution reflects worldwide dengue activity, as well as the popularity of these countries as tourist destinations. Thailand alone was the place of acquisition for 134 cases (28%) of all travel-associated dengue infections. In 2002 there was a proportional rise in the number of infections acquired in South East Asia compared with 2001. Most patients were European-born travellers (94%), but immigrants born outside Europe and foreign visitors had a 4.3 times higher risk of developing dengue haemorrhagic fever compared with those born in Europe (Wichmann & Jelinek, 2003).

There are many records in the literature of cases of dengue which have been imported into Europe. The possible renewal of dengue transmission in Europe would depend on either *Ae. aegypti* populations being re-established at high density, which appears unlikely, or the possibility that *Ae. albopictus*, a species that has recently been introduced into several European countries where it has become established, begins to transmit the virus.

5

Mosquito-borne diseases of Europe – malaria

Until well after the end of World War II, malaria was endemic in much of southern Europe. The Balkans, Italy, Greece and Portugal were particularly affected though seasonal epidemics or outbreaks occurred as far north as Scandinavia, e.g. Finland in 1944. The area of malaria distribution in Europe was greatest at the end of the nineteenth and beginning of the twentieth century. At that time, the northernmost limit of malaria in Europe ran from central England to southern Norway, central Sweden, central Finland and Northern European Russia along the 64° N parallel. The marshlands of coastal southern and eastern England had unusually high levels of mortality from the sixteenth to the nineteenth century attributed to an endemic disease known as 'marsh fever' or 'ague'. The 'marsh fever' was, in fact, malaria. Malaria in these coastal marshlands had a striking impact on local patterns of disease and death (Dobson, 1994). The first noticeable decline of malaria was seen in Europe during the nineteenth century due to new agricultural practices and changed social conditions. The final disappearance of the disease in Europe and North America was due more to the changed ecological conditions than to the use of DDT (de Zulueta, 1994).

Soon after the end of World War II intensive control measures were initiated in the malaria endemic areas of southern Europe and, by 1970, malaria transmission had been virtually eradicated from the continent, an achievement which contributed to the economic development of the some of the worst affected areas in south-east Europe. The last indigenous cases occurred in Macedonia in 1974 and the World Health Organization (WHO) declared malaria as eradicated from Europe in 1975.

Table 5.1 *Number of autochthonous malaria cases registered in the countries of WHO European Region (the other 37 countries did not register any cases)*

Country	1990	1991	1992	1993	1994	1995	1996	1997	1998	1999
Armenia	0	0	0	0	1	0	149	567	542	329
Azerbaijan	24	113	27	23	667	2840	13135	9911	5175	2311
Bulgaria	0	0	0	0	0	11	7	0	0	0
Georgia	0	0	0	0	0	0	3	0	14	35
Greece	0	0	0	0	0	0	0	0	n/a	1
Italy	0	0	0	0	0	0	0	2	0	0
Kazakhstan	0	0	1	0	1	0	1	0	4	1
Kyrgyzstan	0	0	0	0	0	0	1	0	5	0
Republic of Moldova	0	0	0	0	0	0	2	0	0	0
Russian Federation	7	0	0	1	1	4	10	31	63	77
Tajikistan	175	294	404	619	2411	6103	16561	29794	19351	13493
Turkey	8675	12213	18665	47206	84321	81754	60634	35376	36780	20905
Turkmenistan	0	13	5	1	1	0	3	4	115	10
Uzbekistan	3	1	0	0	0	0	0	0	0	7
Total	8884	12634	19102	47850	87403	90712	90506	75685	62049	37169

Source: Sabatinelli *et al.*, 2001.

Recrudescence of autochthonous malaria in Europe

Autochthonous malaria transmission, albeit on a limited scale, is again occurring in some countries of Europe as shown in Table 5.1.

Small-scale autochthonous malaria has actually occurred in many countries of Western Europe in recent years. Most cases have been airport malaria which is transmission by an infected mosquito arriving aboard an aircraft from a malaria-infected country; other cases have resulted when a local vector which had fed on a malaria-infected migrant or returning traveller then transmitted the disease. In view of the developed public health systems in the countries of Europe, it is unlikely that malaria transmission will be renewed in Europe other than as occasional cases. This does not, however, exclude the necessity of a continuing surveillance of human cases of malaria or of potential vector populations. As will be discussed later, the possibility that global warming may allow transmission to resume in Europe, if even on a limited scale, cannot be excluded.

Malaria transmission disappeared from Belgium in 1938 but there have been several autochthonous cases over the last two decades; all of them were associated with 'airport malaria', the most serious of them a cluster of six cases which occurred in 1995 (Van den Ende *et al.*, 1998).

Malaria was eradicated from Bulgaria in 1965. However, in 1995–1996, 18 autochthonous cases of *Plasmodium vivax* were found in southeast Bulgaria, near the border with Greece and Macedonia. It was thought likely that the cases originated from an African migrant who had travelled through the area on his way to Greece. *Anopheles maculipennis* is common in the area and may have been the vector (Vutchev, 2001).

Malaria disappeared spontaneously in France in 1943 except for Corsica. It was eradicated from that island by residual insecticide applications to houses carried out in 1959–1960 (Danis *et al.*, 1996). Both as 'airport malaria', and by local transmission, autochthonous malaria has frequently occurred in several parts of France. In Corsica in 1970–1971, there was an outbreak of 30 cases of *P. vivax; An. labranchiae* was still present and local transmission was confirmed by seroepidemiological studies (Ambroise Thomas *et al.*, 1972). Nine unexplained malaria cases occurred in 1969–1978 in northern and central France and appeared to have been transmitted by indigenous mosquito vectors (Gentilini & Danis, 1981). In 1991, a farmer who had never travelled outside of the Fréjus area of southern France and who had no history of blood transfusions, fell ill; examination confirmed that he had *P. falciparum*. No breeding of *Anopheles* mosquitoes was found within a radius of 3 km. There was no international airport close by but a military camp and a public camping ground were close; it is possible that one of their inhabitants carried an infected mosquito in their luggage on return from an endemic area (Marty *et al.*, 1992).

In Germany, malaria transmission was at one time common in the regions of Friesland and Schleswig Holstein; transmission in these regions disappeared by 1950 and from Berlin in 1951. In 1994, two cases of *P. falciparum* occurred in Berlin in patients who worked at a sewage plant and who had no history of travel to malaria-endemic areas or of blood transfusions. Airport malaria seemed unlikely as the patients worked 20 km away from Schonfeld international airport. The most likely origin of the cases was thought to be importation of an infected mosquito in luggage of a colleague returning from holiday or of a resident of an international migrant camp located 1500 m from the sewage plant (Mantel *et al.*, 1998). Two German children with no history of travel or blood transfusion contracted malaria during a stay in a hospital in the city of Duisburg; a child from Angola with chronic *P. falciparum* was also hospitalized at the same time. *An. plumbeus* was found breeding nearby and as the temperature was relatively high during the period the conditions were suitable for autochthonous transmission which appears to have occurred within the hospital as a nosocomial infection (Kruger *et al.*, 2001).

In Greece in 1997–1999 four presumably autochthonous malaria cases were diagnosed in Evros Province, in northern Greece; a further nine persons out of

1102 tested had specific antibodies against plasmodial parasites (Kampen *et al.*, 2002). Shortly thereafter, two German tourists fell ill and were diagnosed as infected with *P. vivax* after having spent a one-week holiday in Greek Macedonia south of Thessalonica. The infections most probably were transmitted by the same mosquito as the patients developed symptoms simultaneously. The most probable explanation is infection from an asymptomatic carrier of *P. vivax* gametocytes, who infected the mosquito sometime in the beginning of June before the visit of the couple.

Malaria-infected travellers from endemic areas are also very frequent in Italy. These include returning tourists and business travellers, migrants and immigrants who have returned from visits to their families in endemic countries. Sporadic autochthonous cases have appeared in Italy; Baldari *et al.* (1998) described a case of introduced malaria; a woman in Maremma who had no history of travel to a malaria-endemic area fell ill with *P. vivax*. House-to-house investigation identified a 7-year-old girl who had had a feverish illness a few days after her arrival in Italy from India, and who, 3 months later, still had *P. vivax* in her blood; she and her mother had antimalarial antibodies. The child's father later developed a high fever and parasitaemia. The index case of malaria was possibly the result of transmission by local anophelines, most probably *An. labranchiae*.

An unusual case of what was termed 'baggage malaria' was seen in a 2-year-old infant who contracted *P. falciparum* malaria in the absence of the usual risk factors for imported malaria. The clinical history suggested a primary infection acquired during the winter in his home village near Brescia, Italy (100 km from Milan international airport). Among families living within 500 m of the child's house, four reported having travelled to malaria-endemic areas and seven had been visited by travellers from the tropics in December–January. Also, in a village 3 km away there was a large community of migrants from sub-Saharan Africa (600 subjects of a total population of 8000). This suggests that infected mosquitoes can reach areas far distant from international airports having been packed in baggage in countries with a high density of infected *Anopheles* (Castelli *et al.*, 2001).

The importation of active cases of malaria into Italy constitutes a special risk due to the high population densities of *Anopheles* vectors of malaria in many areas of the country especially in rice-field areas.

Large areas of Russia, including the Moscow region, were endemic for malaria until the implementation of a DDT vector control programme combined with active case detection, in 1945. Malaria transmission was prevalent in the regions of Russia where average daily temperatures were above 15 °C for more than 30 days a year. Since 1966 the number of imported cases of malaria has increased and led to renewed local transmission. During 2000 a total of 763 cases of malaria were registered in the Russian Federation, 47 of which were autochthonous (Sokolova & Snow, 2002), including locally transmitted cases of *P. vivax* in the area of Moscow (Makhnev, 2002). The incidence of autochthonous malaria cases in the region of Moscow, has risen from 112 in 1997 to 214 in 2001 and has necessitated the implementation of an active vector control programme.

Malaria was considered as eradicated in Spain by 1964 and all cases thereafter were imported. However, a case of *P. ovale* has been reported by Cuadros *et al.* (2002) from central Spain in a woman who had never travelled outside of the country and had no other risk factors for malaria. This case is the first locally acquired *P. ovale* infection in Europe. There were international airports 4 and 18 km from the patient's home, possibly within the flight range of a mosquito introduced aboard an aircraft. Transmission may have also occurred after one of the local potential vectors, *An. labranchiae* or *An. atroparvus*, had bitten a migrant worker.

The most serious problems of resurgent malaria is in the newly independent states of Eastern Europe; due to a number of factors including an influx of refugees from malaria-endemic areas, the breakdown in health services, especially vector control services, in most of these states and the failure to carry out adequate malaria surveillance, there has been a return of malaria to countries from which it had previously disappeared. While a considerable effort is now being made to control transmission by vector control, the incidence of autochthonous malaria for the period 1996–2000 remains quite high in some of these countries (Table 5.2).

Most of the resurgent cases of malaria in Eastern Europe are due to *Plasmodium vivax* but there have been an increasing number of cases of *P. falciparum*. In view of the mobility of human population, the increased number of autochthonous malaria cases constitutes a threat to the affected areas of Europe, particularly the Balkans.

The problem of imported malaria in Europe

Between 10 000 and 12 000 cases of imported malaria are notified in the European Union each year. Inasmuch as at least an equal number of cases are not properly diagnosed or reported, the actual number of imported cases may be as high as 20 000 per year. The WHO European Regional Office for Europe reports that there were a total of 15 528 cases of imported malaria in Europe in the year 2000. The majority of the cases were imported from Africa and most of these were from West Africa. A study by Muentener *et al.* (1999) on the reliability of the reporting of imported malaria cases in Europe showed that there was gross under-reporting and marked heterogeneity in the type and availability of national data. Only Finland, France and the Netherlands have estimated the level of under-reporting (respectively 20%, 55% and 59%) (Legros *et al.*, 1998). The largest numbers of cases are recorded in continental France and the UK among tourists, businessmen, military personnel etc. These imported cases constitute both a serious public health problem for the recipient countries and often a grave medical problem for the infected patients. Most of the cases are *P. falciparum* and the most common area of origin is Africa. The number of imported cases has been steadily increasing; there has been an eight-fold increase in imported cases since the 1970s: 1500 cases were reported in 1972, 13 000 cases in 1999. France, Germany, Italy and the UK are the west European countries with the largest numbers of cases (Sabatinelli *et al.*, 2001). Within the decade (1989–1999), 680 people have died from imported cases of *P. falciparum* infection in the European region; death is often a result of delayed diagnosis as most physicians are not often confronted with malaria and may have relatively little familiarity with its symptoms. The increasing frequency of drug-resistant strains of *P. falciparum* malaria among imported cases complicates treatment and has probably led to an increase in mortality among the imported cases.

Table 5.2 shows the number of cases of imported malaria from 1996 to 2000 in the most affected countries though imported cases are reported from virtually every country of Europe. The situation in some of the countries from which significant surveillance data are available, will be discussed below. Malaria is a notifiable disease in all European countries.

Austria

During the period 1990–2000, a total of 924 cases of malaria were reported as being introduced into Austria or an average of 84 cases of malaria per year. Of this number, 12 patients died. Almost half the patients had taken no chemoprophylaxis (333/691); of those travellers who did take prophylaxis, 88 people (30%) had taken insufficient and 29 patients (10%) had stopped their prophylactic regimen before the full course. Among the reasons given for not taking chemoprophylaxis were unpleasant side effects, complicated regimens and incorrect information provided by tour operators, pharmacists or doctors (Eurosurveillance Weekly, 2001).

Belgium

Most malaria cases imported to Belgium are *Plasmodium falciparum* acquired in Africa due to the country's close ties with some countries of Africa;

Table 5.2 *Imported cases of malaria in Europe 1996–2000*

Country	1996	1997	1998	1999	2000
Austria	87	75	80	93	62
Belgium	n.a.	n.a.	334	369	337
Denmark	191	213	174	207	202
France	5109	5377	5940*	6127*	8056*
Germany	1021	1017	1,008	918	732
Italy	760	814	931	1006	986
Netherlands	308	223	250	263	691
Norway	101	107	88	74	79
Russia	601	798	1,018	715	752
Spain	224	291	339	260	333
Sweden	189	183	172	153	132
Switzerland	292	319	339	313	317
UK	2500	2364	2073	2045	2069

*Preliminary.

van den Ende *et al.* (2000) studied the epidemiology of imported malaria in Belgium; the number of reported imported malaria cases remained almost stable between 1988 and 1997 (249 cases in 1992, 320 cases in 1993). In 1997, there were more African patients, less infections from Central Africa and 50% less in residents of Belgium. Fewer patients reported use of malaria prophylaxis. The causative agent shifted from *P. falciparum* to other species. In 1988/1989, there were no deaths, and severe malaria did not increase significantly. Treatment of imported malaria cases was considered inadequate. Kockaerts *et al.* (2001) reviewed the clinical presentations of 101 imported cases at the Leuven University Hospital between 1 January 1990 and 31 December 1999. A serious finding was that in 48 of the patients (47%), malaria was not suspected by the referring physician. Only 13% of the malaria patients had taken correct chemoprophylaxis according to WHO recommendations. Eighty-three per cent of the patients were admitted to the hospital with a median duration of hospitalization of 4 days.

Denmark

The number of malaria cases imported to Denmark has been increasing for some years. In 1999, there were 207 imported, confirmed cases of malaria in Denmark, a figure that is largely unchanged from 1998 (174), 1997 (213) and 1996 (191). There was an increase in the number of cases of *P. falciparum* malaria in travellers from Africa to 130 in 1999 as against 90 in 1998, 119 in 1997 and 97 in 1996. Concerned by the low degree of compliance with prophylactic regimes by Danish travellers, Christensen *et al.* (1996) studied the contributing causes of malaria among them. They found that about 20% of the patients with *P. falciparum* malaria had taken no chemoprophylaxis at all while, in many, inadequate prophylaxis was proscribed or the patients themselves had under-dosed. Ronne (2000) reported on a survey of the malaria prophylaxis compliance among some 4500 Danish travellers. On journeys of less than 3 weeks, about 3%, 6% and 6% stopped taking chloroquine, chloroquine + proguanil and mefloquine, respectively. On journeys of over 3 weeks, about 5%, 12% and 8% stopped the respective prophylactic regimens. Molle *et al.* (2000) gave post-travel questionnaires during hospitalization to malaria patients, and sent questionnaires by mail to their travelling companions. In total, 142 persons replied. Only 32% of the travellers used chemoprophylaxis according to Danish recommendations. Twelve per cent of the travellers did not use chemoprophylaxis at all. The average compliance was 52%. Insufficient drug dosage was reported by 13%, and use of non-recommended drugs by 7% of the travellers. Thirty-seven per cent used inadequate anti-mosquito precautions, a problem which often coincided with irregular use of chemoprophylaxis. The authors concluded that increased attention from physicians in educating travellers is important for optimizing malaria prophylaxis. Gjorup (2001) evaluated all patients treated for malaria at the Department of Infectious Diseases, Rigshospitalet, in 1999–2000; 70% of the patients had *P. falciparum*. Of these, 95% had been infected in Africa and 46% of these patients were Africans living in Denmark. Only 50% of all patients had taken chemoprophylaxis and of these only half had been compliant.

Kofoed & Petersen (2003) reviewed imported cases in 320 permanent residents in Denmark who returned from abroad with malaria from 1997 to 1999. Two hundred cases of *P. falciparum* malaria were notified of which 103 had used chloroquine and proguanil, 16 mefloquine and 3 atovaquone and proguanil as prophylaxis, whereas the rest had taken other drugs or no prophylaxis. Risk increased with increasing exposure and compliance was lower especially for mefloquine users in malaria cases compared with controls. Data showed that the risk of malaria varied from 1 per 140 travellers to Ghana to 1 per 40 000 to Thailand.

France

The number of cases of malaria imported into France shows an increasing trend (Table 5.2); epidemiological data from the French National

Reference Center for Imported Diseases showed that the estimated cases of imported malaria in France increased from 5940 in 1998 to 6127 in 1999 and 8056 in 2000. In 2001 the number of estimated cases fell back to 7 223. The regions from which cases originated were tropical Africa (95%), Asia (2.2%) and Latin America (2.7%). During the 3-year period from 1998 to 2000, there were a total of 13 autochthonous cases of malaria in France involving patients with no history of travel to tropical areas.

Of the known, imported malaria cases in France, 83% were due to *Plasmodium falciparum*, 6% to *P. vivax*, 6.5% to *P. ovale* and 1.3% *P. malariae*. Less than 10% of the 45% of patients claiming use of prophylaxis complied with the correct regime. Many of the imported cases were African migrants or African residents of France returning from a visit to their countries of origin. As more than 20 deaths a year are due to imported malaria and the average length of hospital stay is a costly period of 4 days, the public health importance of imported malaria in France is considerable (Danis *et al.*, 2002). The number of cases of paediatric-imported malaria being seen in France seems to be in constant increase. All paediatric malaria cases diagnosed by a positive thin or thick blood film at the Versailles Hospital, from January 1997 to December 2001, were studied retrospectively. They were 58 cases of uncomplicated malaria and two cases of severe malaria; 85% of the children had travelled to sub-Saharan Africa and 15% to Oceania; 90% of the children were of African origin. *Plasmodium falciparum* was found in 84% of the cases. Chemoprophylaxis was inappropriate in 92% of the cases. No child had profited from preventive measures against mosquitoes (Eloy *et al.*, 2003).

Germany

Some 3 million Germans go for a 3–4 weeks' trip to some malaria-endemic area every year. Around 700–1000 of them are known to fall ill with malaria, two-thirds of whom are infected with *Plasmodium falciparum*. Grossly 2% of patients die (Rieke & Fleischer 1993). A total of 1008 imported cases of malaria were reported in 1998, 918 cases in 1999 and 732 cases in 2000, and in 2001, 1040 malaria cases were introduced into Germany. Most of the infected patients were 24–45 years of age. Studies on German travellers have shown that the percentage of travellers above the age of 60 is increasing and that the number of persons with severe malaria complications is also much higher among persons above this age. The duration of hospitalization increased with age from an average of 5 days for the group below 45 years to 21 days for the elderly of 60 years and above. These findings suggest a higher risk for a severe course of malaria in the elderly and extensive advice on the correct use of precautions against exposure and chemoprophylaxis of malaria is especially important for travellers above the age of 60 (Stich *et al.*, 2003). Muhlberger *et al.* (2003) found that the risk of death due to *P. falciparum* malaria, of experiencing cerebral or severe disease in general, and of hospitalization increased significantly with each decade of life. The case-fatality rate was almost six times greater among elderly patients than among younger patients, and cerebral complications occurred three times more often among elderly patients. Eighty per cent of the cases acquired infection in Africa, 8.5% in Asia and 5% in Central and South America. *Plasmodium falciparum* accounted for the largest number of cases (80%) followed by *P. vivax* (12%). In 1999, 60% of all malaria cases were Germans. Most of them travelled for holidays or study purposes. Twenty deaths, all attributed to *P. falciparum* malaria, were notified in 1999, most of them (19) were German citizens. In 1999, 61% of the patients had not taken chemoprophylaxis at all while travelling abroad. Harms *et al.* (2002) found that among the patients at a travel medicine clinic in Berlin, only 34% of the returnees from malaria-endemic areas had taken chemoprophylaxis; in case of travel to Africa and Asia, chemoprophylaxis corresponded to international standards in only 48% and 23%, respectively.

Of the imported cases, 82% of the infections had been acquired in Africa and 11% in Asia. The predominant parasite was *P. falciparum* (70%), followed by *P. vivax* (12% in 2000, and 16% in 2001, respectively). The majority of infections occurred among tourists, fewer in immigrants or business travellers. Some two-thirds of all patients had not taken any chemoprophylaxis showing that compliance continued to be low. Mortality decreased from 18 in 2000 to 8 deaths in 2001 (Schoneberg *et al.*, 2003).

Italy

A total of 1992 imported malaria cases were reported in Italy in 1999–2000. Epidemiological analysis of malaria cases reported in Italy from 1986 to 1996 shows that, despite an increase in the number of people who travelled to malaria-endemic areas (about 314 000 in 1989 compared with about 470 000 in 1996), the number of cases in Italian travellers remained stable. On the other hand, the prevalence of cases among immigrants is increasing. The number of immigrants to Italy from malaria-endemic countries doubled from 1986 to 1992 and is estimated to number about 1.3 million. By 2002, migrants constituted 2.5% of Italy's population of 50 000 000. In 1997, 60% of the 875 malaria cases registered were among migrants mostly from Africa. About 80% of the migrants failed to obtain pretravel advice and did not use malaria prevention measures when travelling back home despite a relatively high knowledge about malaria.

Many of the children of migrants born in Italy are taken to visit relatives in their parent's country of origin and they have never been exposed to malaria (Scolari *et al.*, 2002). Most of the malaria-infected travellers had taken inappropriate treatments or no prophylaxis at all.

Reports from 1989 to 1996 show that half of Italian patients and only 7% of foreign citizens took chemoprophylaxis while travelling (Sabatinelli and Majori, 1998). Ninety-three per cent became infected in Africa, 4% in Asia and 3% in Latin America. *Plasmodium falciparum* accounted for 84% of the cases, followed by *P. vivax* (8%), *P. ovale* (5%) and *P. malariae* (2%). Deaths corresponded to an annual case fatality rate of 0.3% in 1999 and 0.5% in 2000. In general, imported malaria cases reflect the number of Italian travellers who underestimate the risk of becoming infected on visits to endemic countries and permanent residents of African origin who visit their relatives in their native countries (Romi *et al.*, 2001). In the summer of 2000, 22 imported malaria cases, 21 caused by *P. falciparum*, were observed among illegal Chinese immigrants in northern Italy. The rate of severe disease was high because the patients were not immune and they sought health-care services late in their illness because of their clandestine status. Recognition of the outbreak was delayed because no regional alert system among infectious diseases hospitals was in place (Matteelli *et al.*, 2001).

Netherlands

The number of cases of malaria imported into the Netherlands has been increasing, especially among migrants (Makdoembaks & Kager, 2000). The actual number of cases appears to be seriously under-reported; Van Hest *et al.* (2002) found that the number of cases of malaria diagnosed by laboratories in the Netherlands was much greater than the number of cases notified by physicians and estimated that as many as one third of the cases of imported malaria may go unreported. Reep Van den Bergh *et al.* (1996) concluded that the degree of under-notification of malaria in the Netherlands was at least 59% in the period January 1988 to June 1993 inclusive and has increased by 10% each year; by this estimation at least 3170 travellers returned to the country infected with malaria in the period. Most cases seen were *P. falciparum*. Wetsteyn *et al.* (1997) noted that the degree of compliance with proper chemoprophylaxis by Dutch travellers was actually decreasing, and among migrants returning to Africa or Asia for visits compliance was very poor. The lack of chemoprophylaxis compliance among migrants and hence the number of imported malaria cases among migrants coming to or returning to the Netherlands is changing; in the Netherlands, for the period 1979–88, only 15% of cases were persons originating in a malaria-endemic area, whereas for the period 1991–94, about 40% of cases were in persons from malaria-endemic areas and 8% of cases were in children born to settled immigrants resident in the Netherlands (Schlagenhauf & Loutan, 2003).

Portugal

In 1985–1995, Portugal recorded 964 cases of imported malaria (Muentener *et al.*, 1999); in a survey of 205 cases diagnosed as malaria in the Infectious Diseases Hospital in Lisbon during 1989–1995, 47% of the cases originated from Angola and 95% were due to *P. falciparum*; 19% of the patients developed severe complications and 6 patients died (Proenca *et al.* 1997). The number of imported cases has not exceeded 100 since 1991. However, Hanscheid *et al.* (2003), concerned by the possible misdiagnosis of malaria in the country, checked a cohort of patients at a teaching hospital in Lisbon by blind screening of all samples with <100 000 thrombocytes/μL by thick-blood film microscopy and found 5 unsuspected cases of malaria including 3 of *P. falciparum*. The authors felt that the problem of clinically unsuspected malaria seems to be more common than generally expected.

Romania

In the decade 1985–1995, only 146 cases of imported malaria were reported in Romania; however, in 1999, 32 cases of imported malaria were recorded, most of them *P. falciparum* imported from Africa. The remainder of the cases were *P. vivax* originating in Turkey. The vectors *An. atroparvus* and *An. sacharovi* have disappeared from the previously endemic areas of malaria in the Danube Plain and Dobrudja; this is thought to be due to ecological changes including water management and changed agricultural practices as the increase in mechanized farming has led to reduced numbers of draught animals in the region. Bilbie *et al.* (1978) considered that the development of a new malaria epidemic in this area was now unlikely.

Russia

The eradication of malaria in what was then the USSR was certified by the WHO in 1961. The problem of imported cases of malaria began soon after; Dashkova *et al.* (1978) recorded that among 294 cases of malaria introduced into Moscow, in 1974–1976 from 45 countries, *P. falciparum* was the most common (60.8%), mainly from West Africa, and mostly by residents of the countries of that region. A high percentage (72.9%) of carriers of malaria was found among persons visiting the USSR for the first time. The increase in imported cases of malaria was also noted in Lenningrad (St. Petersburg) among students coming to study, among African students returning from holidays in their home countries and among Soviet citizens returning from missions to malaria-endemic countries (Shuvalova & Antonov, 1982). A more serious source of imported cases of malaria was the Soviet military action in Afghanistan. Between 1981 and 1989, a total of 7683 cases of *P. vivax* malaria were imported into the USSR from Afghanistan, mainly by demobilized military personnel. For 23.8% of these cases the clinical manifestations appeared within a month of returning to the USSR, for 22.5% after 1–3 months, for 20% after 4–6 months, for 2% after >1 year, and for 0.6% after >2 years. In 13 patients the clinical manifestations of malaria appeared 3 years after returning from Afghanistan (up to 38 months). Nearly 69% of the patients did not take malaria prophylaxis at all while in Afghanistan, and 19% took chloroquine irregularly. Only 12.5% of the patients received a full course of prophylactic treatment with primaquine before leaving Afghanistan (Sergiev *et al.*, 1993). From 1992 the numbers of imported and local cases of malaria have further increased in Russia and in the countries included in the former USSR. The increase is attributed to migrations, wars, insufficient supplies of antimalarial drugs and insecticides and a reduction in the numbers of specialists in malaria. *Plasmodium vivax* malaria is present in Dagestan, Azerbaijan, Turkmenistan, Uzbekistan and Tajikistan and accounted for most cases imported into Russia since 1993. *Plasmodium falciparum* malaria is present in parts of Tajikistan adjoining Afghanistan. It has been noted that the number of cases of *P. vivax* malaria with long incubation periods has increased. In 1997, 760 cases of imported malaria, 20 of which were local and 13 of which were secondary to imported cases were recorded in Russia (Baranova *et al.*, 1998).

Moscow itself has been experiencing a rising number of autochthonous cases of malaria. The number of registered malaria cases rose from 112 in 1997 to 214 in 2001. A total of 728 people have been infected between 1997 and 2001 in the area. The presence of seasonal workers from Azerbaijan and Tajikistan is a factor that has led to an increase in local transmission. Under suitable climatic conditions, vectors such as *An. messeae* and *An. maculipennis* mosquitoes will transmit malaria. The greatest number of cases to be acquired locally (94) was reported in 2001 in the Moscow and Moscow Region.

Spain

Over the 10-year period 1985–1995, Spain reported 1927 cases of imported malaria (Muentener *et al.*, 1999). Lopez Vélez *et al.* (1999) noted that the number of Spanish travellers visiting malaria endemic areas, and the number of immigrants from malarial countries arriving in Spain is continually increasing; of the imported cases, one third occurred in immigrants and two-thirds in nationals. Among the Spanish nationals, 44% did not follow any prophylaxis, 29% followed a correct prophylaxis, 27% were considered defaulters, and 39% took self-treatment without cure. In a hospital in Barcelona, 80% of the imported malaria cases seen were due to *P. falciparum*, and only 10% of the patients had taken a correct course of chemoprophylaxis (Bartolome Regue *et al.*, 2002). Most of the imported cases of malaria in Spain originate from sub-Saharan Africa and a large proportion of these among immigrants from Equatorial Guinea. Rotaeche Montalvo *et al.* (2001) reviewed the imported cases of malaria in Spain between 1996 and 1999; the annual number of imported cases is increasing and was 224, 291, 389 and 260 for the years 1996, 1997, 1998 and 1999, respectively. Most cases were recorded in the 21- to 40-year age group and almost twice as many males as females were infected. Tourists were responsible for most of the imported cases, closely followed by immigrants; religious pilgrims, temporary workers and sailors were also responsible for a significant number of cases. Most cases originated in Africa, especially in Equatorial Guinea.

The origin of a locally acquired case of *P. ovale* in central Spain (Cuadros *et al.*, 2002) remains unresolved but emphasizes the possibility of local transmission of malaria occurring with its origin as an imported case.

United Kingdom

There are a large number of yearly cases of imported malaria in the UK. In the decade 1985–1995, 21 919 cases of malaria were imported into the UK (Muentener *et al.*, 1999). In 1999 a total of 2045 cases of malaria were imported and there were 14 deaths, and in 2000 there were 2069 cases. Most of the cases originate in Africa.

Despite the large number of malaria cases imported into the UK, there is evidence that surveillance remains unsatisfactory. Therefore, an active surveillance system was established and data on malaria cases diagnosed between 1 January and 31 December 2000 were gathered from local laboratories, the Malaria Reference Laboratory and local neighbouring health authorities. In total, 320 cases were identified in local residents (42.33 per 100 000). Of these 320, 293 were laboratory confirmed (38.75 per 100 000) and there were 47 notifications on clinical suspicion. Only 6.8% (20) laboratory-confirmed cases were formally notified. Males of African descent aged 25–39 years who travelled to West Africa were most affected, and 92.5% of the cases were of *P. falciparum* infection. The surveillance programme confirmed that formal malaria notifications in the UK are unreliable (Cleary *et al.*, 2003).

Children travelling to malaria endemic countries are at particular risk of contracting the disease. Ladhani *et al.* (2003) carried out a retrospective study of all children with malaria admitted to two East London hospitals between 1996 and 2001 using notifications and hospital discharge data. A total of 211 children, with a median age of 9 years (range, 11 to 179 months) were identified. Children living in the UK who acquired malaria while visiting a malaria-endemic country on holiday accounted for 82% of cases, whereas the rest were children visiting the UK from endemic areas. Three-quarters of children who had travelled to a malaria-endemic area were born in the UK, and 93% were of Black African ethnicity. *P. falciparum* acquired in Africa accounted for 91% of cases. Although 42% of children took antimalarial prophylaxis, only 15% of medications were according to recommended guidelines. Another family member, most often a sibling, was found to have concurrent malaria in 23% (49 of 211) of cases. Fortunately, most children had a low level parasitaemia and uncomplicated malaria, and they responded rapidly to antimalarial treatment.

Childhood malaria also constitutes a problem in other areas of the city of London; Williams *et al.* (2002) examined the changes in the presenting number and species of imported malaria in children in southwest London. A study over a period of 25 years (1975–1999) of all cases of paediatric malaria seen at St George's Hospital was carried out. A confirmed diagnosis was made in 249 children (56% boys; 44% girls; median age 8.0 years). Of these, 53% were UK residents and 44% were children travelling to the UK. A significant increase was noted in the number of cases over the 25 years (1975–1979: mean 4.8 cases/year; 1990–1999: mean 13.7 cases/year). Over the 25 years *P. falciparum* was seen in 77%, *P. vivax* in 14%, *P. ovale* in 6%, and *P. malariae* in 3% of cases. *Plasmodium falciparum* had increased in frequency (1975–1979: *P. falciparum* 50%, *P. vivax* 50%; 1990–1999: *P. falciparum* 82%, *P. vivax* 6%), associated with an increase in the proportion of children acquiring their infection in sub-Saharan Africa. Over the 25-year period there has been no improvement in chemoprophylaxis use or time to diagnosis.

The Gambia is a popular winter holiday destination for UK and other European travellers; in 1999, The Gambia recorded 40 588 arrivals of UK nationals at its borders, 25 393 German nationals, and 9625 Dutch nationals. Between 1997–2002, The Gambia was the source of 385 cases of malaria in the UK including 8 deaths, a case fatality rate of 2%. This represents around 4% of all imported cases of *P. falciparum* malaria and over two and a half times the overall case fatality rate of *P. falciparum* malaria reported in the UK. The Gambia, a very small country, is therefore the source of almost 12% of all deaths from malaria that occur in the UK. Most visitors to The Gambia are non-immune holidaymakers, often unaware of the potential severity of malaria. Without use of appropriate chemoprophylaxis, non-immune travellers to West Africa are at a high risk of contracting malaria (Bradley & Lawrence, 2003).

Moore *et al.* (2004) conducted a case-control study to investigate risk factors for malaria among travellers to The Gambia; the participants had visited The Gambia between 1 September and 31 December 2000, travelling with the largest UK tour operator serving this destination. Forty-six cases and 557 controls were studied. Eighty-seven per cent of all participants reported antimalarial use. The strongest risk factors for disease were: early calendar period of visit, longer duration of stay, non-use of antimalarial prophylaxis, non-use of mefloquine, lack of room air-conditioning, less use of insect repellent, prior visit to another malarial area and accommodation in a certain hotel in the country. These findings are probably applicable to persons visiting most of the highly malaria-endemic countries in Africa.

The question has been frequently raised as to whether the importation of large numbers of cases of malaria into the UK carries the risk that local transmission might again be established. Kuhn *et al.* (2003) analysed temporal trends in malaria in Britain between 1840 and 1910, to assess the potential for re-emergence of the disease. Their results demonstrated that at least 20% of the reduction in malaria was due to increasing cattle population and decreasing acreage of marsh wetlands. Although both rainfall and average temperature were associated with year-to-year variability in death rates, there was no evidence for any association with the long-term malaria trend. Model simulations for future scenarios in Britain suggest that the change in temperature projected to occur by 2050 is likely to cause a proportional increase in local malaria transmission of 8–14%. The current risk is negligible, as the >52 000 imported cases since 1953 have not led to any secondary cases. The projected increase in proportional risk is clearly insufficient to lead to the re-establishment of endemicity.

Behrens & Roberts (1994) made an economic appraisal of the benefits of chemoprophylaxis against malaria; the high incidence of imported malaria (0.70%) and low costs of providing chemoprophylaxis gave a cost–benefit ratio of 0.19 for chloroquine and proguanil and 0.57 for a regimen containing mefloquine; they concluded that costs of treating malaria greatly exceed costs of chemoprophylaxis, and therefore that prevention is highly cost effective. However, a survey of 172 general practitioners practicing in the West Yorkshire area of the UK who have travelled to South Asia showed that of the 145 (84%) responding to the survey, 50 (35%) took no chemoprophylaxis, 28 (19%) did not complete the chemoprophylaxis course, and 67 (46%) were fully compliant (Banerjee & Stanley, 2001).

The low compliance with recommended malaria prophylaxis is due in great part to the fact that many travellers simply do not seek advice. The European Travel Health Advisory Board conducted a cross-sectional pilot survey to evaluate current travel health knowledge, attitudes and practice (KAP) and to determine where travellers going to developing countries obtain travel health information, what information they receive and what preventive travel health measures they employ. A total of 609 responses were collected at the departure gates of three international airports: London Heathrow, Paris Charles de Gaulle and Munich. The study showed that more than one-third of travellers questioned had not sought pretravel health advice and of those who did, over 20% sought advice 14 days or less prior to travel. One-third of the respondents were aged 50 or more, and 20% had planned their trip less than 2 weeks before leaving. Only a minority had taken chemoprophylatic drugs as per the World Health Organization or national recommendations. Respondents often misperceived both the risk of malaria at the destination and recommended preventive measures (Van Herck et al., 2003).

Imported malaria in other countries of Europe

Obviously, the number of imported cases of malaria in other countries of Europe is a function of the number of travellers from those countries visiting malaria-endemic countries and the degree of their compliance with effective prophylactic measures. The number of tourists and persons travelling for business is steadily rising in the countries of Eastern Europe; the degree of compliance is similar to that of the countries described above and will surely result in a steady increase in imported cases.

More effort should be devoted to making travellers aware of the necessity of complying with a correct regime of chemoprophylaxis when travelling to countries where they will be exposed to malaria; in view of the growing problem of P. falciparum drug resistance, physicians and health clinics should have information on the most appropriate antimalarial drugs for chemoprophylaxis and treatment. A large proportion of the cases of malaria introduced by travellers will be resistant to one or, frequently, several, of the antimalarial drugs, thus presenting a therapeutic challenge to health-care professionals. Virtually every review of imported malaria in Europe emphasizes the low percentage compliance with chemoprophylaxis among travellers to malaria-endemic countries and, in particular among those persons who fell ill with malaria on return from travel.

Airport malaria

Local transmission has frequently occurred in Europe in the form of airport malaria; this is the transmission of malaria as a result of the inadvertent transport of live, malaria-infected mosquitoes aboard aircraft arriving from tropical malaria-endemic countries. Some 90 cases of airport malaria have been recorded, most of them in Europe; most of the cases in Europe have occurred in countries which have the most air transport connections with Africa, i.e. Belgium, France, Switzerland and the UK (Gratz *et al.*, 2000). In the summer of 1995 six cases of airport malaria occurred at the international airport of Brussels, Belgium. Of the 6 patients 3 were airport employees and 3 were occasional visitors. One patient died and the diagnosis was made by PCR amplification and DNA sequencing after exhumation. Two different species of *Plasmodium* were detected, and infections occurred on at least two different floors of the airport. An inquiry revealed that the cabins of aeroplanes were correctly sprayed, according to the WHO recommendations, but that the inside of the hand luggage bins, the cargo hold, the animal compartment, the wheel bays and the containers remain possible shelters for infected mosquitoes (van den Ende *et al.*, 1998). In the Netherlands, *P. vivax* malaria has recently been reported in a woman who had never travelled to an endemic area but lives near Schiphol airport, Amsterdam (Thang *et al.*, 2002). Airport malaria poses a particular health risk as it can occur among people living near international airports but who have had no travel exposure to malaria; consequently diagnosis is often delayed which may result in prolonged illness and even death. Many cases of airport malaria occur in countries where malaria transmission has long since been eradicated and may go undiagnosed and unreported.

The transport of mosquitoes, and other vectors, aboard aircraft has also resulted in the introduction of vector species into countries in which they have not previously been found, though this has not occurred in Europe. The number of mosquitoes which may be imported aboard aircraft can be substantial. Karch *et al.* (2001) conducted a survey between August and September 2000 of aircraft arriving at Paris Roissy airport, France. Cabin and cargo areas of 42 aircraft arriving from tropical Africa were examined upon arrival. Several live mosquitoes were found: 2 *Anopheles gambiae*, one in the passenger cabin, engorged with fresh blood and one in the walkway, non-bloodfed; 3 *Culex quinquefasciatus*, 1 in the passenger cabin, 1 in the cargo hold and 1 in the walkway. In Africa for every 100 female *An. gambiae* an average of 3 are infected and able to transmit malaria, which might also apply on aircraft on arrival at an airport. There has been opposition expressed to the spraying of aircraft in the belief that the pyrethroid insecticides recommended by the WHO may constitute a health hazard; this was considered by Aitio (2000) who concluded: 'At present, there is no information that would contradict WHO's advice that aircraft disinfection is necessary when there is a risk of vector spread and that permethrin and phenothrin, when used for this purpose in accordance with WHO recommendations are safe.

Malaria may also be introduced as a result of mosquitoes imported aboard vessels; *P. falciparum* malaria occurred during the summer of 1993 in two inhabitants living close to Marseille harbour. History of blood transfusion and travel outside France were excluded as was airport malaria. Entomological investigations confirmed the absence of *Anopheles* breeding sites in the port area. An hypothesis is that transmission occurred following the introduction of *Anopheles* on a ship coming from tropical Africa (Delmont *et al.*, 1994). Indigenous malaria disappeared from Belgium in 1938 but indigenous malaria was detected in a patient living 3.5 km from the port of Ghent. An infected *Anopheles* mosquito from a cargo load or baggage aboard an incoming vessel could have flown this distance (Peleman *et al.*, 2000).

Autochthonous malaria does not constitute a grave problem in Europe though sporadic cases and even small outbreaks continue to occur. It is likely that normal medical surveillance and reporting would detect any sizeable outbreak relatively quickly ensuring that appropriate chemotherapy and vector control measures were undertaken. However, imported malaria is a serious and growing problem in Europe, often under-reported and incorrectly diagnosed. Such cases can be the source of local outbreaks. The surveillance of malaria should remain a priority.

6

Mosquito-borne filarial infections

Dirofilaria immitis and *D. repens*, the agents of human pulmonary and subcutaneous dirofilariasis respectively, may coexist in areas of southern Europe. They are the only filarial parasites in Europe occasionally transmitted to man. The usual hosts of these nematodes are domestic and wild carnivores. Dogs are the most common host though cats, foxes (*Vulpes vulpes*) and wolves among other carnivores may also serve as hosts. In their canine and feline hosts these parasites are commonly known as canine heartworm. The vectors include mosquito species in the genera *Aedes, Anopheles* and *Culex*.

Canine subcutaneous and cardiopulmonary dirofilarioses are widely extended through the southern countries of the continent and are less frequent or completely absent toward the north. Italy, Spain, France, Portugal and Greece are the countries where epidemiological studies have demonstrated the presence of both *D. immitis* and *D. repens* in domestic (dogs and cats) and wild carnivores (mainly foxes). Italy is the European country where *Dirofilaria* species are most frequently found. The northern limit in which these two species have been reported, is the area of Cherbourg, France (Doby *et al.*, 1986). There is clear evidence that *Dirofilaria* infections are spreading in animal populations (Rossi *et al.*, 1996). Mosquito density and large numbers of microfilaraemic dogs are the most important risk factors for the transmission of the infection to humans. *Dirofilaria repens* infection is the most frequent and widespread dirofilariasis species.

Human infection manifests with either subcutaneous nodules or lung parenchymal disease that may be asymptomatic. The significance of infection in humans is that pulmonary and some subcutaneous lesions are commonly labelled as malignant tumors requiring invasive investigation and surgery before a correct diagnosis is made. The pathology of the condition is associated with aberrant localization of immature worms that do not reach maturity and if they mate microfilariae are rarely formed and so microfilariae are almost always absent in the blood. The most common reported manifestation of human diro-filariasis is subcutaneous nodular disease caused by *Dirofilaria repens*, with more than 400 case reports in the medical literature. There are endemic foci for *D. repens* in southern and eastern Europe, Asia Minor, central Asia and Sri Lanka. *Dirofilaria immitis* infection is usually associated with pulmonary lesions or coin lesions of the lung.

Italy has the highest prevalence of human dirofilariasis in Europe (66% of the total cases), followed by France (22%), Greece (8%) and Spain (4%). Cases of canine and human disease have been described in northern European coun-tries, but these have all been the results of patients having been exposed during a southern European visit. The prevalence of human dirofilariasis due to *D. repens* in Italy is relatively high especially in the area of Piedmont comprising the provinces of Alessandria, Asti, Novara and Vercelli, which is one of the most severely affected areas of the world (Pampiglione *et al.*, 2001). In the Piedmont region, four mosquito species (*Aedes caspius*, *Anopheles maculipennis* s.l., *Culex modestus* and *Cx. pipiens*) play a role in the transmission of canine filariasis (Pollono *et al.*, 1998).

Human dirofilariasis due to *D. immitis* is less common in Europe than *D. repens*-induced disease. Dirofilariasis is endemic in canine populations in several countries from which human cases have also been reported including Hungary, Portugal, Serbia and southern Switzerland (Bucklar *et al.*, 1998). The true preva-lence of human exposure and disease with *D. immitis* is probably underestimated as canine infection is widespread.

A seroprevalence study of humans in Spain, where 33% of dogs are infected with *D. immitis*, revealed that 22% of the human population had antibodies to the parasite. IgG seroconversion was most prevalent in people older than 60 while IgM seropositivity was most commonly observed in those younger than 19 years. The authors concluded that human contact with *D. immitis* in this endemic population begins at an early age (Nissen & Walker, 2002).

In the Ticino region of southern Switzerland dirofilariasis is common among dogs; of blood samples taken from 308 local dogs, 10.7% were circulating microfi-lariae for *D. immitis* and 5.5% for *D. repens*. Infective stages of *D. immitis* were found in local strains of *Aedes geniculatus* and *Culex pipiens*, following engorgement on a microfilaraemic dog; *Ae. vexans* was the most abundant mosquito species in the area (Petruschke *et al.*, 2001).

The public health importance of dirofilariasis

No fatality directly due to dirofilariasis has been recorded in the medical literature. Infection is symptomatic in 38–45% of cases. The primary significance of infection in adults is the confusion and invariable radiological misdiagnosis of a primary or metastatic lung tumour, which usually leads to thoracotomy with open lung biopsy or resection of the lung to obtain the correct diagnosis. Infections can remain unidentified due to poor knowledge of the parasite on the part of physicians. Considering the high rates of seropositivity found in serological surveys for the parasites the number of human cases is probably considerably underestimated. Contact of humans with these parasites is more frequent than shown by the number of clinical cases published, and, as most individuals do not develop symptoms, the infections are not diagnosed.

7

Sandfly-borne diseases

Sandfly-borne diseases -- viruses

The sandfly-transmitted viruses are all in the Bunyavirus group, the Phleboviruses; some 45 viruses are associated with sandflies globally. Some Phleboviruses are transmitted by mosquitoes, e.g. Rift Valley fever, and some by ticks, such as Crimean-Congo haemorrhagic fever. The sandfly-transmitted fevers identified in Europe include Arbia virus, Corfou virus, Naples virus, Radi virus, Sicilian virus and Toscana virus. Arbia virus has been isolated from the sandflies *Phlebotomus perniciosus* and *P. perfiliewi* in Italy, and Corfou virus from *P. major* on Corfu Island, Greece. Neither of these viruses appear to be of public health importance.

The three sandfly-transmitted Phleboviruses in the Mediterranean region causing human disease are Sicilian virus, Naples virus and Toscana virus. Sandfly fever is an acute, self-limited flu-like illness of 2–5 days duration. Onset is sudden with a fever, severe frontal headache, low back pain, myalgia, marked conjunctival secretion, malaise and occasionally nausea. Recovery is complete. No deaths from sandfly fever have been reported although weakness and depression may persist for a week or more after acute illness (Tesh, 1989). The public health importance of sandfly fever lies in the large numbers of non-immune persons, such as tourists or military, who may fall ill with the infection after entering an endemic area.

Sandfly fever in Europe

Naples and Sicilian viruses are largely responsible for the sandfly-borne diseases widely known as 'pappataci fever' or sandfly fever. Both are common throughout southern Europe, the Balkans, the eastern Mediterranean including Cyprus, along the Black Sea coast and extending eastwards. The two viruses frequently overlap. In Europe they are found mainly in areas where the sandfly *Phlebotomus papatasi* occurs (Tesh *et al.*, 1976). After the end of World War II and with the widespread application of DDT residual house-spraying for control of malaria vectors in much of southern Europe, Sicilian and Naples viruses virtually disappeared (Nicoletti *et al.*, 1996). However, with the cessation of house-spraying, transmission resumed and the incidence of these two viruses is now high in most endemic areas; illness from these viruses is a significant public health problem although clinical data are sparse. In a survey of the Adriatic littoral of Croatia, 23% of the persons surveyed were positive for the Naples virus (Borcic & Punda, 1987). Surveys in Greece in 1981–1988 showed antibodies to Naples virus in 16.7% and to Sicilian virus in 2% of the population. In Cyprus high antibody prevalence rates of 57% and 32% to Naples and Sicilian viruses, respectively, show that the sandfly fevers pose a significant public health problem in the country (Eitrem *et al.*, 1991a). The importation of sandfly fevers by tourists and soldiers returning from endemic areas is a growing problem. Many Swedish tourists returning from Cyprus and Spain were antibody positive in 1987; serum from a group of 95 tourists indicated that only 20% of the real number of sandfly fever infections had been correctly diagnosed by physicians (Eitrem *et al.*, 1991b).

Toscana virus was isolated from *P. perniciosus* in Tuscany in Italy in 1971 (Nicoletti *et al.*, 1996). It has been associated with acute neurological disease; clinical cases of aseptic meningitis or meningoencephalitis caused by Toscana virus are observed annually in Central Italy during the summer. Acute meningitis is perhaps the most frequent central nervous system infection in Tuscany. Valassina *et al.* (2000) found that a high percentage of such cases were due to infections with Toscana virus and concluded that Toscana virus is the most common viral agent involved in acute infections of CNS in children in central Italy. Braito *et al.* (1998) determined that the virus is responsible for at least 80% of acute viral infections of the CNS in children in Siena during the summer.

Toscana virus is widely distributed in Italy; a serological survey was carried out on 360 serum samples from a high-risk population occupationally exposed to Toscana virus in two regions of Italy, Tuscany and Piedmont; the results indicated a high seroprevalence of Toscana virus of 77.2% in the forestry workers, particularly in the Tuscany region. The distribution and prevalence of the virus correlated with the ecological niches specific for the survival of the sandfly vector (Valassina *et al.*, 2003b). Francisci *et al.* (2003) reported the circulation of Toscana virus in the Umbria region of Italy for the first time; from 1989–2001, specific antibodies were found in 36.6% of cases of aseptic meningitis, in 6.06% of meningoencephalitis and in 16% of healthy subjects. Because of the serious illness that may result from infections with Toscana virus, it is an important public health problem. Cases have been detected among tourists returning home from holiday in Italy (and other endemic areas).

Toscana virus has also been reported from France, Portugal, Cyprus and Spain. In Cyprus, 20% of the antibodies collected from 479 healthy people were positive for Toscana virus (Eitem et al., 1991a). The frequency of neuropathic infection due to Toscana virus infection increases in the summer months, peaking in August in the endemic Mediterranean areas of Italy, Portugal, Spain and Cyprus (Valassina et al., 2003b). In Spain, Mendoza-Montero et al. (1998) studied 1181 adults and 87 children from different regions of Spain and found high prevalence rates of 26.2% compared to 2.2% for Sicilian virus and 11.9% for Naples virus. Echevarria et al. (2003) studied a collection of 88 sera and 53 cerebrospinal fluid samples from 81 patients with acute aseptic meningitis of unknown aetiology, residing in Madrid or the southern Mediterranean coast of Spain. Anti-TOSV (Toscana) IgG was also investigated in 457 serum samples from healthy individuals from the south of Madrid. Specific IgM in serum and anti-TOSV IgG were detected in 7 patients, 3 were residents of the Mediterranean region and 4 came from the Region of Madrid. The overall prevalence of antibodies among the healthy population studied was 5%. These results confirm the role of Toscana virus as an agent causing illness on the Spanish Mediterranean coast and extend the findings of Toscana virus to the central region of the country; they suggest that the virus might be producing infection and neurological disease in every area of Spain having dense populations of the sandfly vectors.

Sandfly-borne diseases – Leishmaniasis

Leishmaniasis is a protozoan disease whose clinical manifestations are dependent both on the infecting species of Leishmania and the immune response of the host. Transmission of the disease is by the bite of a sandfly infected with Leishmania parasites. Infection may be restricted to the skin in cutaneous leishmaniasis (CL), limited to the mucous membranes in mucosal leishmaniasis, or spread internally in visceral leishmaniasis (VL), also called kala-azar.

Leishmaniasis is found in more than 100 countries, mainly in tropical and subtropical areas but is also widespread in southern Europe. The overall prevalence of leishmaniasis is 12 million cases worldwide, and the WHO estimates that the global yearly incidence approaches 2 million new cases (WHO Fact Sheet no. 116, 2000). In Mediterranean Europe, VL was traditionally a childhood disease whereas today the disease strikes immunocompromised patients. In these atypical forms of VL, diagnosis and treatment are particularly difficult and the infection presents a serious problem. Surveillance and accurate diagnosis of VL are of the utmost importance as classic VL is usually fatal if untreated.

In Europe, VL is found in Albania, Bosnia, Croatia, Cyprus, France (southern regions: Nice, Marseille, Montpellier, Toulon, Avignon, Alpes-Maritimes), Greece, Hungary, Italy, Macedonia, Malta, Montenegro, Portugal, Romania, Spain, Serbia and Turkey. *Leishmania infantum*, (*Leishmania donovani infantum* of some authors) is the causative organism of VL, or kala-azar, in Europe and in most of the Mediterranean basin. The main reservoir host is the dog and, to a lesser extent, foxes (*Vulpes vulpes*) and rodents. The distribution of *L. tropica*, the agent of urban, or anthroponotic, cutaneous leishmaniasis is limited in the Mediterranean region, being restricted to the Middle East, Tunisia and Greece (Gradoni *et al.*, 1994). The global distribution of leishmaniasis is shown in Figure 7.1.

Distribution and incidence of visceral leishmaniasis

The following section reviews the distribution and reported incidence of VL in the populations of the most affected countries of Europe. It must be emphasized that reporting of the infections is not obligatory in most countries and data are therefore taken almost entirely from the literature rather than national health services reports.

Albania

Leishmaniasis is an important public health problem in Albania. From 1984–1996 a total of 1230 cases were reported (1136 were of VL and 94 were of CL). Following the cessation of indoor insecticide applications for malaria control, the number of observed cases of VL has increased each year, probably due to the increase of sandfly and dog populations along with poor hygienic and sanitary conditions, especially in the outskirts of cities. Leishmaniasis constitutes a widespread endemic infection, affecting 32 of 36 districts in the country. About 3 cases/100 000 inhabitants are identified each year, and infection has spread to new districts (only 23 districts were affected before 1992). The ratio between cutaneous and visceral leishmaniasis cases is 1:12, and between urban and rural areas 1:2. About 40% of cases are in the age group 0–4 years (Kero & Xinxo, 1998). It is estimated that there are some 200 new cases of VL per year.

Velo *et al.* (2003) carried out a retrospective analysis of recent VL cases in 2001. From January 1997 to December 2001, 867 VL cases were recorded in 35 of 36

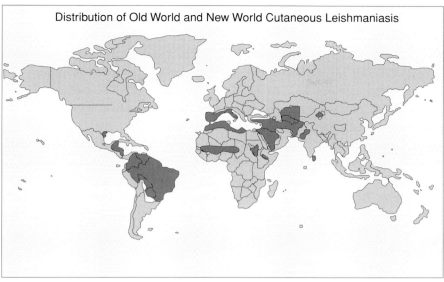

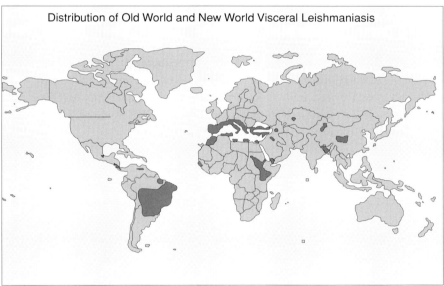

Figure 7.1 The global distribution of cutaneous and visceral leishmaniasis. Public Health mapping Group Communicable diseases (CDS). World Health Organization, October 2003.

Albanian districts, an average of 173 cases/year and a cumulative morbidity of 2.8/10 000 population. The incidence is increasing and has almost doubled during the past 10 years. A high proportion of patients (67.6%) were children below 5 years of age. The increase in VL morbidity in Albania is 20–40-fold higher than in other southern European countries.

France

The incidence of both human and canine leishmaniasis is increasing in France; infections are widespread throughout the south of the country and most cases occur on or near the Mediterranean coast. However, an autochthonous case of visceral leishmaniasis has been reported from Brest in an 18-month-old child who had never left the city (Castel *et al.*, 1982). The reservoir host is the dog and the prevalence of canine *Leishmania* infection is particularly high in the Provence and Cevennes regions of the south. All ages of humans may be affected by this disease. In the southeast of France between 1965 and 1975, 972 cases of canine leishmaniasis and from 1968 to 1975, 89 cases of human VL were reported. The sandfly vector, *Phlebotomus perniciosus*, is found in the suburbs of towns, the hills in the centre of Marseilles or in villages surrounding Toulon.

In serological surveys in the Marseilles region, the number of seropositive dogs detected rose from 240 to 2278 between 1976 and 1986. At that time there were from 40 to 50 cases of human leishmaniasis yearly in the region and the visceral form was more common than the cutaneous form; it was also noted that there were an increasing number of cases in immunocompromised adults rather than among children (Marty & Le Fichoux, 1988). Nevertheless, the incidence of infantile VL is also increasing in the Mediterranean region and both teenagers and young children may be affected by the disease. Although the number of cases of VL is relatively small, the occurrence of autochthonous cases persists; there were 22 cases of human VL in France in 1999, 30 in 2000 and 31 in 2002, mostly in the southeast.

Germany

Although Germany was thought to be free of autochthonous leishmaniasis transmission, there is now a concern about the possibility that leishmaniasis transmission is occurring in Germany. The new findings of *Phlebotomus mascittii*, an experimental vector, in Baden-Wurttemberg in the south of the country during 1999–2001 and *P. perniciosus* (a proven vector of *L. infantum*) in Rhineland-Palatinate in 2001, may explain the sporadic cases of autochthonous leishmaniasis in animals and humans and indicate transmission via a sandfly vector in Bavaria (Naucke & Schmitt, 2004).

Leishmaniasis has not been a notifiable disease in Germany. In September 2000, a national advice and reference centre was therefore created at the Institute of Tropical Medicine in Berlin to monitor the frequency, origin, and type of leishmaniasis seen in Germany, to provide advise physicians and to improve information for travellers to disease-endemic areas. Within 2 years, 70 cases of leishmaniasis were reported from Germany (43 cutaneous or mucocutaneous/mucosal; 27 visceral). For 58 case-patients (35 cutaneous or mucocutaneous/mucosal; 23 visceral), data were available on the age, sex, residence, travel destination, possible exposure location, reason for travel, duration of stay, duration and type of symptoms, concomitant diseases or therapies, type of diagnosis and treatment received. A total of 18 of the 23 VL patients were German tourists; three were immigrants from Angola, Iran and Togo; and two were visitors from Italy and Portugal. Six cases of VL occurred in children aged 2 to 11 years of age. Four German tourists and two immigrants had long-known HIV infection (median duration 3 years, range 8 months–6 years). Of the visceral cases, 78% were contracted in the European Mediterranean area and Portugal, and most of the infections were caused by a species of the *L. donovani* complex, probably *L. infantum*. Thirteen infections (22%) were acquired on the Mediterranean islands of Ibiza, Ischia, Majorca, Malta, Corfu or Sicily.

This distribution reflects the fact that the Mediterranean countries – Spain, Italy and the Mediterranean islands in particular – are the favourite vacation areas for Germans. Annually, Germans take 18 million vacations to the European Mediterranean area (including 8 million to Spain and 6 million to Italy) with a median duration of 2 weeks. Sixty per cent of travel to Italy and 90% of travel to Spain are to *Leishmania*-endemic areas. The maximum northern latitude for sandfly survival and hence transmission of leishmaniasis may move further to the north, even beyond Germany, due to global warming. Should this occur, imported cases may serve as a source of infection for sandfly vectors. Dogs imported from disease-endemic areas, or dogs that contract the infection when accompanying their owners for vacation, are a potential source of infection (Harms *et al.*, 2003).

Greece

Before World War II, VL was common throughout Greece; after initiation of malaria control activities in 1946 by indoor application of residual insecticides, there was a significant decline in the incidence of VL. When malaria vector control spraying was stopped, sandfly densities remounted and VL incidence then increased. From 1962–1992, 1005 cases of human kala-azar from the Athens basin were reported to the Greek Ministry of Health. However, VL seems to be present in practically all areas of the country, both continental and insular.

There are three major foci of VL in Greece; Athens with 33.4% of the cases, Epirus with 20.6% and the Peloponnese with 14.8%. In the Athens region there are two zones of transmission; one a high zone occupying the foothills of the mountains surrounding or traversing the Athens basin and the other a low urban zone consisting of the plains. Ninety per cent of the cases occurred near quarries in the foothills bordering the city and hills within the metropolitan area. The presence of stray dog reservoir hosts of *L. infantum* and a susceptible human population provide favourable conditions for transmission (Tselentis *et al.*, 1994).

The role of dogs in the transmission of VL to man was recognized very early in Greece. From December 1910 to March 1911 more than 200 stray dogs were screened for *Leishmania* infection; 8.2% of the dogs from Athens, 7.5% from Piraeus and 1.7% from the provinces were found positive. Early studies carried out between 1912 and 1939 reported that 5–20% of all dogs were infected in Athens and other areas of Greece (in Sideris *et al.*, 1996). Tzamouranis *et al.* (1984) typed 3 human and 19 canine leishmanial stocks and showed them to be identical with *L. infantum* thus verifying that the parasites which cause human and canine visceral leishmaniasis in Greece are the same organism and that dogs are the reservoir hosts for human infection. A survey was carried out in 1986–1994 to investigate the prevalence of canine visceral leishmaniasis in apparently healthy dogs in Athens. Of 1638 dogs examined, 366 (22.4%) had anti-*L. infantum* antibodies at titres greater than, or equal to 1/200 which were considered positive; 53 (3.2%) had antibody titres of 1/100 and these were considered uncertain; and 1219 (74.4%) dogs were seronegative. The rate of asymptomatic, chronic infection with *L. infantum* in dogs in the greater Athens area is high (Sideris *et al.*, 1999).

Human leishmaniasis is a persistent and important public health problem in Greece and is likely to remain so as long as a large reservoir of asymptomatic infected dogs is present in the country.

Italy

There has been a marked increase in the incidence of canine leishmaniasis in Italy. Canine leishmaniasis due to *L. infantum* is a disease of great veterinary importance and a serious public health problem. In humans, *L. infantum* causes visceral and cutaneous leishmaniasis and the distribution of VL overlaps that of canine leishmaniasis. The dog is the only domestic reservoir host of the infection. Canine leishmaniasis is endemic, particularly in the central and southern regions of Italy, including islands. Until 1983, all regions of northern Italy except Liguria and some territories of Emilia Romagna were considered free of the infection. From the early 1990s new stable foci of canine leishmaniasis have appeared, most of them located within classical endemic areas including territories of Emilia Romagna, Tuscany, Umbria, Marche and Abruzzi regions. But the most relevant aspect, from an epidemiological point of view, has been the appearance of stable foci in northern Italy, in the Veneto and Piedmont regions. In these two foci, entomological surveys showed the presence of *P. perniciosus* and a second vector, *P. neglectus*, which may play a role in the diffusion of the infection. Autochthonous human VL cases also have occurred in these areas. There is a real risk that canine leishmaniasis could rapidly spread through northern latitudes of the country and beyond (Capelli *et al.*, 2004).

In the first half of the twentieth century, VL was a common infantile syndrome in coastal territories of the Campania region of Italy. After World War II, the incidence dropped to a few cases/year for 3 decades; in the late 1980s the disease re-emerged among both children and adults. By now the VL focus of the Naples-Caserta area is probably responsible for the highest number of infantile cases among any VL foci described in southern Europe. Considering the limited efforts being given to canine reservoir host control, rapid diagnosis and treatment of patients remain the first-line control measures aimed at reducing the health impact of the disease (di Martino *et al.*, 2004).

Malta

Visceral leishmaniasis in Malta was recognized at the beginning of the twentieth century. In 1946, when human diseases became compulsorily notifiable, the leishmaniasis figures were 1264 visceral cases, 36 cutaneous cases and 5 unspecified (Headington *et al.*, 2002).

During the period 1947–1962, a total of 901 cases of kala-azar were notified (Cachia & Fenech, 1964). Visceral leishmaniasis was the only form of the disease recorded in Malta until the early 1980s, when CL was recognized. Five cases of cutaneous infection were reported in 1997 and 23 cases of CL and 3 of VL in January–October 1998 but there may be considerable under-reporting of the disease. As dogs are the reservoir hosts of the infection, the incidence of canine leishmaniasis is an important indicator of the risk of leishmaniasis to the human population. In Malta, figures of between 18–47% have been reported for the prevalence of canine leishmaniasis. However, in a recent survey, as many as 62% of 60 dogs were seropositive for *L. donovani* by polymerase chain reaction (PCR) (Headington *et al.*, 2002).

A centre for the immunodiagnosis of leishmaniasis has been established on Malta and several hundred dog sera have been examined by immunofluorescence antibody (IFA). Isolation and isoenzyme characterization of *Leishmania* from man (VL and CL cases), dogs and sandflies have shown that leishmaniasis in Malta is due to viscerotropic and dermotropic strains of the same species, *L. infantum* (Gradoni & Gramiccia, 1990). In a further study on leishmaniasis and sandflies of the Maltese islands, 22 *Leishmania* stocks were isolated from human visceral and cutaneous cases, dogs and sandflies. The commonest Mediterranean *L. infantum* zymodeme, MON 1, was the cause of human and canine visceral leishmaniasis; *L. infantum* MON 78, which has so far been isolated only in Malta, was the agent of human CL. Both zymodemes were isolated from the same sandfly species, *P. perniciosus* (Gradoni *et al.*, 1991).

Although the incidence of CL has recently increased, the overall numbers of cases of human leishmaniasis have markedly decreased since the 1960s. Prior to 1963, almost all cases were among children aged <10 years. Although leishmaniasis in Malta has been primarily a disease of children, the proportion of adult cases is increasing (Fenech, 1997). The case records of all 81 paediatric patients with VL between 1980 and 1998 were analysed. The annual incidence of VL has declined for all cases of VL, and declined significantly for paediatric cases. From 1994–1998, the overall incidence of VL was 0.9 per 100 000 total population and the paediatric incidence was 2.5 per 100 000 population. The decreased incidence is attributed to the eradication of stray dogs (Grech *et al.*, 2000).

Portugal

During the period 1910–1950, 1616 cases of VL had been reported in Portugal; the year of highest incidence was 1950 when 448 cases of VL were reported. In the same year a malaria vector control programme was started with the residual application of insecticides to homes; as a result, by 1970 only 20 cases of VL were reported. After cessation of malaria vector control, there has been an increase with 20–30 cases a year of VL now being reported. Sporadic infection is present in the whole of the country though more than 80% of the cases occur in the north (Abranches *et al.*, 1993). Kala-azar has always been endemic in the sub-region southwest of Lisbon (45 cases from 1961–1978). In the Setubal peninsula the disease has always had a low incidence (10 cases from 1961–1978). The Alcacer do Sal sub-region was, in the past, the second most active endemic zone in the country but after 1950 there was a sudden fall in prevalence (only two cases reported from 1961–1978) (Abranches *et al.*, 1982). Autochthonous VL was first reported in the Algarve region in the 1980s.

Abranches *et al.* (1991) examined the prevalence of canine leishmaniasis in 1823 dogs from the Lisbon metropolitan region. The breeds most affected were Doberman and German shepherd. Young adults (12.2%) and older dogs (14.7%) had higher prevalences of infection; 53.8% of the dogs with significant antibody titres (greater than or equal to 1:128) showed no symptoms, suggesting that canine leishmaniasis has a prolonged asymptomatic period. There is also evidence that there is a wide spectrum of clinical diseases in dogs and that the prevalence of canine leishmaniasis infection is higher than was presumed (Cabral *et al.*, 1993). The area of the city of Evora appears to be a new focus for canine leishmaniasis in Portugal. In a serological survey to determine the prevalence of anti-*Leishmania* antibodies in the local human and canine populations residing in Evora town and 14 adjacent villages, blood samples collected from

885 children and 3614 dogs were tested. None of the children screened showed an antibody level indicative of VL; however, antibody rates ranging from 0.7% to 6.9% were obtained in dogs residing in Evora and 11 adjacent villages (Semiao Santos *et al.*, 1995). Cabral *et al.* (1998) studied 49 asymptomatic mixed breed dogs from a region of high incidence of VL in Portugal for the presence of *Leishmania*-specific cellular immunity. Of the 49 dogs, 20 had demonstrable parasite-specific cellular immunity. The study showed that the infection rate of canine leishmaniasis is higher than previously thought using serological tests alone. Visceral leishmaniasis has, in the past, primarily involved children: between 1993 and September 1997 there were 87 cases of VL, of which 57.5% were in children <15 years old; 36.8% of all cases were in the age group 1–4 years. In 1996 and 1997 there was an increase in the number of adult cases, mainly due to cases among persons with HIV immunosuppression. In co-infected VL/AIDS patients, the most affected risk-group was that of intravenous drug users. *Leishmania infantum* MON-I was the most common zymodeme, but other zymodemes have been isolated from co-infected cases. Three VL endemic areas are known, with canine leishmaniasis prevalence between 7.0 and 11.3%. The relatively high number of positive serological reactions with low titres (as found in the population of Alto Douro endemic region) suggests previous contact with the parasite or asymptomatic infections (Santos Gomes *et al.*, 1998).

Spain

Leishmaniasis is a compulsory notifiable disease in Spain and relatively complete data are available; nevertheless, there are indications that the actual incidence is considerably higher than that notified. Arnedo Pena *et al.* (1994) studied the sandfly vectors and canine reservoir hosts of VL in Castellon where the incidence of leishmaniasis is among the highest in Spain. They found that there was under-reporting by the hospitals in the region.

The infection is endemic over much Spain with foci of transmission near Madrid and in Catalonia, in the east as well as in provinces and regions of Alicante, Almeria, Aragon, Castellon, Granada, Guadalajara, Huelva, Huesca, Malaga, Murcia, Navarra, Salamanca and Zaragoza and on the Balearic Islands. Leishmaniasis was first discovered in Spain in 1912 on the Levante coast. It has been estimated that before 1985 approximately two-thirds of leishmaniasis cases were in patients less than 15 years of age. However, since then, almost 80% of cases have been in immunodepressed patients, 60% being HIV positive. With the increase in the number of cases of VL in Spain, patients with unusual cutaneous manifestations are also being seen.

The epidemiology of canine leishmaniasis was studied in the Madrid Autonomous Region by Amela *et al.* (1995); 591 dogs were screened, revealing a prevalence of 5.25% with no difference between rural and periurban areas. Age-specific prevalence exhibits a peak at 2–3 years and another at 7–8 years. There was a 70% greater risk of infection among pet than among working dogs. An epidemiological survey was carried out in the leishmaniasis endemic area of Axarquía region in southern Spain between February and May 1992. The prevalence of *Leishmania* was determined in 1258 individuals by the leishmanin skin test (LST) and in 344 randomly selected dogs by immunofluorescence. The LST showed that 533 (42.4%) of the individuals tested were positive. There was an increase in positivity with age, ranging from 21% at 4 years to 58% at 14 years. Seroprevalence in dogs was 34.6% (119 of 344) and longhaired breeds showed a significantly lower seroprevalence (20.9%) than shorthaired breeds (37.6%). A discrepancy in disease prevalence between two rural districts was thought to result from the design of the houses; houses in the area with greater prevalence (54.8%) were associated with stables which provided an ideal habitat for sandflies, whereas houses in the area with lower prevalence (25.1%) rarely had stables attached to them (Morillas *et al.*, 1996). In the region of the Alpujarras (southern Spain, in Granada province), nine villages were chosen at random in five bioclimatic zones and leishmaniasis in the canine and human populations was studied, concentrating especially upon schoolchildren. A total of 615 dogs were screened, which was almost 100% of the canine census. Of the screened dogs, 33 showed an antibody titre > or = 1/160 when tested by immunofluorescence assay (seroprevalence of 5.3%). Among the human population, infection from *Leishmania* was studied using the LST. Of 1286 people tested, 568 (44.16%) were positive. Most of the subjects were schoolchildren (878; practically 100% of the pupils), of whom 288 (32.8%) tested positive. A close relation was seen to exist between the percentage of positive LSTs and age. Finally, a close relationship was also observed between canine seroprevalence and percentage of schoolchildren who tested positive in the LST (Acedo Sanchez *et al.*, 1996). A high prevalence of canine leishmaniasis was also found in a survey over a period of 10 years (1985–1994) in the Priorat, a rural region in the northeast of Spain in Catalonia province. Seroprevalence throughout the region was 10.2% (8–12%). Forty per cent of the dogs had a low level of anti-*Leishmania* antibodies, whereas only 50% were seronegative. Only one-third of the seropositive dogs had evident symptoms of the disease. Annual incidence of the disease was 5.7% (Fisa *et al.*, 1999).

A high prevalence of leishmaniasis was also found among 100 dogs on the island of Mallorca. The prevalence of the infection, 67%, was calculated using all animals seropositive and/or positive by PCR with any tissue. The results showed that the majority of dogs living in an area where canine leishmaniasis is endemic are infected by *Leishmania* and that the prevalence of infection is much greater than the prevalence of overt *Leishmania*-related disease (Solano Gallego *et al.*, 2001).

An epidemiological study was carried out in 1997–1998 in the leishmaniasis-endemic Alicante region of southeast Spain to estimate the magnitude of and the factors related to subclinical *L. infantum* human infection. In the main focus of leishmaniasis in the region, 11.5% of the children and 52.8% of the adults reacted as positive to the skin test. Among the adults, the response was significantly greater for males and for those who had resided in the area for 15 years or more. In the region, 3.7% of 14-year-old students reacted to the skin test, with no gender differences. The main factors related to a positive skin test result were having a parent or sibling recovered from leishmaniasis and living in the rural periphery of the region as opposed to the metropolitan area. Results indicate a high frequency of subclinical leishmaniasis in the region. The authors postulated that the decline in childhood visceral leishmaniasis in southern Europe in the second half of the twentieth century is related to social changes, which gave rise to a less frequent exposure at a young age and lowered susceptibility to disease through nutritional and immune improvements (Moral *et al.*, 2002). Cascio & Colomba (2002) reviewed childhood Mediterranean visceral leishmaniasis in France, Italy and Spain and observed that in comparison with the past, when VL was observed more frequently in children, the current ratio of child to adult cases is approximately 1:1.

Martin Sanchez *et al.* (2004) found considerable polymorphism of *L. infantum* in southern Spain. Over a period of more than 10 years they isolated and classified 161 *Leishmania* strains from cases of human visceral, cutaneous and mucosal leishmaniasis in immunocompetent subjects, from cases of visceral leishmaniasis in immunocompromised individuals with HIV, from dogs with leishmaniasis (visceral and cutaneous), from *Rattus rattus* (black rat) and from sandflies. The strains were all *L. infantum*, the only species endemic in Spain, and corresponded to 20 different zymodemes.

Cutaneous leishmaniasis

Although *L. infantum* usually causes visceral disease, cutaneous lesions may be the only manifestation along the Mediterranean coast in Spain, France, Italy and Greece where dogs are the main reservoir host. A large number of cases of cutaneous leishmaniasis in the Old World go unreported or undiagnosed.

Cutaneous leishmaniasis has been reported from Albania, Austria, Bosnia Herzegovina, Bulgaria, Croatia, Cyprus, France, Greece, Italy, Malta, Monaco,

Portugal (with the Azores and Madeira), Romania, Spain (with the Canary Islands), Macedonia and Yugoslavia. The causative organisms are *Leishmania infantum, L. major* and *L. tropica*. Koehler *et al.* (2002) reported a case of CL due to *L. infantum* in a horse in southern Germany. Since neither the infected horse nor its dam had ever left their rural area, the presence of autochthonous CL infection in Germany cannot be excluded.

Cutaneous leishmaniasis is endemic in Greece. Cases collected between the years 1975 and 1979 were analysed. Prevalence is highest in the Ionian Islands and Crete. The disease most commonly affects individuals 10 to 20 years of age. Exposed parts of the body are most commonly involved, particularly the face. The highest incidence is mid-winter (Stratigos *et al.*, 1980).

Human CL has been known in Spain since 1914, when it was described in the Alpujarras region (Andalusia). The causative agent was characterized in 1986 for the first time by isoenzyme electrophoresis and identified as a zymodeme of *L. infantum*.

Leishmaniasis/HIV

In the last two decades, leishmaniasis, especially VL, has been recognized as an opportunistic disease in the immunocompromised, particularly in patients infected with HIV (Choi & Lerner, 2001). As the AIDS pandemic spreads to rural areas and human VL becomes more common in suburban areas, there is a growing degree of overlap between the geographical distributions of the two diseases and, in consequence, an increasing incidence of *Leishmania*/HIV co-infection. Cases of the co-infection have been reported from 35 countries around the world but most of the cases to the present, have been recorded in south-western Europe. A total of 1911 cases have been detected in Spain, France, Italy and Portugal (Desjeux & Alvar, 2003) with Spain reporting the greatest number of cases. In southern Europe 25–70% of adult VL cases are related to HIV and 1.5–9% of AIDS cases suffer from newly acquired or reactivated VL. Of the first 1700 cases of co-infection reported to the WHO up to 1998, 1440 cases were in southwestern Europe (Paredes *et al.*, 2003). The distribution of leishmaniasis/HIV co-infection is shown in Figure 7.2.

The incidence of *Leishmania*/HIV co-infection is expected to continue to increase in eastern Europe but to fall in south-western Europe as increasing numbers of HIV-positive patients in the latter region are given the new, highly active antiretroviral therapy (HAART). In 1996, HAART was started in France and resulted in a significant decrease in the incidence of leishmania/HIV co-infections (del Giudice *et al.*, 2002). A marked decrease in the annual incidence

Figure 7.2 Countries reporting leishmaniasis/HIV co-infections (dark shading).

of VL relapses among the patients attributable to HAART was also reported from Catania, Italy (Russo *et al.*, 2003). The appearance of relapses may differ in Spain; in Madrid, HAART appears to have decreased the annual incidence of VL among local AIDS cases, from 4.81 cases/100 to just 0.8 case/100 (p < 0.0005), a first episode of VL now appearing only when there is obvious HAART failure. Unfortunately, it does not seem to be very good at preventing VL relapses; within 24 months of antileishmanial treatment, 70% of patients who were receiving HAART had relapses. Visceral leishmaniasis also seems to hamper the immunological recovery of the HIV-positive, although HAART appears to have little effect on the clinical manifestations of VL (Lopez Vélez, 2003).

Leishmania/HIV co-infection is changing the epidemiology of VL. HIV infection modifies the traditional zoonotic/anthroponotic patterns of *Leishmania* transmission. The poor therapeutic outcome, the higher rate of relapses, and the polyparasitic nature of VL in HIV-infected persons, as well as the atypical manifestations of the disease that make diagnosis difficult and impaired access to health-care resources of co-infected patients, make HIV-infected individuals prone to enlarge the number of human reservoir hosts in areas where transmission of leishmaniasis is already anthroponotic. In addition, these characteristics help create a new focus of anthroponotic transmission in areas where the spread of leishmaniasis has traditionally been zoonotic (Paredes *et al.*, 2003). The levels of transmission of the parasites causing such leishmaniasis were previously dependent on the conventional zoonotic cycle, in which sandflies transmitted the parasites from infected canids to other canids or humans. The *Leishmania/*HIV co-infection, how-ever, has led not only to marked increases in the sandfly transmission of the parasites from immunodepressed individuals directly to other humans but also, probably, to artificial transmission between immunodepressed intravenous-drug users, as the result of needle sharing (Molina *et al.*, 2003). In fact, Cruz *et al.* (2003) looked for parasites in syringes discarded by intravenous drug users (IVDUs) using two different PCR techniques. *Leishmania* species were detected in 65 (52%) of 125 syringes collected in southern Madrid, Spain, in 1998, and in 52 (34%) of 154 collected in southwestern Madrid in 2000–01. They found shared restriction fragment length polymorphisms in 12 of 65 positive samples of syringes tested, suggesting that syringe sharing can indeed promote the spread of leishmania clones among IVDUs.

Between 1988 and 1998, 258 *Leishmania* strains from patients infected with HIV were characterized by iso-enzyme electrophoresis at the Istituto Superiore di Sanita (ISS) in Rome. Most (227) of the isolates came from 80 Italian patients with VL, the rest from cases of *Leishmania/*HIV co-infection in other Mediter-ranean countries. Every strain was found to be *L. infantum*. Other changes in VL epidemiology include the visceralization in HIV-positive individuals of variants considered to be dermotropic in the immunocompetent, and the appearance of new zymodemes among the HIV-positive human population (Chicharro *et al.*, 2003). Marty *et al.* (1994) reported that in the department of Alpes-Maritimes of France, cases of VL are increasing in adults and 40% of the adult cases are in HIV positive patients.

The number of HIV-positive patients infected by leishmaniasis remains unknown both globally and in Europe especially inasmuch as subclinical cases of VL appear to be common in AIDs patients. Pineda *et al.* (1998) studied this ques-tion in southern Spain and found that VL is very prevalent among HIV-1-infected patients and that a high proportion of cases were subclinical. Like other oppor-tunistic infections, subclinical VL can be found at any stage of HIV-1 infection, but symptomatic cases of infection appeared mainly when a deep immunosup-pression is present. The disease is associated with male sex and intravenous drug use.

The sandfly vectors of leishmaniasis in Europe

Despite the influence of man on the ecology of the leishmaniasis-endemic countries with the destruction of forests and expansion of crop and grazing land, studies show that phlebotomine sandfly populations continue to thrive even in areas under cultivation. In the Mediterranean area of southern

Europe, five species are known to transmit VL, *Phlebotomus ariasi, P. neglectus, P. perfiliewi, P. tobbi* and *P. perniciosus.* The sandfly vectors of cutaneous leishmaniasis are *Phlebotomus perfiliewi* in Italy, *P. ariasi* in the Cévennes of southern France and Spain, and *P. perniciosus* in France, Italy, Malta and Spain.

In the foci of southern France, *P. ariasi* is the vector of *Leishmania infantum.* The species is distributed throughout the western Mediterranean but its area of main vectorial importance appears to be restricted to southern France west of the Rhone valley. To the east it is limited to eastern France and the Piedmont and Liguria in Italy but in the west it is present in Portugal and Spain. While mainly exophilic, it occasionally enters houses and feeds on man and dogs. It has a relatively large flight range of up to 2200 m.

Phlebotomus perniciosus is found from Portugal to Turkey, particularly in coastal areas and on islands with a semi-arid Mediterranean climate. Populations of this species are associated with sporadic cases of VL and possibly CL (Martinez Ortega, 1984). It is endophilic and zoophilic. In France *P. perniciosus* was first found infected in the region of Nice in 1991 (Izri *et al.*, 1992). It is the confirmed vector in several areas of Italy including on Sardinia and Sicily. It is found naturally infected in Malta where it is the established vector of VL and represents a large proportion of the sandflies caught (Leger *et al.*, 1991). It is probably the vector of VL in the Algarve, Portugal and in much of Spain including the Balearic Islands; in the province of Salamanca it is the main vector of VL (Encinas Grandes *et al.*, 1988). Although the species is present in southern Switzerland where it has been found feeding on humans, no leishmaniasis has been transmitted in that country (Grimm *et al.*, 1993) though there is concern that this could take place in the warm, southern canton of Ticino.

Phlebotomus papatasi occurs throughout the Mediterranean region from Portugal in the west and eastwards to southern Russia and to India. It is endophilic and readily feeds on man but reaches high densities only in urban areas. It is mainly important as a vector of CL in the eastern Mediterranean, the near east and Russia.

Phlebotomus perfiliewi has been found in southeast France, in Monaco and four neighbouring French communes (Izri *et al.*, 1994) but reaches its highest densities in Italy and the countries of ex-Yugoslavia. It is a vector of both VL and CL. In Italy, population densities of *P. perfiliewi* may be very dense and it is the vector of CL. In 1948, DDT sprayings of houses in some villages of the Abruzzi region for malaria vector control brought about a dramatic decrease in *P. perfiliewi* densities followed by a drop in the incidence of CL. It feeds on dogs, horses, sheep and birds; it has been incriminated as a vector of VL in Italy (Maroli & Khoury, 1998) and is probably the vector of CL in the Adriatic area. It is considered the vector of CL in Sicily and was present in high densities in collections in Agrigento and

Palermo provinces; of 2410 specimens collected, 77.5–9% were *P. perfiliewi*, 12.78% were *P. perniciosus*, 0.74% *P. major* and 2.07% *Sergentomyia minuta* (a sandfly that feeds on reptiles not mammals) (Maroli *et al.*, 1988). Infection rates in *P. perfiliewi* are generally much lower than those in *P. perniciosus*.

A study in 1997 investigating how climate may be influencing the distribution of *Phlebotomus* and leishmaniasis in Europe, focused on Italy due to unexplained increases in VL cases in that country; the abundance of *P. perniciosus* and *P. perfiliewi* were different, *P. perfiliewi* being distributed in regions with relatively cold winter temperatures and high summer temperatures while *P. perniciosus* was found in regions with comparatively warm winters, but mild summers. Models predicted that *P. perfiliewi* would be positively affected by temperature increases, extending its distribution, while *P. perniciosus* would decrease in currently inhabited areas but spread further into Switzerland. Human disease models predict a dramatic increase in the incidence of VL and a slight increase in CL. Climate change would most likely also affect parasite development and the infection rate in dogs and thus overall disease incidence (Kuhn, 1999).

The species and ecology of sandflies present in the coastal district of Bar, in Montenegro, an endemic focus of VL, were investigated in 1996–1999. *P. perfiliewi* was recorded for the first time in Montenegro (Ivovic *et al.*, 2003). In southeast Serbia *P. perfiliewi* was the most abundant phlebotomine, making up some 80% of the total. Naples sandfly fever and Jug Bogdanovac viruses were also isolated from the species (Milutinovic *et al.*, 1989).

Phlebotomus sergenti is present in Albania, Cyprus, France, Italy, Portugal and Spain and is a vector of CL (*L. tropica* and *L. major*) in some of the areas where it is found in high densities.

Conclusions on the public health importance of leishmaniasis in Europe

To ensure better monitoring of leishmaniasis it would be desirable if the reporting of human or canine cases of leishmaniasis were made obligatory in all countries of Europe. Because of the recent increases in the incidence of VL, surveillance activities should be strengthened as a basis for the implementation of control measures. While HAART may result in a decline of leishmaniasis/HIV co-infections, VL may continue to increase as a result of the persistent sandfly vector populations present in endemic areas. In any event, human (and canine) leishmaniasis remain a serious problem in southern Europe and the infections may be spreading to more northern countries as has been reported in Germany.

Global warming as a result of climate change would likely result in further spread.

Relatively neglected in the past, more attention should be given to the control of the sandfly vectors; while control is generally difficult it can be achieved (Gradoni *et al.*, 1988), and the use of pyrethroid-impregnated dog collars has been shown to be an effective method of controlling sandflies in urban areas of Italy (Maroli *et al.*, 2002).

8

Ceratopogonidae – biting midge-borne diseases

Members of this family of biting midges can be severe nuisances; they are also the vectors of two important animal diseases in Europe, bluetongue virus and African horse sickness. *Culicoides* biting midges are among the most abundant of haematophagous insects, and occur throughout most of the inhabited world. Across this broad range they transmit a great number of assorted pathogens of humans and domestic and wild animals, but it is as vectors of arboviruses, and particularly arboviruses of domestic livestock, that they achieve their prime importance. To date, more than 50 such viruses have been isolated from *Culicoides* species and some of these cause important diseases, particularly in animals (Mellor *et al.*, 2000). As regards human infections in Europe, there has only been a single isolation of Tahyna virus from *Culicoides* species in Czechoslovakia (Halouzka *et al.*, 1991). The group has no apparent importance as vectors of human disease in Europe; however, *Culicoides* are vectors of human diseases such as Oropouche virus in Brazil and vesicular stomatitis in the Americas, and surveillance of the activity of this group is necessary in Europe.

Haemoglobins in the midge family Chironomidae are potent human allergens and have been identified as causative allergens in asthmatic patients. Eriksson *et al.* (1989) concluded that Chironomidae might be allergens of clinical importance in asthma and rhinitis, in that cross-allergy exists between chironomids and shrimp and that cross-allergy also might occur among chironomids, crustaceans and molluscs.

9

Dipteran-caused infections – myiasis

Myiasis is the infestation of live human and other vertebrate animals with dipterous larvae, which, at least for a certain period, feed on the host's dead or living tissue, liquid body substances or ingested food (Zumpt, 1965). The animal importance of myiasis is well known and well described in the veterinary literature; the frequency of human myiasis is greater than generally realized in Europe or North America. Inasmuch as infestations are self-limited, the presence of larvae may go unnoticed and unreported unless there is a secondary bacterial infection. The diagnosis of myiasis is easily overlooked in Europe because of its relative rarity.

Most myiasis-causing flies belong to one of three major families: Oestridae, Sarcophagidae or Calliphoridae, although representatives of other families, such as Muscidae and Phoridae, also are known to cause myiasis.

An aetiological classification of myiasis-causing flies separates them into three forms of myiasis: obligate, in which it is necessary for the maggots to feed on living tissues; facultative, where flies opportunistically take advantage of wounds or degenerative necrotic conditions as a site in which to incubate their larvae; and accidental, which occurs when egg-stage flies are ingested on contaminated food or come in contact with the genitourinary tract. In general obligate myiasis of humans is tropical in origin, whereas facultative and accidental myiasis can occur anywhere in the world. In the furuncular form, boil-like lesions develop gradually over a few days. Each lesion has a central punctum, which discharges serosanguinous fluid. The posterior end of the larva is usually visible in the punctum and its movements may be noticed by the patient. The lesions are often extremely painful. There is an inflammatory reaction around the lesions that may be accompanied by lymphangitis or lymphadenopathy. These symptoms rapidly resolve after the larva has been removed. The flies causing furuncular myiasis in humans are the human botfly, *Dermatobia hominis*, the rabbit or rodent botflies, *Cuterebra* spp., the tumbu fly, *Cordylobia anthropophaga*, Lund's fly, *Cordylobia rodhaini*, *Wohlfahrtia* spp. and the warble flies, *Hypoderma* spp.

Another common clinical form is a 'creeping eruption'. The larva lies ahead of the vesicle in apparently normal skin. This form of myiasis is produced by *Gasterophilus* larvae, the horse botflies.

Wound or traumatic myiasis is the infestation of animal and human wounds by dipterous larvae. It may be benign, as when secondary species confine their activities to diseased and dead tissue, or it may be malign, as when the obligate and primary species attack living tissue. The three major species of obligate parasites encountered in wound myiasis are the New World screwworm, *Cochliomyia hominivorax*, the Old World screwworm, *Chrysomya bezziana* and Wohlfahrt's wound myiasis fly, *Wohlfahrtia magnifica*. One must also distinguish between autochthonous myiasis acquired in Europe and introduced cases of myiasis which will be considered below.

The cause of autochthonous cases of myiasis reported in Europe have included, among others, the drone fly *Eristalis tenax* in Belgium; the lesser house fly, *Fannia canicularis* in Bosnia Herzegovina; the blowflies *Calliphora erythrocephala* and *C. vicina* in Denmark; the sheep botflies, *Hypoderma bovis* and *H. lineatum*, and the greenbottle fly, *Lucilia sericata* in France; *Fannia scalaris*, the latrine fly and *C. vicina*, in Germany; *Calliphora vomitoria*, the blue bottle fly, the sheep botfly *Oestrus ovis* and *Wohlfahrtia magnifica* in Italy; *C. vicina* and *O. ovis* in Spain. Unfortunately cases of myiasis may occur among the elderly or injured during hospitalization. Daniel *et al.* (1994) described a case in the Czech Republic in which 50 larvae of *Lucilia sericata* were discovered in the oral cavity, nose, paranasal sinuses and enucleated eye-socket. Timing indicated that the eggs were laid while the patient was hospitalized. The development of myiasis was facilitated by the mental and physical debility and dependency of the patient, numerous and deep facial necrotic wounds and a lengthy period of hot weather which led to prolonged open window ventilation of his room. When myiasis occurs in a patient after hospitalization the disease is termed nosocomial myiasis. In another case in the Czech Republic, Minar *et al.* (1995) described a case of nosocomial myiasis of cavities caused by larvae of the fly *L. sericata* in a patient hospitalized at the resuscitation unit after a motor car accident. On the fourth day in hospital, in the nasal and oral cavity of the patient 1st and 2nd instar larvae were detected in his wounds. The eggs of the fly had been laid in hospital. They noted that 7 other cases of human myiasis had been reported in the country up to that time and 3 of them had also been nosocomial myiasis. Mielke (1997) found records in the literature of 23 cases of myiasis as hospital infections, which have occurred in a number of countries.

Doby *et al.* (1985) collected information on all the human cases of human hypodermosis in France (myiasis due to *Hypoderma bovis* and *H. lineatum* – the common cattle grub) and found that there were 266 cases either reported in the literature, or as unpublished records.

Maggot debridement therapy (MDT) is the medical use of live maggots (fly larvae) for cleaning non-healing wounds.

In maggot debridement therapy (also known as maggot therapy, larva therapy, larval therapy, biodebridement or biosurgery), sterile fly larvae are applied to the wound within special dressings. Medical grade maggots have three primary actions: they clean the wounds by removing dead and infected tissue ('debridement'), they disinfect the wound (kill bacteria) and they speed the rate of healing. In 1996, the International Biotherapy Society was founded in Wales. Today, over 3 000 therapists are using maggot therapy in 20 countries. Approximately 30 000 treatments were applied in the year 2003.

In January 2004, the US Food and Drug Administration (FDA) issued licence '510(k) #33391', allowing the production and distribution of 'Medical Maggots' as a medical device. In February 2004, the British National Health Service (NHS) permitted its doctors to prescribe maggot therapy. Patients no longer have to be referred to one of a few regional wound-specialty hospitals to get maggot treatments.

10

The flea-borne diseases

Plague

The best known of the human infections transmitted by fleas is plague, whose causative organism is *Yersinia pestis*. The reservoir hosts are various species of rodents and historically the black rat (*Rattus rattus*) was considered the main reservoir host during the great epidemics that ravaged Europe from the sixth century to until 1720, when the last major outbreak occurred in Europe, in Marseille. The pandemics of plague from the thirteenth century resulted in the deaths of tens of millions of people in Europe. Though plague remains endemic in many foci in Africa, Asia and the Americas, there are now no foci in Europe other than the natural foci of plague in the North Caucasus.

There are, however, a number of other flea-borne diseases of man endemic in Europe, some of which are emerging diseases while others, like murine or endemic typhus, are well known.

Flea-borne rickettsial diseases

Murine typhus

Murine typhus (endemic typhus), whose causative agent is *Rickettsia typhi*, has a worldwide distribution. Rodents, most often rats (*Rattus norvegicus*, *R. rattus*) are the reservoir hosts. Transmission from rodent to rodent or to humans is by contamination with rickettsia-infected flea faeces or tissue during or after blood-feeding by the flea by the faeces or tissues being rubbed into skin abrasions or into delicate mucous membranes. Murine typhus is usually a benign acute, febrile disease, characterized with headache and rash and a fever that persists for about 12 days. Mortality is low and more common in elderly patients.

Misdiagnosis is frequent and the infection is probably much more common than reported (Azad *et al.*, 1997).

In Europe, the main reservoir host is the brown rat *Rattus norvegicus* and the most common flea vector is the Oriental rat flea, *Xenopsylla cheopis* and, less frequently, the cat flea *Ctenocephalides felis*. Murine typhus has been reported from Bosnia, Croatia, the Czech Republic, France, Greece, Italy, Macedonia, Portugal, Romania, Russia, Serbia, the Slovak Republic, Slovenia and Spain and is probably present in other countries as well.

Surveys in the northwestern part of Bosnia–Herzegovina found that among 231 human sera tested the rate for *R. typhi* was 61.5% (Punda Polic *et al.*, 1995). Foci of murine typhus were known to exist on the Dalmatian Islands of Croatia; in a serological survey in 1985, from 63.3% to 68.1% of the inhabitants were found positive for *R. typhi* with immunity being detected from the age of 10 years on (Dobec & Hrabar, 1990).

In Evia Prefecture, Greece, 92% of the 53 rat sera tested by IFA were positive for anti-*R. typhi* antibodies (Tselentis *et al.*, 1996) and human cases were common in the Island's hospital. In a follow-up study in Evia, the transmission cycle of the disease in Greece was elucidated for the first time. In 1993, 53 *R. norvegicus* were trapped in localities where the cases of murine typhus had occurred. Some 300 fleas, probably all *X. cheopis*, were found on the rats. Eight of the fleas were positive for *R. typhi* by polymerase chain reaction (PCR). Anti-*R. typhus* antibodies were found in the sera of 48 (91%) of the rats at high titres. The commensal rat population in the general area of study was both ubiquitous and abundant. It was concluded that murine typhus was obviously endemic in the area and probably of much greater public health significance than the number of reported cases indicates (Chaniotis *et al.*, 1994). Daniel *et al.* (2002) found that the prevalence of antibodies to *R. typhi* was 2% of 1584 human sera collected in northern Greece in an area which had never been surveyed before for this infection.

A high antibody prevalence for murine typhus has also been found among rat populations in Spain and Portugal; a death from murine typhus has been reported of an English tourist after her return to the UK from the Costa del Sol in Spain (Pether *et al.*, 1994). Murine typhus is a re-emerging disease on the Canary Islands of Spain. Cases in the Las Palmas hospital showed a rather distinct clinical pattern characterized by higher incidence of complications, especially renal damage (including acute failure and urinalysis abnormalities). The greater clinical severity could be related to different strains of *R. typhi* or other cross-reactive species (Hernandez Cabrera *et al.*, 2004). In Soria province, central Spain 9 of 73 dogs (12.3%) checked carried antibodies against *R. typhi*. The gender, age and breed of the dog, and whether it was used for hunting, shepherding, guarding or simply as a pet, apparently had no significant effect on the probability of it being seropositive. Being infested with fleas or having a history of such infestation was, however, significantly associated with seropositivity (Lledo *et al.*, 2003).

A long-term, 17-years study was carried out in the south of Spain, 1975–1995 on 104 patients with persistent fevers. Murine typhus was the cause in 6.7% of 926 cases of fevers of intermediate duration, characterized by a febrile syndrome lasting from 7 to 28 days. Most patients had a benign course of the diseases but some developed severe complications (Bernabeu Wittel *et al.*, 1999).

Murine typhus appears to be re-emerging in Portugal; the disease was fairly common in Portugal until the 1940s, when several cases were diagnosed in the Lisbon area. In 1993, antibodies against *R. typhi* were detected for the first time in serum from a patient in the Madeira Autonomous Region, an archipelago with two major islands: Madeira and Porto Santo. The patient was an HIV-positive woman with no clinical features of murine typhus. In November and December 1996, antibody titres against *R. typhi* exceeding 1:128 were detected in sera from several patients at Madeira Central Hospital; antibodies to *R. typhi* were also found in the local rats, *Rattus norvegicus* and *R. rattus*. The outbreak of 1996 was probably facilitated by conditions on the island in that year favourable to an increase of the rat population; the high level of rain allowed an increase of vegetation, food and shelter and hence the rats (Bacellar *et al.*, 1998).

Northern Europe seems to be largely free of murine typhus although the tale of the Pied Piper of Hamelin in Germany has already been referred to earlier. Small numbers of cases of the disease have been reported from France and from Germany though the validity of the diagnostic methods was uncertain (Traub *et al.*, 1978).

The ELB agent

A flea-borne disease, the 'ELB agent' caused by *Rickettsia felis*, was reported in 2000 in a couple in Düsseldorf, Germany with the symptoms of a high fever and rash and confirmed by PCR testing; this is the first report of this agent in Europe. The cat flea, *Ctenocephalides felis*, transmits the infection, first identified in the USA in 1990. Studies have shown that cat fleas can maintain the infection for up to 12 generations by vertical (transovarial) transmission (Wedincamp & Foil, 2002). In the cases in Germany, the agent was probably contracted from fleas on the couple's dog (Richter *et al.*, 2002). The same infection has also been identified in a cat flea in southwestern Spain (Marquez *et al.*, 2002). ELB agent was also identified in two patients with fever and rash in France (Raoult *et al.*, 2001) and has also been isolated from the cat flea in that country (Rolain *et al.*, 2003a). In a survey in the UK, DNA extracted from fleas was analysed using genus- and species-specific PCR and amplicons were characterized using DNA sequencing. Fifty per cent of flea samples were PCR positive for at least one pathogen and 21% were positive for *R. felis* (Shaw *et al.*, 2004).

In view of the wide distribution of the cat flea through Europe, the infection will almost certainly be found to be common in more countries; its diagnosis should be considered with patients with a murine typhus-like disease with rash and fever following fleabites.

Cat scratch disease

The causative agent of cat scratch disease is *Bartonella henselae*. While the disease itself was first described in France in 1950, it was only in 1992 that the bacterial agent was described. Infection with this agent can give rise to bacillary angiomatosis in humans, a vascular proliferative disease most commonly associated with long-standing HIV infection or other significant immunosuppression. *Bartonella henselae* has also been associated with bacillary peliosis, relapsing bacteraemia and endocarditis in humans. Cats are healthy carriers of *Bartonella henselae*, and can be bacteraemic for months or years. Cat-to-cat transmission of the organism by the cat flea, *Ctenocephalides felis*, with no direct contact transmission, has been demonstrated (Chomel, 2000). *Bartonella henselae* is widespread in Europe. About 10% of pet cats and 33% of stray cats harbour the bacterium in their blood. In immunocompetent patients, *B. henselae* is responsible for human cat scratch disease, characterized by lymph node enlargement near the entry site of the bacteria (Piemont & Heller, 1998).

In Europe, *B. henselae* and clinical cat scratch disease have been reported from Belgium, Bulgaria, Croatia, Czech Republic, Denmark, France, Germany, Greece, Italy, the Netherlands, Poland, Russia, Spain, Sweden, Switzerland and the UK.

Cats may harbour three *Bartonella* species, *B. henselae, B. clarridgeiae* and *B. henselae* genotype/serotype Marseille (type II) which has been isolated from cats and cat fleas in France (La Scola *et al.*, 2002). Despite the serious clinical infections sometimes seen in infected patients, both *B. henselae* and *B. quintana* (an infection transmitted by human lice to humans causing trench fever) have been found in many healthy persons in an urban area of Greece; of 500 individuals, 99 (19.8%) and 75 (15%) were IgG seropositive to *B. henselae* and *B. quintana* respectively. A high percentage (12.4%) of cross-positivity between the two species was seen. The prevalence of both *Bartonella* species was high. Cat owners had significantly higher antibody titres only to *B. henselae* and not to *B. quintana* (Tea *et al.*, 2003).

The prevalence of *B. henselae* in cat populations was evaluated by serological and bacteriological tests in northern Italy. A total of 769 stray cats from three urban and three rural areas were sampled between January 1999 and December 2000. Of these, 540 cats were tested by serology; 207 (38%) were seropositive (Fabbi *et al.*, 2004). Considering the high percentage of cats positive for *B. henselae*, the finding by Massei *et al.* (2004) that the infection is common among children in central Italy and occurs early in life is not unexpected. In most cases, however, the infection is asymptomatic, and resolves spontaneously. This was not necessarily the case in Poland where Podsiadly *et al.* (2002) reported that *B. henselae* and *B. quintana* infections resulted in illnesses with symptoms of severity ranging from mild lymphadenopathy to systemic disease. Most cases of the disease were in children 8–16 years old. The number of detected and reported *B. henselae* infections in Poland is low but may be underestimated. Cat scratch disease is the most frequently clinically and serologically identified bartonellosis in Poland.

In Italy, *Bartonella* DNA was amplified and sequenced from four *Ixodes ricinus* ticks (1.48%) removed from humans in Belluno Province. This report indicates that the role of ticks in the transmission of *Bartonella* species should be further investigated (Sanogo *et al.*, 2003).

Cat scratch disease is an emerging infection in Europe; its incidence in humans and its prevalence in cat populations are now recognized as high. Its public health importance may grow as the infection is more frequently diagnosed.

11

The louse-borne diseases

Of the 200 or so species of sucking lice, only three infest man: the body louse, *Pediculus humanus*, the head louse, *P. capitis* and the pubic louse, *Pthirus pubis* (also known as the crab louse). Only the body louse is a vector of disease. In Europe, infestations by body lice have, until recently, been uncommon for some time; head lice infestations on the other hand are extremely common, mainly among children, while infestations by pubic lice appear to be increasing to the extent that they may be considered a marker for sexually transmitted diseases. Head lice infestations are found virtually only on the hairs of the head; body lice attach to garments where they feed on the skin and attach their eggs to clothes, most frequently along the seams of clothing. Pubic lice, as indicated by their name, are generally found only in the pubic areas though many infestations of eyelids and eyebrows have been reported.

Louse-borne rickettsial diseases

Epidemic or louse-borne typhus

Epidemic typhus or louse-borne typhus due to *Rickettsia prowazekii* is transmitted by the human body louse, *Pediculus humanus*. The infective agent is transmitted from person to person through contaminated faeces of the human body louse which is the only known vector. Until World War II, the disease was periodically responsible for an enormous number of cases and a great many deaths especially among armies, refugees and inmates of camps of all types. Russia suffered heavily from epidemic typhus during the World War II with as many as 20 million cases (Rydkina *et al.*, 1999). In 1943 an outbreak of typhus in Naples, Italy was controlled by mass dusting of the population with DDT powder;

there were outbreaks of louse-borne typhus in Bosnia–Herzegovina between 1946 and 1949 and these were also successfully controlled by DDT. There were scattered cases up until the 1960s, but there had been no important outbreak in Europe since that time until an outbreak occurred in Russia in 1997 in a hospital in Lipetsk. A total of 23 patients and 6 members of the staff suffered symptoms of typhus with 22 shown to be seropositive for *R. prowazekii*. The clothes of the patients were infested with body lice and these tested positive for typhus. It is thought that poor public health practices following breakdown of the hospital's heating system led to body louse infestations (Tarasevich *et al.*, 1998). This outbreak demonstrates that epidemic typhus has the potential for recrudescence even in areas where it has been absent for a number of years. In view of the recrudescence of body lice populations in several countries of Europe (Gratz, 1997; Rydkina *et al.*, 1999) the risk of the recrudescence of louse-borne infections such as louse-borne typhus and trench fever is increasing.

In areas where louse-borne typhus was once epidemic, Brill–Zinsser's disease may occur; this disease is actually a recrudescence of an earlier infection with epidemic typhus and may occur many years after the initial illness. It may affect people long after they have recovered from epidemic typhus. The disease may be due to a weakening of the immune system such as by ageing, surgery or illness. Brill–Zinsser disease tends to be extremely mild and there is no mortality. However, as the disease arises from rickettsia retained in the body, lice that feed on patients may acquire infection and transmit the agent. *Rickettsia prowazekii* may even be isolated from the blood by animal inoculation. The disease is sporadic, occurring at any season and in the absence of infected lice. There have been recent reports of the disease in France in 3 patients, all of whom had a history of body louse infestations at one time (Stein *et al.*, 1999). Cases are also reported in eastern Europe, generally among older persons with a history of louse-borne typhus. Twenty-five cases of Brill–Zinsser's disease were reported from 1 January 1980 through 31 December 2000 in Croatia. Fifteen of these cases (60%) had a history of a primary attack of epidemic typhus during or after World War II. During the course of the disease in the patients aseptic meningitis was verified in 21 (84%) patients, rash in 17 (68%), liver lesion in 14 (56%), pneumonitis in 7 (28%), myopericarditis in 7 (28%) and 5 (20%) had renal lesion (Turcinov *et al.*, 2002). In populations infested with body lice, a person falling ill with this disease may thus trigger a recrudescence of louse-borne typhus.

Trench fever

Trench fever, whose causative organism is *Bartonella quintana* (previously *Rochalimaea quintana*), affected as many as a million soldiers during World War I;

body louse infestations were very common among the troops in the trenches facilitating the transmission of the infection. After the end of the war, the disease apparently disappeared after 1918; it later reappeared to infect substantial though smaller numbers of soldiers and camp inmates during World War II. Like epidemic typhus, it is transmitted from person to person through contaminated faeces of the human body louse, which is the only known vector. Infection can give rise to fever, cutaneous bacillary angiomatosis and endocarditis. Mortality is rare but the disease may be the cause of prolonged disability.

The disease apparently disappeared again after World War II. However, in the early 1980s, trench fever reappeared, generally among homeless, often HIV-infected men, in Europe and North America (Foucault *et al.*, 2002) and has since been reported in Australia, Burundi, France, Germany, Mexico, Peru, Portugal, Russia, the UK and the USA. Infections with this agent are associated with poor personal sanitation and lousiness; *Bartonella quintana* has been isolated from lice on some of the patients (Roux & Raoult, 1999; La Scola *et al.*, 2001). The current epidemiology of this infection is not well understood other than that the disease appears to mainly affect the homeless. Drancourt *et al.* (1995) searched for this organism in 3 alcohol-dependent homeless men with endocarditis in France. *Bartonella quintana* was isolated from 1 patient in the blood-agar culture and from the other 2 patients in the endothelial-cell culture. As part of a survey for trench fever among homeless people in Marseilles, France, *B. quintana* was isolated from 15 of 161 body lice and was detected in 41 of 161 lice by PCR (La Scola *et al.*, 2001). Rolain *et al.* (2003b) compared the growth and the number of bacteria per erythrocyte in vitro in laboratory-infected red blood cells from alcohol-dependent patients versus normal blood donor erythrocytes. They found that erythrocytes from alcohol-dependent patients contain significantly more bacteria per cell than erythrocytes from blood donors suggesting that there is a link between alcoholism and infections of *B. quintana* that may be due to the macrocytosis of erythrocytes.

During the last decade, pediculosis increased markedly in Russia; in a study carried out by Rydkina *et al.* (1999), lice were collected at the Moscow Municipal Disinfection Center, where the homeless wash and delouse themselves, as well as disinfect or change their clothes, and examined for the presence of the agents of typhus, trench fever and relapsing fever; 19% of the persons examined at the shelter were infested with body lice. The study showed that 12.3% of studied lice samples were *B. quintana*-positive.

The resurgence of a disease long thought to have disappeared is, in itself, serious. It is also serious that a resurgence of populations of body lice which serve as vectors of trench fever is occurring in many countries. If body lice infestations are found, especially among homeless and alcohol-dependent individuals, a serological examination should be made to determine if *B. quintana* is present to ensure that appropriate treatment is provided if necessary.

Louse-borne relapsing fever

The aetological agent of louse-borne relapsing fever, otherwise known as epidemic relapsing fever, is *Borrelia recurrentis*. The vector is the human body louse, *P. humanus*. At one time the infection was widely spread in Europe but after outbreaks, which occurred during and immediately after World War II, the disease has disappeared from Europe. It remains common in Africa and has recently caused major epidemics in Burundi and Ethiopia. Inasmuch as the reappearance of body lice infestations in Europe has resulted in the reappearance of trench fever, the possible resurgence of louse-borne relapsing fever, a disease with a high mortality, cannot be excluded.

Body louse infestations

The body louse, *P. humanus*, is, as noted above, the vector of epidemic typhus, epidemic relapsing fever and trench fever. Until shortly after World War II, body lice infestations were common in Europe. With the advent of DDT and other modern insecticides which provided easy and persistent control, body louse infestations and the diseases associated with them virtually disappeared.

The presence in many large cities of sizeable populations of homeless persons with low levels of personal sanitation has resulted in what appears to be a growing recrudescence of body lice infestations. Such increases have been reported in the Czech Republic, France, the Netherlands and Russia; the return of body lice has been associated with the reappearance of trench fever in some countries. During 1993 and 1994, body lice were found 41 times on 31 patients at a clinic for the homeless in Utrecht in the Netherlands (Laan & Smit, 1996).

Head louse infestations

Head lice, *Pediculus capitis*, are not regarded as vectors of disease although Robinson *et al.* (2003) considered that they are potential vectors of *R. prowazekii*; there is, in fact, little evidence that head lice have a vectorial role for epidemic typhus or any other infection. However, their feeding activity irritates the scalp, and can cause intense itching; secondary infections may result if the skin is broken by repeatedly scratching the area.

Infestations are extremely common, especially among school children where lice are easily passed from one child to another, and infestation rates of

10–20% or even higher are frequently seen in school classes. All socioeconomic levels may be infested. A study on the prevalence of head lice in schools in the UK was made by questionnaires addressed to the parents of students at 235 primary schools; a total of 21 556 of 43 889 (49%) questionnaires were returned by parents. Overall 438 children had head lice at the time of the survey, giving a prevalence of 2.03%; 8059 had had lice at some time in the last year giving an annual incidence of 37.4% (Harris *et al.*, 2003). Courtiade *et al.* (1993) conducted a survey of four schools in the Bordeaux area of France. Between January 1990 and March 1991, 48.7% of children had had at least one episode of head lice, at rates of 38.8–62.6% depending on the schools. For 30.5% of children, this was their first infestation; 95% of the cases were detected by parents, and the prevalence was 60% in girls and 40% in boys. The highest prevalence was noted at a suburban school in an area where 17% of the parents were unemployed. Similar increases in the prevalence of head lice infestations are found in virtually all the countries of Europe (Gratz, 1997).

Large sums of money are expended for the control of this ubiquitous pest in most countries of Europe; a UK government review estimated that £25 million is spent each year in the UK on head louse treatment. Hundreds of thousands of insecticide preparations are purchased annually for head lice control (Gratz, 1997). The widespread occurrence of insecticide resistance to malathion, carbaryl and the pyrethroids impedes control. The systemic ivermectin offers promise as a control agent.

Although head lice are not vectors of disease, the fact that such large sums of money are spent on their control, detracts from resources that would be available for other health-care activities. Head lice must, therefore, be considered a public health problem.

Pubic louse infestations

As is the case with head lice, pubic lice or crab lice, *Pthirus pubis*, are not vectors of disease. Accurate information on the prevalence of infestations of this species of human louse is, understandably, difficult to obtain but from articles originating mainly from sexually transmitted disease (STD) clinics, it is apparent that infestations are quite common. Pubic lice are usually found on the hairs of the pubic region and transmission from one person to another is, for the most part, through sexual contact. Infestations may cause irritation and, at times, secondary infections.

Physicians finding pubic lice should consider the possible presence of other venereal infections as there is a high degree of correlation between the presence of *P. pubis* and other venereal infections. Furthermore, the possibility of sexual abuse having occurred should be considered in children with pubic louse infestations.

Valle (1988) studied 235 homosexually active men in Finland in 1983. Of this group, 88.5% reported at least one STD, the most common being pubic lice (64.7%) followed by gonorrhea (42.9%). Opaneye *et al.* (1993) found that 37% of the patients at a clinic in Coventry, UK, who were positive for *P. pubis* also had another STD. In the USA, Pierzchalski *et al.* (2002) compared the rate of *Chlamydia* and gonorrhea infections in adolescents with and without pubic lice in order to evaluate pubic lice infestation as a predictor for concurrent *Chlamydia* or gonorrhea infection. The study demonstrated that pubic lice infestation predicted *C. trachomatis* infection (odds ratio = 3.31) and pubic lice infestation was, therefore, predictive of a concurrent *C. trachomatis* infection in this population. The authors recommended that adolescents infested with pubic lice should be screened for other STDs, including *Chlamydia* and gonorrhea infections.

In Spain, Lopez Garcia *et al.* (2003) reported that phthiriasis palpebrarum (infestation of the eye lashes) caused by *P. pubis*, is an uncommon cause of blepharoconjunctivitis; therefore, this condition is easily misdiagnosed. The number of such cases in their hospital department has increased in recent years. They also recommended that affected children be searched for signs of child abuse.

12

Tick-borne diseases of Europe

Tick-borne viruses

By 1972 some 68 different viruses had been recorded from more than 80 tick species, 20 of which were believed to cause disease in man or domestic animals (Hoogstraal, 1973). Since the publication of Hoogstraal's review, many other viruses have been isolated from ticks though their role as causative agents of human, or animal, disease is often unknown or uncertain. Many areas of Europe remain poorly surveyed and more viruses are likely to be found.

Tick-borne encephalitis TBE

Tick-borne encephalitis (TBE) is the most important and widespread of the arboviruses transmitted by ticks in Europe; TBE is a member of the family Flaviviridae. Tick-borne encephalitis should be considered a general term encompassing at least three diseases caused by similar flaviviruses, whose range spans an area from the British Isles (Louping ill), across Europe, as central Europe tick-borne encephalitis transmitted mainly by *Ixodes ricinus*, to far-eastern Russia, as Russian spring-summer encephalitis, which is transmitted mainly by *I. persulcatus*. The three diseases differ in severity, with Louping ill being the mildest and Russian spring-summer encephalitis the most severe. Man is infected by the bite of infected ticks and, much more rarely, by ingestion of milk from infected domestic animals (Dumpis *et al.*, 1999). The tick is both a vector and a reservoir host; once infected, it remains infected throughout its life, through its metamorphosis and transmits the virus to its progeny. Small rodents (field-mice or voles) are the prime vertebrate hosts.

Tick-borne encephalitis is often the cause of serious acute central nervous system (CNS) disease which may result in death or long-term neurological sequelae for a considerable period after recovery from the initial infection. The case fatality rate in those developing symptomatic disease is 0.5–2% for the western subtype and 5–20% for the eastern subtype. About 40% of the infected patients are left with a residual post-encephalitic syndrome. The course of the disease is more severe in the elderly than in young people. The mortality of the central European form of TBE is 0.7–2% (Ozdemir *et al.*, 1999) and may be even higher in severe infections. For the far eastern form, the mortality rate may be as high as 25–30%. There is no specific treatment for the disease. In infected persons, the post-TBE recovery lasts about 1 year and sequelae occur in about one-third of the cases (Laiskonis & Mickiene, 2002). Haglund & Gunther (2003) considered the percentage of sequelae as even higher; they observed that a defined post-encephalitic TBE syndrome exists, causing long-lasting morbidity that often affects the quality of life and sometimes also forces the individual to a change in lifestyle. The sequelae render high costs for individual patients and society. Three clinical courses may be identified: one with complete recovery within 2 months, occurring in approximately one-quarter of patients, one with protracted, mainly cognitive dysfunction, and one with persisting spinal nerve paralysis with or without other post-encephalitic symptoms. Up to 46% of patients are left with permanent sequelae at long-term follow-up, the most commonly reported residuals being various cognitive or neuropsychiatric complaints, balance disorders, headache, dysphasia, hearing defects and spinal paralysis.

Tick-borne encephalitis is a notifiable disease in the entire region except for Denmark and Iceland. A voluntary reporting system in Sweden has since July 2004 been replaced by mandatory reporting. Estonia, Finland, Lithuania, Norway and Sweden only report laboratory-confirmed cases. Northwest Russia reports both laboratory-confirmed cases and cases with a typical clinical picture without laboratory confirmation but with an epidemiological link to risk factors (seasonal forest visits, goat milk, etc.). In western and central European countries, TBE is particularly common in forest and mountainous regions of Austria, Estonia, Latvia, the Czech Republic, Slovakia, Germany, Hungary, Poland, Switzerland, western Russia, Ukraine, Belarus and northern Yugoslavia. It occurs at a lower frequency in Bulgaria, Denmark, France, Romania, the Aland archipelago and neighbouring Finnish coastline, and along the coastline of southern Sweden, from Uppsala to Karlshamn. Thousands of cases occur each year from late spring to early autumn; the total number of annual cases in western European countries has averaged at least 3000 over the last 5 years although as will be seen, the incidence may be still higher. The International Scientific Working Group on Tick-Borne-Encephalitis (2003) estimates that at least 10 000 cases of TBE are referred to hospitals each year, yet the incidence of TBE is so far not fully

Table 12.1 *Tick-borne encephalitis in Europe, most recent year available*

Country	Most recent year	Number of cases reported	Incidence/100 000
Austria	2003	87	1.09
Czech Republic	2003	–	5.9
Finland	2001	>40	–
Germany	2003	276	–
Hungary	2001–2003	63 (annual average)	–
Latvia	2003	–	15.7
Lithuania	2003	763	22
Norway	2003	1	–
Poland	2003	339	0.89
Slovakia	2003	74	1.38
Slovenia	2003	272	13.6
Finland	2003	107	–

Eurosurveillance Weekly Archives 2004 8(29)

recognized. The reason for this is that TBE produces clinical features similar to those of many other types of meningitis and/or encephalitis. The group also considered that until recently TBE was believed to be a rather limited problem in a few well-defined endemic areas; however, this notion has now been revised (Table 12.1).

Outbreaks often follow periods when voles (*Clethrionymus* or *Microtus* species), the principle reservoir hosts, and ticks are numerous (between May and June and between September and October). Russian summer-spring encephalitis occurs in the spring and summer months in eastern Russia, Bulgaria and Iran.

The incidence of TBE in central and eastern Europe has increased sharply, supposedly due, among other things, to climate warming and changes in leisure activities. The probability of contracting the disease following a viral infection is about 30%. Some 70% of victims manifest a biphasic course with two fever peaks. Some 10% will have lasting, sometimes severe, neurological deficits, and 1–2% will die. Diagnosis is based on confirmed exposure to ticks in a high-risk area, a tick bite within the previous three weeks, clinical symptomatology, infected cerebrospinal fluid and FSME-specific IgM and IgG antibodies in the serum. Causal treatment is not possible, but the disease can be effectively prevented by prophylactic measures and vaccination (Lademann *et al.*, 2003). The commercially available vaccines in Europe consist of highly purified inactivated whole TBE virus.

Many of the following country summaries are taken from the website euro-surveillance.weekly@hpa.org.uk, others from the author's database. The graphs of TBE incidence (Figure 12.1) are also taken from the web site.

Austria

Meningoencephalitis is a notifiable disease in Austria. There were 87 cases of TBE in Austria in 2003: an incidence rate of 1.09 per 100 000. In 2002, there were 51 cases, and in 2002, 60 cases. The regions most affected by TBE are in the south: Steiermark (Styria) and Karnten (Carinthia). All of these cases were in unvaccinated people, or people who had not had the vaccine according to the recommended schedule.

Austria is the country with the highest coverage of TBE vaccination (86% of the total population); this has led to a dramatic reduction in the annual number of clinical cases, proving that vaccination is an effective means for the prophylaxis of TBE (Heinz & Kunz, 2004). Until the early 1980s, TBE was a frequent cause of CNS infectious disease in Austria. In 1958, the proportion of TBE in the total number of CNS viral diseases in Austria was 56%. Thus, before the start of the vaccination programme, it was the most important and most frequent disease of this type in adults, with several hundred cases being reported each year. From 1981, voluntary vaccination was encouraged by intensive media campaigns. As a result the number of hospitalized cases due to TBE declined significantly from 1981 to 1990 with important savings in health-care costs (Schwarz, 1993). A study of the rate of vaccination among school children in Austria showed that the prevalence of at least one TBE vaccination was 91.4% for 7-year-olds, 97.3% for the 10-year-olds and 97.1% for the 13-year-olds. The prevalence of basic TBE immunization was 84.0%, 91.7% and 92.3% respectively. The lowest vaccination rates were found in families with four or more children and for those children with mothers of the lowest educational level (Stronegger *et al.*, 1998). The coverage rate for very young children, and people over 65, is under 70%. The low coverage in older people represents a challenge for prevention of TBE in Austria.

Belarus

There have been dramatic increases in TBE incidence in Belarus since the early 1990s. Tick-borne encephalitis incidence is highest in June and has been observed predominantly in people between 30 and 39 years of age (Korzan *et al.*, 1996).

Croatia

In 1984/1985, Galinovic Weisglass & Vesenjak Hirjan (1986) examined 243-paired sera of persons who had been in contact with ticks. Of this group,

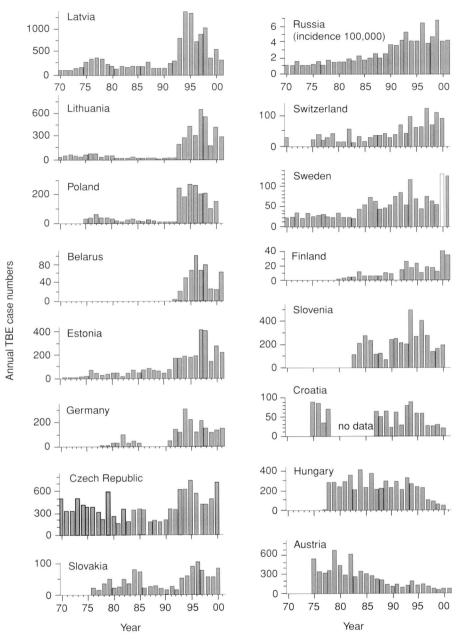

Figure 12.1 Annual case number of tick-borne-encephalitis in European countries, 1970–2001. NB – the data for Russia are expressed as incidence/100 000 population.

51.85% were positive by complement fixation (CF) and 53.18% by haemagglu-tination inhibition (HI) tests. The largest number of positive cases was in the 15–49 age group, many of whom were from the Zagreb area. Over a period of 20 years, 92 patients were treated for TBE in the University Hospital in Osijek in eastern Croatia, most of them forestry workers. In eastern Croatia TBE is a relatively common disease, appearing almost every year. In a small number (9.8%) of cases the clinical picture was aseptic meningitis and in the majority of patients (90.2%) it presented as an acute meningoencephalomyelitic form. The course was relatively severe in the majority of the patients analysed, with disturbances of consciousness (32.6%) and transitory neurological signs (61.9%). Three patients died (3.3%) in the early phase of the disease (Anic *et al.*, 1998).

Czech Republic

Tick-borne encephalitis was isolated for the first time in Europe in the Czech Republic and occurs throughout the country in natural foci where the virus cycle is easily maintained. The landscape consists of forests and farmlands. There is an abundance of game and the presence of roe deer (*Capreolus capreolus*) plays a key role in TBE natural foci (Danielova, 2002). Laboratory-confirmed cases of TBE have been reported in the Czech Republic since 1979. In 2003, the approx-imate incidence of TBE was 5.9/100 000 population. Incidence is higher south of Prague near the city of Ceske Budejovice and near the town of Pilsen in the west. Recently, TBE foci have been identified in the northern part of the province of Bohemia. In the east of the country, incidence is high near Olomouc. Clinical cases of TBE are notified from April until November every year.

Since 1970, the incidence of TBE has changed twice: during the 1980s, it fell by about 30% compared with previous levels, but in 1993 incidence doubled to its present level, about 50% above the pre-1980 level. The increase was character-ized by: (a) a higher number of cases in areas well known for TBE occurrence; (b) re-emergence in areas where TBE human cases had not been observed, or only sporadically, for a long time; (c) emergence of TBE in places unknown previously including at high elevations (Daniel *et al.*, 2004). No single factor adequately explains the rising incidence of the disease in the Czech Republic. Changing weather patterns in the past few years is a possible factor. The aver-age annual temperature in the Czech Republic increased slightly from 1970, but then much more markedly from 1989 and rainfall patterns have also changed, possibly affecting tick survival and development rates (Beran, 2004). Changes in the geographical distribution of *Ixodes ricinus* have been observed, with ticks appearing at higher altitudes in mountains than previously as a possible effect of global warming (Zeman & Benes, 2004).

Vaccination is recommended for persons who intend to camp or hike in endemic areas.

Denmark

Tick-borne encephalitis is not a notifiable disease in Denmark. The only area where it was considered that there is a risk of acquiring TBE is on the island of Bornholm. People who live on Bornholm permanently, or have a summer holiday home on the island, are advised to be vaccinated if they leave the designated paths in woods or scrub land. Tourists and school parties are not generally required to have a vaccination. However, Skarphedinsson *et al.* (2005) conducted a study of the distribution and prevalence of tick-borne infections in Denmark by using roe deer (*Capreolus capreolus*), as sentinels. Blood samples from 237 deer showed that 8.7% were positive for TBE-complex virus. The authors considered that these findings show that a marked shift has occurred in the distribution of TBE-complex virus in Denmark, supporting studies that predict alterations in distribution of TBE due to climatic changes as previously TBE has been found only on the island of Bornholm.

Estonia

Tick-borne encephalitis is considered as a severe public health problem in the country. There has been a rising trend of reported cases of TBE in Estonia; in 1990 only 37 cases were reported and by 1997 there were 404 cases; in 1998 there were 387, in 1999, 185 and in 2000, 272 cases. Occupational exposure in forests is a risk factor for infections.

Finland

Tick-borne encephalitis is a notifiable disease in Finland. The number of TBE cases has risen from an annual 10–20 in the 1990s to over 40 cases annually in 2001 (population 5.2 million). The incidence of identified cases is highest (i.e. over 100/100 000/year) on the island of Aland, situated between Finland and Sweden. Approximately one out of five Alanders is infected during his or her lifetime, although TBE infections are rare in children and adolescents. In addition to the Alanders, approximately 10 Swedes annually fall ill with TBE after visiting Aland. Foci of TBE exist elsewhere in Finland such as in the Turku archipelago, in some areas of the southeast around Kokkola, on Isosaari and in coastal regions close to Helsinki (Han *et al.*, 2001).

The National Public Health Institute recommends vaccination against TBE for all those over 7 years of age who reside in, or intend to spend long periods in, the known endemic areas. The vaccine is not, however, currently part of the Finnish national immunization programme.

France

In France, natural foci have been detected in the Alsace in the Illkirch forest, near Strasbourg. Tick-borne encephalitis was determined as a cause of a case of encephalitis in 1968 and was isolated from *Ixodes ricinus* in 1970 (Hannoun *et al.*, 1971). An epidemiological survey carried out from 1970 to 1974 led to the conclusion that the Alsatian TBE focus is stable as virus activity was detected every year; the focus is an extended one; virus was isolated from both *I. ricinus* and various rodents in five of six study sites, as well as in several control sites (Perez Eid *et al.*, 1992). Eight new cases of TBE were observed in Alsace between 1985 and 1990. The clinical presentation in these patients and the 2 earlier cases, was a pure meningitis syndrome in 4 cases and meningo-encephalitis in 6 cases, very severe in 3 of them. All patients recovered rapidly, and only 3 had slight sequelae. In a seroprevalence survey conducted in 1989 among 619 professional foresters of eastern France, 8% were found to be seropositive, which suggests that the disease is often unrecognized (Collard *et al.*, 1993).

Christmann *et al.* (1995) noted that between the first report of the presence of TBE in France in 1968 and 1994, a total of 21 cases were reported in the literature or through personal observations (by 1996 this had risen to some 30 cases). All the patients were infected in the Alsace where the rate of infection is probably underestimated. Because 2 cases of TBE were diagnosed in the Nancy region, a seroepidemiological survey was conducted in the Lorraine (Meurthe and Moselle, Moselle, Vosges, Meuse) in 1966 on 1777 persons, of whom 21% had experienced tick bites. Tick-borne encephalitis serology (IgG) was positive in 19 subjects; 9 sera were positive on Western blot (0.76). No IgM positive serum was found. Seroprevalence was higher in subjects with a history of tick bites; the authors concluded that the infection was not a public health problem in the Lorraine (Schuhmacher *et al.*, 1999). The number of cases and tick population densities has continued to rise in France since the mid-1990s (George & Chastel, 2002).

Germany

Tick-borne encephalitis is a notifiable disease in Germany. In 2003, 276 cases of TBE were reported (in 2002, 239 and in 2001, 256 cases), mainly in southern Germany in the federal states of Baden-Wurttemberg (42%) and Bavaria (38%).

Counties in Germany are classified according to three levels of TBE risk. A county is classified as a high-risk area if at least 25 TBE cases occurred within a 5-year period between 1984–2003, and as a risk area if at least 2 cases occurred in a single year, or at least 5 cases occurred within a 5-year period between 1984–2003. Areas are declared to be TBE endemic based on elevated TBE seroprevalence in non-immunized forestry workers. In 2003, three new districts were identified as risk areas; 74 of Germany's 440 counties are currently classified as TBE risk areas, and 9 as high risk areas. They are located in Baden-Wurttemberg (30), Bavaria (45), Hesse (4), Thuringia (3) and Rhineland-Palatinate. A further five counties in Baden-Wurttemberg are classified as TBE endemic based on sero-prevalence studies. There was a change in the definition of risk areas of TBE in Germany in 1998. In 1998, 63 country and town districts were TBE risk areas, in 2001, 79 and in 2002, 86. There were new risk districts within Bavaria and Baden-Wurttemberg and in Thuringia, Hesse and the Rhineland-Palatinate. The TBE incidence in Bavaria and Baden-Wurttemberg has been stable at a high level for years; outside these areas it has steadily been climbing (Odenwald, Thuringia) (Suss *et al.*, 2004).

The German Standing Committee on Vaccination (STIKO) recommends TBE vaccination for persons exposed to ticks in risk areas.

Greece

Little epidemiological information is available on human infections with TBE in Greece. The virus has been isolated from goats in the north of the country and dogs have been found positive at low levels. Blood samples for serosurveys were collected from healthy farmers, wood cutters and shepherds in a rural area; 1.7% carried antibodies to TBE (Antoniadis *et al.*, 1990). The virus circulates in the country but is not a public health problem.

Hungary

Tick-borne encephalitis has been a mandatory notifiable disease in Hungary since 1977. Samples from patients with aseptic meningitis and encephalitis have been tested for TBE since 1958. The average yearly incidence between 1977 and 1996 was 2.5/100 000 population, with the highest incidences between 1981 and 1990. From 1997 to 2000, a decrease in the number of the diagnosed TBE cases was seen, with an incidence of 0.5/100 000 in 2000 (Voss, 2000). Since 2001, the incidence has been increasing again. In the last 3 years, the yearly average of the reported cases was 63. The high-risk areas are in western and northern Hungary, the known natural foci.

Vaccination for forestry and agricultural workers was introduced in 1977. Since 1991, TBE vaccine has been available for all through purchase at pharmacies, and employers must ensure the vaccination of employees at risk. No detailed data on TBE vaccination coverage are currently available, although a rough estimate is that 3–5% of the population has been vaccinated, mostly people living in high-risk areas (Lakos *et al.*, 1996).

Italy

The first clinical cases of TBE were reported in 1975 and 1978 in northern Italy; the two cases in 1978 were both mild forms. Both patients lived or frequented rural areas where there are rodents likely to be natural reservoir hosts, and sheep were hosts of *Ixodes ricinus* (Amaducci *et al.*, 1978).

Since then, two more endemic areas and five more TBE foci have been recognized in northeast Italy, in the provinces of Trento (Trentino-Alto Adige) and Belluno (Veneto). The disease initially appeared to be very rare, only 18 cases having occurred in the period 1975–1991, but this has changed in recent years, with 84 new cases diagnosed from 1992–2001. Thus, a total of 102 indigenous TBE cases were recorded in 1975–2001. By comparing the median number of TBE cases/year recorded in the period 1975–1991 and in 1992–2001, an almost eight-fold increase in the incidence of the disease can be estimated. Tick-borne encephalitis predominates in males (M/F ratio = 72.5%), can occur in any age (age range of 16–73 years) and tends to be acquired during outdoor leisure activities. Unlike other countries, however, the incidence is seasonally distributed throughout the year, with the exception of January and March, and displays a biphasic peak in July and October. Only 58% of Italian TBE patients recalled a recent tick bite. No fatalities have occurred; however, five patients required respiratory support in an intensive care unit and persistent motor deficits affected 4 subjects (Caruso 2003).

The low prevalence of TBE in reservoir hosts and man does not warrant carrying out vaccination campaigns.

Latvia

Tick-borne encephalitis is an important public health problem in Latvia. The disease has been notifiable since 1955. Since 1963, there has been a steady increase in the incidence with the greatest number of cases occurring in the Riga district, including cases from inside the city itself (Babenko *et al.*, 1975). During the period of 1993–2002 there was a significant increase in the incidence of TBE. The highest increase of morbidity was in 1994 and 1995, and to a lesser extent in 1998, then this decreased during subsequent years. Both *Ixodes ricinus* and *I. persulcatus* are vectors; *I. ricinus* is found in the western and central part of Latvia and rarely in small numbers in the east while *I. persulcatus* dominates in the east. The increased incidence of the infection in 1994–1995 was primarily in the area where *I. ricinus* is the vector. The annual field-collected adult tick infection rate with TBE from 1993–2002 for *I. ricinus* adults varied between 1.7–26.6% and for *I. persulcatus* between 0–37.3%. Infection levels in ticks removed from humans was much higher and from 1998–2002 surpassed 30%. A study of the presence of the virus in adults and nymphs of *I. ricinus* in 2002 showed a high TBE prevalence of 43% (Bormane *et al.*, 2004).

In 1994, a campaign to vaccinate children against TBE began in areas of high risk. There are five rural areas where child TBE incidence level exceeded the mean level in the country (20/100 000 children); in areas with the highest incidence, the levels exceeded the mean by six-fold. These areas were a vaccination priority and 75% of children in these rural districts are now covered. Vaccination in the two highest risk groups of infected territories was completed in 1998. Altogether, children have been vaccinated in more than 100 rural districts. The immunization coverage for the whole population of Latvia is about 5%, but results of a survey of TBE prophylaxis awareness (1000 respondents) suggested the percentage of vaccinated adults was higher: 15% people on low incomes and 26% of all respondents reported that they had been vaccinated (Lucenko *et al.*, 2004).

Lithuania

Tick-borne encephalitis is a severe public health problem in Lithuania; from 171 to 645 serologically confirmed cases occurred each year between 1993 and 1999. A sharp increase in TBE cases was observed between 1992 and 1993 and continued in the 1990s. The increase was greatest in 1993–1995. In 2003, the epidemiology of TBE in Lithuania was very unusual. The incidence rate (763 cases, 22/100 000 population) was double the average incidence over the last 10 years, and was the highest annual rate recorded since notification began at the end of the 1960s. This rate was also the highest of all the Baltic countries in 2003 with four cases of death in 2003. The highest incidence, about 80% of all notified cases, is recorded yearly in the northern and central part of the country. People from rural areas are 1.7 times more affected than people in urban areas. About 40% of all cases of TBE were in retired and unemployed people, who constitute a particular risk group more likely to go into forests to collect mushrooms and berries as an additional source of income. Incidence is about 2–3 times higher in adults than in children; typically, 20% of all cases of TBE in Lithuania are in people over 60 (Asokliene, 2004).

Clinically, infections in the country are often severe. Mickiene *et al.* (2002) examined 250 patients with CNS symptoms during a 1-year-period; TBE presented as mild (meningeal) in 43.6% of patients and as moderate or severe (encephalitic) in 43.6% and 12.8% of patients, respectively. Paralytic disease was observed in 3.8% of the subjects, and cranial nerve injury was observed in 5.3%. One patient died of TBE. Permanent CNS dysfunction after a year was found in 30.8% of patients; in 8.5% of all TBE cases, severe disabilities required adjustment of daily activities. A progressive course of TBE was noted in 2 patients. The risk of incomplete recovery was significantly higher among patients with the encephalitic form of TBE.

Tick-borne encephalitis is, therefore, an important pathogen in CNS infection in the Kaunas region of Lithuania, and it causes long-lasting morbidity in one-third of cases. It remains geographically restricted in Lithuania – certain areas are well-known 'hot spots' of infection. A vaccine is available and has been recommended to people living or visiting high-risk areas.

Norway

All cases of encephalitis are notifiable in Norway, including TBE. In 2003, one case of TBE was reported. There have only been 8 cases of TBE acquired in Norway reported. The first case was identified in 1998. All cases were acquired within a limited area on the southern coast, and four cases were diagnosed in the municipality of Tromoy. A study done among regular patients in a health centre in Tromoy showed a seroprevalence of 2.4% with TBE antibodies (4 cases). This area probably represents a small focus of the disease in Norway. In addition, 2 cases of imported TBE were reported since 1994. These were acquired in Sweden and Austria.

Vaccination is not recommended as protection against transmission within Norway.

Poland

Tick-borne encephalitis is a notifiable disease in Poland. Since 1993, the number of reported cases at country level has ranged from 100 to 350 cases per year. In 2002 the number of reported cases was 126 (incidence 0.33/100 000), and in 2003 the number was 339 (incidence 0.89/100 000). Eighty per cent of cases occurred in two northeastern provinces of Poland adjacent to Lithuania and Belarus. A second focus of the disease is in the southwestern part of Poland, in districts adjacent to the Czech Republic.

Vaccination, using a 3-dose schedule, is recommended for groups living in endemic areas, and tourists visiting endemic places. Certain risk groups (foresters, soldiers, timber industry employees) are immunized in regular campaigns paid for by their employers.

Russia

Tick-borne encephalitis was first observed in the taiga zone of Russia in 1937 and was characterized by a severe illness and high mortality. Endemicity extends along the southern band of the forest zone of non-tropical Eurasia from the Atlantic and Mediterranean to the Pacific. A greater part of the range is in

Table 12.2 *Incidence in Russia of tick-borne encephalitis*

| Year | Incidence | | Deaths |
	Total	Per 100 000	
1992	6310	4.25	112
1993	7520	5.08	101
1994	5593	3.78	89
1995	5982	4.04	111
1996	10298	6.97	166
1997	6702	4.60	102
1998	7520	5.10	118
1999	9955	6.79	134

Russia and 80–90% of the TBE cases occur in Russian territory. Its main reservoir hosts and vectors are *Ixodes persulcatus* and *I. ricinus*. The range of the virus and the location of natural foci within it are closely associated with the distribution pattern of these ticks.

There were two epidemic peaks in 1956 and 1964 with 5163 and 5205 cases respectively (4.1–4.5/100 000). From 1965–1971 morbidity gradually decreased from 3658 cases in 1965 to 1226 in 1971 due to the control of *I. persulcatus* carried out by DDT applications to vegetation. Since 1972 TBE incidence has been increasing, in part because DDT applications have been stopped completely. This was followed by a return of high tick population densities. In 1990 and 1991 there were 5486 and 5225 cases respectively. There were further increases with 6301 cases in 1992 and 7893 cases in 1993 and 9548 in 1996 (6.45/100 000) (Korenberg, 1997). In the period from 1995–1999, the number of people contracting TBE increased 5.7 times in comparison with the period from 1971–1987. The rise in morbidity of TBE in Russia over the last years is not only explained by the discontinuation of the use of acaricides but also by the changing ecological situation in the country; this includes urbanization of new territories, increased contact between the population and ticks in the natural nidi of TBE through visits to the forests for relaxation, berry and mushroom picking, tourism, etc. At the same time, the use of specific prophylaxis has sharply decreased. Based on their analysis of data from 1992–1999, a sharp increase in TBE morbidity was detected in the Russian Federation (from 6310 to 10 298 cases). In addition to an overall increase, the number of deaths also rose (Vorobjeva *et al.*, 2001). These authors also presented a review table (Table 12.2).

Children up to 14 years of age constitute one-fifth to one-third of all reported cases. From 1995–1999, 27 children under 14 died of TBE in Russia.

Tick-borne encephalitis is a serious public health problem in Russia and the incidence is increasing in many foci. Vaccination rates are unsatisfactory and, as noted above, a well-functioning tick control programme has disappeared.

Slovakia

Tick-borne encephalitis is a compulsory notifiable disease in Slovakia. Typically, foci occur in forested areas with an average temperature of 8 °C and a yearly rainfall of 800 mm. The density of ixodid ticks within the foci is relatively high. The natural foci of TBE are confined to southern slopes of the Carpathian mountains and to the Danube basin. The foci appear most frequently in western parts of the country, but also in middle and eastern Slovakia. The morbidity of TBE, which is actually increasing, ranges from 0.6 to 1.6/100 000 inhabitants. The TBE virus infection rate of the ticks within Carpathian foci may be as high as 2.6%; the number of infected ticks within the Pannonian foci at the Danube never exceeds 0.1%. Four isolates of TBE have been obtained from brains of small rodents, the long-tailed field mouse *Apodemus sylvaticus*, and the bank vole *Clethrionomys glareolus* (Eleckova *et al.*, 2003).

The number of reported cases has ranged from 54–101 cases/year in the last 10 years. In 2002, the number of reported cases was 62 (incidence 1.15/100 000), and in 2003, the number of reported cases was 74 (incidence 1.38/100 000). Some of the reported cases were caused by drinking raw goat and sheep milk (home production). Longitudinal monitoring of TBE virus in ticks and vertebrate hosts (including humans) between 1964–1997 identified 37 endemic foci.

Vaccination is recommended for people living, or working in, endemic areas, and for tourists visiting endemic areas. The cost of vaccination for those who work in TBE endemic foci is reimbursed by health insurance.

Slovenia

Tick-borne encephalitis is endemic in northern Slovenia and is a notifiable disease. In 2003, 272 cases of TBE were reported, an incidence of 13.6/100 000. In 2002 there were 262 cases and in 2001, 260 cases.

Control efforts are directed towards early diagnosis and awareness campaigns. A vaccination campaign coordinated by the National Institute of Public Health is in place throughout the country, from late autumn to spring, annually. Immunization against TBE is recommended and offered by general practitioners and epidemiologists to anybody who spends time outdoors in the endemic areas, including short-term visitors.

Vaccination is obligatory for those in military service and other professionally exposed persons, including forestry and agricultural students. The cost of vaccination is covered by health insurance for students only. Coverage in these groups, and in students, is very high (98%). Coverage in the general population is below 10%.

Sweden

Tick-borne encephalitis is included in voluntary laboratory reporting for infectious disease surveillance in Sweden. A human case of TBE was first described in Sweden in 1954 and the virus first isolated in 1958 from a patient and from *Ixodes ricinus*. By the late 1980s and early 1990s, around 50–70 TBE cases were reported annually. The majority of the infected patients are diagnosed through hospitals. Since the end of the 1990s, around 100 cases have been reported annually. During the same period, the disease attracted increased public attention. It is, therefore, difficult to say whether there has been a real increase in the number of cases, or increased diagnosis due to a higher clinical awareness. Cases are being reported more frequently in recent years from areas where, previously, only occasional cases had been detected.

In 2003, 107 cases of TBE were notified (in 75 men and 32 women). Most of the infections were acquired in the counties of Stockholm (56%), Sodermanland (15%) and Uppsala (6%). In the county of Vastra Gotaland (to the south of Lake Vanern), 5–10 cases are notified annually. Sporadic cases occur in the rest of Sweden every year.

Lindgren *et al.* (2000) examined the possibility that a reported northward expansion of the geographic distribution of *I. ricinus* and an increased tick density between the early 1980s and mid-1990s in Sweden was related to climatic changes. The annual number of days with minimum temperatures above vital bioclimatic thresholds for the tick's life-cycle dynamics were related to tick density in both the early 1980s and the mid-1990s in 20 districts in central and northern Sweden. The winters were markedly milder in all of the study areas in the 1990s as compared to the 1980s. Their results indicate that the reported northern shift in the distribution limit of ticks is related to fewer days during the winter seasons with low minimum temperatures. This finding has implications for a likely northward shift of the distribution of TBE transmission as well.

Vaccination is recommended for high-risk groups residing in endemic areas, and for people visiting endemic areas during the summer.

Switzerland

Tick-borne encephalitis has been known in Switzerland since 1969 and is endemic in areas of northern Switzerland and the principality of Liechtenstein; TBE foci have not, so far, been found in the far western part of Switzerland, despite the prevalence of *Ixodes ricinus* in that region. There has been a steady increase of cases reported each year during the last 20 years. In addition, the highest incidence is now found in different geographic regions than previously.

There are two main endemic areas, one covering most parts of the Swiss midlands, the other one in the upper Rhine valley. The number of yearly reported new infections in Switzerland ranges between 60 and 120 cases, most of them with neurological manifestations. From 1984–2004 a total of 1370 cases of TBE were reported in Switzerland; and from 1999 the incidence was 1.4/100 000, but was higher in the towns (Zimmmerman & Koch, 2005). Seventy-five per cent of the diagnosed patients were hospitalized (Schwanda et al., 2000).

In the last 10 years the number of reported cases has risen from 30–70 cases per year to 60–120 cases per year. This increase was mainly caused by the increase of cases in the cantons of Thurgau, Aargau and St. Gallen, all located in the northeastern part of the country. In the canton Thurgau with 223 000 inhabitants, between 1990 and 2000, incidence increased from one to 26 registered TBE-cases per year, an incidence of up to 12 cases per 100 000 inhabitants per year and the canton became the area of highest risk in Switzerland for TBE. Between 1970 and 1980 there were only 3 cases in the canton and between 1980 and 1990 not a single case was reported (Krech, 2002). In 1996, 6 cases were diagnosed, in 1997, 15, in 1998, 10, in 1999, 26, and in 2000, 12 cases, almost all requiring hospitalization (Baumberger et al., 1996).

UK

Louping ill (LI) disease of sheep has been recognized in Scotland for centuries. It causes encephalitis and is transmitted by *Ixodes ricinus*. The LI virus has been shown to be most closely related antigenically to strains of the Western European subtype of tick-borne encephalitis virus (Stephenson et al., 1984). A very similar infection transmitted by *I. ricinus* has been reported from Bulgaria, Greece, Norway, Spain and Turkey (Gould, 2000).

Louping ill virus is transmissible to humans. Humans can develop any one of four clinical syndromes: either an influenza-type illness, a biphasic encephalitis, a poliomyelitis-like illness or a haemorrhagic fever following infection with LI virus. Transmission can take place by tick bite, exposure to aerosolized infective material, or through skin abrasions or wounds. Non-laboratory-acquired infections most frequently result from handling infected carcasses in abattoirs. The potential for oral transmission of LI virus to humans also exists where milk for human consumption is obtained from goats or sheep that are in the acute phase of the infection.

Human infection was first reported in 1934 and more than 31 human cases have now been described in the UK. Humans, though susceptible to infection with LI virus, are considered an accidental or tangential host of the virus. The occurrence of LI is closely related to the distribution of the primary vector, the sheep tick. There may well have been more cases but it appears that many cases are subclinical and, therefore, not recognized or reported. The risk of contracting the infection is higher for persons who kill infected sheep. A vaccine is available for sheep which should also reduce human disease among persons most exposed.

Ukraine

Tick-borne encephalitis is endemic in the steppe regions in the western part of the country. In the Crimea the most active and potentially dangerous foci are located in forests of mountainous areas of the peninsula coinciding with the distribution of *Ixodes ricinus*. Some 40–50 cases a year are reported. Among the residents of the Crimean mountain forest zone, 13.9% were found to have immunity to TBE, indicating wide contact of the population with the pathogen. Tick-borne encephalitis morbidity had a pronounced seasonal character with a prevalence of mild clinical forms (Evstaf'ev, 2001).

Tick-borne encephalitis is rare or absent from Portugal and Spain and little information is available from other possibly endemic countries.

Conclusions on the public health importance of TBE in Europe

Tick-borne encephalitis is a serious public health problem throughout much of its endemic areas. While use of an effective vaccine has greatly reduced the number of infections in those areas where vaccination rates are high, such as Austria, there are many areas where vaccination rates remain low and infections persist. Randolph (2001) believes that there will be an eventual decline in the distribution and incidence of TBE due to global climatic changes. On the other hand, Lindgren (1998), in a model for Sweden based on climate change, concludes that there will be an increase in TBE incidence in Stockholm County, a high-endemic region in Sweden, during the next 50 years. According to this simplified model, the annual vaccination rate needs to increase by three to four-fold during the next half century to prevent the projected increases in TBE incidence in the region. Zeman & Benes (2004) concluded that, overall, the fluctuations of TBE incidence and TBE transmission altitude ceiling are synchronous processes that correspond with temperature changes. Although the dependence of TBE on temperature is not a direct one and various factors could be involved, an impact of climate warming on the vertical disease distribution in the Czech Republic and Central Europe is evident.

Whatever the effect of climate, both the distribution and incidence of TBE infection are increasing in most endemic countries as a result of increased human exposure to ticks and despite vaccine availability. There is a need for improved surveillance and vaccination programmes in the most seriously affected endemic areas.

Crimean–Congo Haemorrhagic fever

The disease Crimean–Congo Haemorrhagic fever (CCHF) is caused by a *Nairovirus*, a group of related viruses forming one of the five genera in the Bunyaviridae family of viruses. All of the 32 members of the *Nairovirus* genus are transmitted by argasid or ixodid ticks, but only three have been implicated as causes of human disease: Dugbe and Nairobi sheep viruses, and CCHF, which is the most important human pathogen. Crimean–Congo Haemorrhagic fever was first observed in the Crimea by Russian scientists in 1944 and 1945. At that time it was established by studies in human volunteers that the aetiological agent was filterable and that the disease in man was associated with the bite of the tick *Hyalomma marginatum*. The agent was detected in the larvae and in adult ticks, as well as in the blood of patients during fever. This agent, presumably a virus, was not maintained in the laboratory and was lost. Congo virus was first isolated in Africa from the blood of a febrile patient in Zaire in 1956 and later shown to be identical to the infection detected in Crimea.

In Europe CCHF has been reported from Albania, Armenia, Azerbaijan, Bulgaria, France, Greece, Hungary, Kosovo, Macedonia, Moldova, Portugal, Russia, Turkey and Ukraine either by the occurrence of human cases, by isolations from ticks, or in serological surveys.

A severe disease in humans, CCHF has a mortality of approximately 30% or higher. There is no specific treatment and general supportive therapy is the mainstay of patient management. When CCHF patients are admitted to the hospital, there is a serious risk of nosocomial spread of infection and a number of such outbreaks have occurred. Patients with suspected or confirmed CCHF must be isolated and cared for using strict barrier nursing techniques.

The most recent outbreaks in Europe have been 8 cases in Albania in 2002 (Papa *et al.*, 2002), and in Kosovo in 2001 where the WHO reported 69 suspected cases, out of which 6 died (WHO, 2001a). From 8 May to 28 July 2002, 12 confirmed cases of CCHF were registered in Kosovo and 3 deaths (WHO, 2002b); CCHF has been persistent in Kosovo including an outbreak in 1995 with 119 cases of which 16 were fatal in Peja. On 26 July 1999 the Ministry of Health of Russia reported that CCHF in the Stavropol region between the Black Sea and the Caspian Sea had been confirmed. A total of 65 cases were reported, with 6 deaths (3 of which were among children). The outbreak appeared to be tick-transmitted. From 1975 to 1996, 279 cases were reported in Bulgaria with a fatality rate of 11.4%. Since 1997, a total of 124 cases of CCHF have occrred in Bulgaria, 27 of them fatal (Papa *et al.*, 2004). No human cases of CCHF have been reported from Greece.

The virus may infect a wide range of domestic and wild animals which become infected from the bite of infected ticks. Several tick genera may be infected, but the most efficient and common vectors for CCHF are *Hyalomma* species. Transovarial and venereal transmission have been demonstrated amongst some ticks species. The most important source for acquisition of the virus by ticks is believed to be infected small vertebrates on which immature *Hyalomma* species feed. Once infected, the tick remains infected through its developmental stages, and the mature tick may transmit the infection to large vertebrates, such as livestock. Domestic ruminant animals, such as cattle, sheep and goats, are viraemic for around 1 week after becoming infected. Humans who become infected acquire the virus from direct contact with blood or other infected tissues from livestock or may become infected from a tick bite. The majority of human cases are involved with the livestock industry, such as agricultural workers, slaughterhouse workers and veterinarians (WHO, 1998).

In western Europe and the Middle East, the most frequent vector is *Hyalomma marginatum*. In Moldavia, CCHF has been isolated from *Ixodes ricinus, Dermacentor marginatus* and *Haemaphysalis punctata* (Chumakov *et al.*, 1974) but the actual vectorial importance of these species is uncertain. However, Markeshin *et al.*, (1991) found *D. marginatus* infected with CCHF virus in the Crimea and considered it to be a vector; CCHF has also been isolated from *Rhipicephalus bursa* in Greece (Papadopoulos & Koptopoulos, 1978).

Although an inactivated, mouse brain-derived vaccine against CCHF has been used on a small scale in eastern Europe, there is no safe and effective vaccine widely available for human use. Prevention of CCHF must depend on the control of tick populations in endemic areas or the avoidance of tick bites by personal protection measures such as insecticide-impregnated clothing or topical repellents.

Bhanja virus

A *Bunyavirus* and member of the Bhanja antigen group, the virus is distributed mainly in southern Europe and the Balkans as well as Africa and Asia. Its northernmost record in Europe is an isolation from *Dermacentor marginatus* in the Czech Republic (Hubalek *et al.*, 1988). Antibodies to the virus have been detected in Romania in 1986 in ticks, humans and domestic animals (Ungureanu *et al.*, 1990). The first report of human illness due to Bhanja virus was in Dalmatia, Croatia in 1977 where an antibody rate of 31% was found on the island of Brac and the virus was isolated from *Haemaphysalis punctata* (Vesenjak Hirjan *et al.*, 1991).

Antibodies to Bhanja virus have also been detected in animals and humans or isolations made from ticks in Bulgaria, Italy, Kosovo, Portugal, the Slovak Republic and Spain. Hubalek (1987) reviewed the isolations or antibody findings of Bhanja virus and reported that Bhanja virus has been isolated in 15 countries of Asia, Africa and Europe, and antibodies against it have been detected in 15 additional countries. Vectors include ixodid ticks and species of six genera (*Haemaphysalis, Dermacentor, Hyalomma, Amblyomma, Rhipicephalus* and *Boophilus*) have yielded the virus. Bhanja virus has rarely been isolated from vertebrates but antibodies have been detected in a wide range of mammals, mainly ruminants, in birds (Passeriformes, Galliformes) and in reptiles. Antibodies to the virus were isolated from brown bears in Croatia (Madic *et al.*, 1993).

Natural foci of the Bhanja virus infections are associated with pastures of domestic ruminants infested by ticks in the regions of tropical, subtropical and partly temperate climatic zones. The high incidence of this virus in animals implies that the risk of human exposure is high. Its overall public health importance is, at present, small with no evidence of human clinical disease despite its common presence in human populations.

Thogoto virus

Thogoto virus is an influenza-like tick-borne member of the family Orthomyxoviridae; it is widely distributed in Africa and in Europe has been found in Italy (Sicily) and Portugal. Antibodies were found in human sera in Portugal (Filipe *et al.*, 1985) and it has been isolated from *Rhipicephalus bursa* in Sicily (Albanese *et al.*, 1972). Dobler (1996) classes this virus among those known to be a cause of neurological disorders although overall it does not appear to be of public health importance in Europe.

Dhori virus

Also a member of the Orthomyxoviridae, both Thogoto and Dhori viruses share structural and genetic properties with the influenza viruses. Dhori virus has been isolated from *Hyalomma marginatum* in Portugal (Filipe & Casals, 1979); antibodies were also found in humans in Portugal (Filipe *et al.*, 1985). Four strains of Dhori virus were isolated from *H. marginatum*, and from a hare in the middle zone of the Volga River delta, in the Astrakhan region in 2001, which was the first isolation of Thogoto virus from wild vertebrates (L'vov *et al.*, 2002b). The potential clinical importance of infection by this virus was shown when

5 laboratory workers were accidentally infected during the preparation of cultural agents. Clinically, Dhori infection is characterized by an acute course with marked general toxicity and a febrile period of 2–4 days. However, like Thogoto virus, this agent does not appear to be of public health importance in Europe at present. Dhori virus is closely related to Batken virus (Frese *et al.*, 1997) and they may indeed be identical.

Tribec virus

Tribec virus, an *Orbivirus*, was first isolated from *Ixodes ricinus* in Slovakia and has been reported from Belarus, Estonia, France, the Czech Republic, Hungary, Italy, Moldavia, Norway, Romania, Russia and the Ukraine either by isolation or as antibodies (Hubalek & Halouzka, 1996). In the Tribec mountains the virus circulates among rodents, goats and different stages of *I. ricinus*. Some 20 patients with CNS infection in Czechoslovakia had antibodies to Tribec (Libikova *et al.*, 1978) but its overall public health importance is very limited.

Tettnang virus

Three strains of this virus were isolated from *Ixodes ricinus* in the then Czechoslovakia; the isolates were lethal for suckling mice (Kozuch *et al.*, 1978). In Czechoslovakia, a virus, identified as Tettnang virus, was isolated from the cerebrospinal fluid (CSF) of an 18-month-old child with pharyngitis accompanied by an encephalitic reaction (Malkova *et al.*, 1980). Although the agent may have been the cause of disease, its restricted distribution and the few reports of illness indicate that the virus is not of public health importance.

Eyach virus

This *Coltivirus* virus, family Reoviridae, is closely related to the American Colorado tick fever virus. Eyach virus has been isolated from *Ixodes ricinus* in Baden-Wurttemberg, Germany (Rehse Krüpper *et al.*, 1976), and from *I. ricinus* and *I. ventalloi* in France (Chastel *et al.*, 1984). Eyach virus as well as Erve virus, also isolated in Western France in 1981–1982 from the long-tailed field mouse, or wood mouse *Apodemus sylvaticus*, are able to infect human beings and have been responsible for severe neurological disorders (Chastel, 1998). Newly developed serological techniques for the detection of this virus will help clarify its epidemiological status (Mohd Jaafar *et al.*, 2004). Eyach virus has also been found in the Netherlands and the Czech Republic. The virus is widespread in hares and probably in rodent populations.

A number of other viruses have been isolated from ticks in Europe whose importance as causative agents of disease is, at present, unknown or uncertain. Further studies will undoubtedly find still other viruses as many areas of the continent remain unstudied.

Tick-borne bacterial infections

Since the identification of *Borrelia burgdorferi* as the agent of Lyme disease in 1982, 11 tick-borne human bacterial pathogens have been described throughout Europe. These include five spotted fever rickettsiae, the agent of human granulocytic ehrlichiosis, four species of the *B. burgdorferi* complex and a new relapsing fever caused by a *Borrelia* species (Parola & Raoult, 2001). The species of *Borrelia* causing Lyme disease are closely related to the *Borrelia* species causing relapsing fever.

Tick-borne relapsing fever

Several species of spirochaete are the causative agents of tick-borne relapsing fever (TBRF) in Europe. Globally TBRF can be caused by some 15 different *Borrelia* species; these should be distinguished from louse-borne relapsing fever caused by *Borrelia recurrentis*, a disease generally associated with a higher mortality. The human body louse transmits an epidemic form and is always associated with *B. recurrentis*, whereas a soft-bodied (argasid) tick transmits endemic relapsing fever and is caused by several different *Borrelia* species. Tick-borne relapsing fever is a serious disease; if appropriately treated, it has a mortality rate of less than 5%. If acquired during pregnancy, tick-borne relapsing fever poses a high risk of foetal loss (up to 50%).

The *Borrelia* agents of TBRF are transmitted by the soft ticks of the family Argasidae, primarily *Ornithodoros* species. The reservoir hosts are rodents and other small mammals. The infection is widespread in Africa where *O. moubata* and *O. erraticus* are the vectors of *Borrelia duttonii* and *B. crocidurae*. The wild reservoir hosts of the infectious agents are usually rodents. In Europe, *B. hispanica* is the causative agent of relapsing fever in Spain although cases of disease are now only rarely detected from patients (Anda *et al.*, 1996). There have been no cases of TBRF in Portugal since the 1960s (Sofia Nuncio, personal communication). The infection has also been reported from Greece and Cyprus (Goubau, 1984). The vector in Europe is *O. erraticus*. In Sweden, Fraenkel *et al.* (2002) found a *Borrelia* closely related to the relapsing fever borrelia species i.e. *B. miyamotoi*. This is the first report of a *B. miyamotoi*-like borrelia in *Ixodes ricinus*, a hard tick in Europe.

Tick-borne relapsing fever is now of minor public health importance in Europe and is only seen as an imported disease; however, a survey near the Rhine valley of Germany has shown that relapsing fever-like spirochaetes infected 3.5% of questing *I. ricinus*. These spirochaetes differed genetically from American and Asian analogues and their public health importance is still unknown.

Lyme disease

Lyme borreliosis (LB) or Lyme disease (LD) is the most commonly reported tick-borne infection in Europe and North America and, indeed, the most commonly reported vector-borne disease in these regions. The disease is a multi-system disorder, which can affect a complex range of tissues including the skin, heart, nervous system, and to a lesser extent the eyes, kidneys and liver. As will be seen, the incidence of the disease is clearly increasing in many parts of Europe.

The first record of a clinical condition associated with this infection was recorded in Breslau, Germany in 1883, where a physician named Alfred Buchwald described a degenerative skin disorder now known as acrodermatitis chronica atrophicans (ACA). In a 1909 meeting of the Swedish Society of Dermatology, a Swedish physician, Arvid Afzelius, presented his research on an expanding, ring-like lesion he had observed though he only published his observations 12 years later (Afzelius, 1921); Afzelius speculated that the rash came from the bite of an *Ixodes* tick. The condition known as erythema chronicum migrans, or more commonly, erythema migrans (EM), occurred widely in Europe and was often found associated in *Ixodes ricinus* though the causative agent remained unknown.

In 1972 an outbreak of a disease occurred in the town of Lyme, Connecticut, USA as an epidemic form of arthritis; Steere *et al.* (1977) described the infection, observed that 'Lyme arthritis' was a previously unrecognized clinical entity, whose epidemiology suggested transmission by an arthropod vector, possibly ticks (Steere *et al.*, 1978a), specifically by the blacklegged or deer tick *Ixodes scapularis* (Steere *et al.*, 1978b). Burgdorfer (1982) later determined that a spirochaete was the causative agent of EM and it was given the name *Borrelia burgdorferi*. Lyme disease in the USA has now spread throughout almost all of the country.

To date *Borrelia burgdorferi* can be divided into at least 10 to 12 species: *B. burgdorferi* sensu stricto, present in Europe and in the USA but absent from Asia; *B. garinii, B. afzelii*, the genospecies *B. valaisiana* and *B. lusitaniae* in Eurasia; *B. japonica, B. tanukii* and *B. turdae* restricted to Japan; and *B. andersonii* and *B. bissettii* in the USA. Other tick-transmitted *Borrelia* have been identified in the USA (*B. lonestari*), and in Japan and Sweden *B. miyamotoi* has been identified.

Of the different species, only *Borrelia burgdorferi* s. str., *B. garinii* and *B. afzelii* are undoubtedly involved in clinical cases of LD. There is evidence resulting from isolations made from patients and PCR and serological data, that the division of *B. burgdorferi* s. l. into genospecies has clinical relevance. Thus, *B. burgdorferi* s. str. is most often associated with arthritis, particularly in North America where it is the only known cause of LD, *B. garinii* is associated with neurological symptoms and *B. afzelii* with the chronic skin condition, acrodermatitis chronica atrophicans (ACA). Overlap between species in relation to clinical manifestations occurs and all cause the symptom erythema migrans, though there is evidence in Europe that this sign occurs more frequently in *B. afzelii* infections than in those caused by *B. garinii*. *B. valaisiana* has so far only been associated with erythema migrans. The least information is available for *B. bissettii*, a species mostly encountered in California. Rare cases of human disease due to it have been reported in Europe.

In Europe the *Borrelia burgdorferi* s. l. complex is represented by five distinct genospecies: *B. burgdorferi* s. str., *B. afzelii*, *B. garinii*, *B. valaisiana* and *B. lusitaniae*. These taxonomic entities differ in their associations with vertebrate hosts and in provoking distinct clinical manifestations in human patients (Derdakova *et al.*, 2003). In a small proportion of untreated cases serious sequelae may occur, but LD alone does not cause death.

DNA has been amplified from ticks which have been stored under museum conditions for nearly a century. Spirochaetal DNA was detected by PCR in six ticks; the oldest was collected in 1884. *Borrelia garinii*, which predominates in modern ticks in the region in Germany, infected three of these older ticks, and the presently infrequent *B. burgdorferi* s. str. infected two ticks. These data indicate that residents of Europe have been exposed to diverse LD spirochaetes at least since 1884, concurrent with the oldest record of apparent human infection (Matuschka *et al.*, 1996).

Many mammal species have been found to be reservoir hosts of LB. The majority of these are rodents, the most important probably being field mice (*Apodemus* species), voles (*Clethrionymus* species) and squirrels (*Sciuris* species). Ungulates are involved in the epidemiology of borreliosis as maintenance hosts for the ticks. Among birds, pheasants (*Phasianus colchicus*) and blackbirds (*Turdus merula*) are reservoir-competent and infect ticks with *B. garinii*, and in the case of blackbirds, with *B. valaisiana* as well.

The incidence of Lyme disease in Europe

As only a few countries have made LD a compulsorily notifiable disease, case rates provide only an approximate estimation of the incidence in Europe. In most countries reporting is mainly from diagnostic laboratories reporting

Table 12.3 *Estimated* Lyme borreliosis *annual incidence in selected European countries*

Country	Incidence per 100 000 population	Annual number of cases
UK*	0.3	200
Ireland	0.6	30
France	16.0	7200
Germany	25.0	20 000 (++)
Switzerland*	30.4	2000
Czech Republic*	39.0	3500
Bulgaria	55.0	3500
Sweden (south)	69.0	7120
Slovenia	120.0	2000
Austria	130.0	14 000

*No published figures available. (Based on Report of WHO workshop on Lyme Borreliosis Diagnosis and Surveillance, Warsaw, Poland, 20–22 June, 1995, WHO/CDS/VPH/95. (1996) 141–1.)

patients with positive tests results. Human seroprevalence studies probably represent the best method of obtaining epidemiological data throughout Europe.

Data presented at a WHO workshop show that there is a gradient of increasing incidence from west to east with the highest incidences in central-eastern Europe (Table 12.3).

Although the clinical entity erythema migrans was known in Europe since the early part of the twentieth century, the incidence of infection appears to have been relatively stable during most of the century. In the last decades, the incidence of the disease has been increasing in most of its endemic areas. Korenberg (1998) contends that the observed increase is due to improved diagnosis, but it is generally believed that the increase is a real one. Barbour (1998) noted that LB infections were more often acquired in suburban residential and recreational areas and consequently the rise in cases of LD and the other *Ixodes* tick-borne infections is, in part, the consequence of reforestation and increased deer populations in developed countries. Human recreational activities are expanding in these same areas increasing the chance of being bitten by tick vectors. George & Chastel (2002) found that the number of persons suffering from tick-borne diseases has notably increased in the French region of Lorraine since the mid 1990s and that greater awareness of the pathology is insufficient to explain the increase; proliferation of ticks is a major factor in the increased incidence and is mainly due to a modification of the ecosystem.

The human impact on the environment has increased both the habitat suitable for ticks and their wildlife hosts, allowing tick populations to multiply. This probably accounts for a real emergence of LD in both the USA and Europe. The following section details information for some European countries with a high incidence of LD.

Austria

Tick densities are high in Austria and the incidence of LD disease is the highest in Europe, around 120/100 000 population with about 14 000 new cases reported per year. The IgG seroprevalence of healthy blood donors in Graz/Styria was 13% (Aberer *et al.*, 1999). The prevalence of antibodies to *Borrelia burgdorferi* is generally higher in TBE-endemic areas (7.7%) as compared to TBE non-endemic areas (3.8%). There was a significant increase in positive antibodies against *B. burgdorferi* with age, exposure and the number of tick bites remembered by test persons (Pierer *et al.*, 1993). The distribution of cases peaks in July and August but clinical cases are seen throughout the year. The infection is endemic in all Austrian states.

In 1997, 1163 *Ixodes ricinus* were collected in three different regions (15 localities) in Styria and examined for spirochaetes. The mean infection rate was 20.8%. Among 310 adults, 24.2% were positive and among 853 nymphs, 19.6% were positive. All 15 collection areas harboured infected nymphs with a positivity rate ranging from 5.8% (3/52) to 32.1% (18/56). Species identification by polymerase chain reaction-restriction fragment length polymorphism (PCR-RFLP) analysis revealed 16 strains of *B. garinii*, 10 *B. afzelii* and 2 *B. burgdorferi* s. str. One isolate showed a mixed population of *B. garinii* and *B. afzelii*. In collection areas, all three *Borrelia* species were present in the tick population (Stunzner *et al.*, 1998).

Because of the importance of rodents in the epidemiology of Lyme disease, Khanakahg *et al.* (2003) studied the prevalence of *Borrelia* species in rodents in Lower Austria. They found that heart specimens of 226 animals were *Borrelia* PCR positive (24%) and of the rodent species the yellow-necked mouse, *Apodemus flavicollis* (43%), and the bank vole, *Clethrionomys glareolus* (38%) were positive. *Borrelia afzelii* was most frequently identified, followed by *B. burgdorferi* s. s. and by mixed infection of *B. afzelii* and *B. burgdorferi* s. str. *Borrelia garinii* was detected only once in the common vole, *Microtus arvalis*.

Belgium

Lyme disease is widely endemic in Belgium. The incidence ranges from low near the coast to medium in the central part of the country and high in the wooded southeastern part of the country, with an overall incidence of 14.1/100 000. The disease is reportable and there is a rising trend in its incidence; in 1991, 137 cases were reported, and there were 1442 cases in 2000 and 795 in 2001.

There is an ongoing study in Belgium carried out through the general practitioner network (163 practices) which aims at estimating the number of consultations for tick bites and the incidence of LD in Belgium; 18/10 000 inhabitants consulted for a tick bite in 2003, and EM has been observed in 7.8/10 000 and a clinically diagnosed LD has been reported in 12.4/10 000.

In addition to infections caused by *Borrelia burgdorferi*, at least two separate genospecies have been described in Belgium. A relationship between infection by strains belonging to different genospecies and clinical outcome is suspected. Anthonissen *et al.* (1994) described 9 cases of Lyme arthritis attributed to infection by *B. burgdorferi* s. str., 18 cases of neuroborreliosis attributed to *B. garinii* and 1 case of ACA attributed to a strain of *B. afzelii*. Where *B. garinii* is the infecting agent, half of the reported cases of LD in the country have neurological manifestations (Machurot *et al.*, 2001).

A total of 489 ticks, collected in four locations of an endemic region of southern Belgium, were examined for the presence of the spirochaete; 23% of the ticks were found to be infected. *Borrelia garinii* was most prevalent (53% of infected ticks), followed by *B. burgdorferi* s. str. (38%) and *B. afzelii* (9%). Of the infected ticks, 40% were infected with a single species, 40% were infected with two species, and 5% were infected with all three species. For 15% of the ticks, the infecting species could not be identified (Misonne *et al.*, 1998).

Bulgaria

Although the reporting of cases of Lyme disease is not mandatory, about 500 cases are reported annually to the authorities.

Borrelia burgdorferi was first isolated in Bulgaria from adult and nymphs of *Ixodes ricinus* collected in a park in the centre of Sofia; some 18% of the ticks were infected (Tomov *et al.*, 1994). The finding of infected ticks in parks in urban areas has been reported from several cities in Europe and illustrates the urbanization of both the vector tick and the infective agent.

Angelov *et al.* (1990) carried out a serological survey of 467 at-risk individuals in Bulgaria and made observations on 134 patients with LD. Erythema migrans was the only disease symptom in most patients; 76.9% of the patients were from the city and had probably contracted the infection from ticks on visits to their dachas (holiday villas). In a group of animal farmers 15.4% had specific antibodies as did 17.8% of forestry workers. In wild and domestic animals and ticks in the LD-endemic area of the country, antibodies were found in 44.5% of sheep, 58% of goats, 74.5% of dogs and 31.6% of rodents. Of the *Ixodes ricinus* examined, 35.6%

of adults, 7.5% of nymphs and 0.6% of larvae captured from various biotopes and 17.5% of adults taken from deer and other wild animals were positive for *Borrelia burgdorferi* (Angelov *et al.*, 1993).

Christova & Komitova (2004) studied LB in Bulgarian patients analysing the clinical data and epidemiological characteristics of 1257 LD patients between 1999 and 2002. The most affected age group was 5–9 years, followed by 45–49 years, 50–54 years and 10–14 years. Most patients (68%) lived in a rural area or were bitten by ticks during activities in a rural area. Lyme borreliosis cases occurred throughout the year with two peaks – one in June and a second smaller one in September. The most common clinical manifestation was EM in 868 (69.1%) of the patients. Neuroborreliosis was the second most common presentation of LD, diagnosed in 19% of the patients. Lyme arthritis was found in 8%. Heart and ocular manifestations were noted in 1.1% and 0.9% of the patients, respectively.

The regions of Veliko, Tarnoyo and Burgas were surveyed in 1989–1990. In the Veliko region 31.5% of 215 engorged ticks of five species and 18.5% of 205 unfed ticks of four species were positive for *Borrelia*. Further, 34.2% of engorged and 21.6% of unfed *Ixodes ricinus* (79.2% of the total ticks taken) were positive for *Borrelia* as were a few unfed *Rhipicephalus bursa* (16.7%). In the Burgas region 6.2% of 193 engorged ticks of nine species and 4.5% of 285 unfed ticks of six species were positive. The dominant species was *R. sanguineus* of which 6.7% of fed and 4.5% of unfed were positive. Only a few fed *I. ricinus* were positive in this area (Georgieva *et al.*, 1993).

Both *Dermacentor marginatus* and *Hyalomma punctata* were found infected with *Borrelia burgdorferi* in an area of low endemicity of LB and may be secondary vectors (Angelov *et al.*, 1996).

Croatia

Croatia has the highest seroprevalence of LB in southern Europe (Santino *et al.*, 1997); the first case in Croatia was described in 1986. Compulsory reporting began in 1991. An annual average of some 150 cases, varying from 93 (1992) up to 335 (1996) has been reported. About 92% of the cases occur between May and August. Lyme disease occurs mostly inland and only sporadically on the Adriatic coast (Borcic *et al.*, 1999) and generally does not occur in southern Croatia; it has been only sporadically reported in the area south of Zadar. During the period 1989–91, 1529 cases of LD were reported. Erythema migrans was seen in 91% of the patients and other clinical symptoms in only 8.5%. Clinical manifestations vary, depending on the type of *Borrelia* involved.

A study of antibody prevalence in the endemic region in northeast Croatia comprised 265 subjects: 97 living in the rural region, 51 forest rangers working in the region and 117 members of the army who spent a certain period of time in the region. Positive titres of specific antibodies were found in 11.3% of the local inhabitants, 25.5% of forest rangers and 6% of the soldiers (Zivanovic *et al.*, 1991).

In a study of the prevalence of antibody to *Borrelia burgdorferi* in Croatia in various population groups, 263 individuals were examined: 93 healthy subjects (travellers); 100 controls separated into two subgroups, 50 from a high-risk zone (endemic) and 50 from a low-risk zone (non-endemic); and 70 members of a high-risk population (forestry workers). IgG antibodies were detected in 9/93 (9.7%) of the general population; 22/50 (44.0%) in the control group from the high-risk zone and 4/50 (8.0%) from the low-risk zone; and 30/70 (42.9%) among forestry workers, a group which appears to be at high risk to *B. burgdorferi* infection throughout Europe (Burek *et al.*, 1992).

Rijpkema *et al.* (1996) investigated *Borrelia burgdorferi* s. l. in *Ixodes ricinus* collected in a LB-endemic region of northern Croatia. Ticks were collected at five locations and analysed by PCR. *B. burgdorferi* s. l. DNA was detected in 56 out of 124 ticks (45%). Four genomic groups were identified: *B. afzelii* (n = 26), *B. garinii* (n = 5), *group VS116* (n = 5) and *B. burgdorferi* s. str. (n = 1). Mixed infections of *B. afzelii* with group VS116 (n = 10) and *B. afzelii* with *B. burgdorferi* s. str. (n = 1) were also detected. Eight ticks contained *B. burgdorferi* s. l., which could not be typed.

Czech Republic

The presence of LB in what is now the Czech Republic was confirmed in the mid 1980s; Jirous (1987) presented evidence that the newly found agent was very common in the country.

In 1988 Pokorny (1990) collected nearly 3000 *Ixodes ricinus* from Prague. Infection rates with *B. burgdorferi* varied from 1.9% to 22.0%. Positive ticks were found virtually in the centre of the city showing a risk of urban transmission. In a later survey in Prague, Basta *et al.* (1999) again studied *B. burgdorferi* s. l. and its individual genospecies in an urban park; the incidence of *B. burgdorferi* s. l. was 9.2% in 1995, 3.4% in 1996, 4.5% in 1997 and 2.8% in 1998. The *B. garinii* to *B. afzelii* ratio was 1.4:1 and it did not differ significantly throughout the study period. *Borrelia burgdorferi* s. str. was not detected. *Borrelia garinii/B. afzelii* co-infection was found in 5.7% of positive ticks. Among some 1000 *I. ricinus* collected in 13 localities in the city of Brno, infection rates varied between 0.9–18.6% (Pokorny & Zahradkova, 1990). Zeman *et al.* (1990) analysed the co-infection of TBE and LB in the Central Bohemian region based on clinical cases; a divergence was seen in the dispersion patterns of the diseases. Whilst TBE infections occurred in a few limited areas and its clinical cases tended to aggregate into well-defined clusters,

LD cases were scattered more or less randomly over nearly all the region without forming such marked clusters and had little topographical correlation with TBE. Between 1988–1989, 529 cases of LD were reported in the West Bohemian region; 342 (64.6%) patients reported contact with a tick (Pazdiora *et al.*, 1991).

Hubalek *et al.* (2003) monitored host-seeking *Ixodes ricinus* for borreliae in South Moravia each May from 1991–2001(150 nymphs, 100 females and 100 males each year). The mean annual percentage of infected ticks was 16.8% in nymphs, 24.9% in females and 26.1% in males. Annual incidence of LD in humans of the area in the same period (range, 8.7–41.7 per 100 000) correlated with the frequency of nymphs infected with >50 borreliae. The infection rate in *I. ricinus* correlated significantly with the North Atlantic Oscillation winter index of the last year (in nymphs) or of the year before last (in adults).

In 2000–2001, Drevova *et al.* (2003) made a study of the knowledge of young people about ticks and tick-borne diseases. Of 2763 respondents from 6 -26 years of age, more than 98% knew about the existence of ticks and almost 93% of children and 97% of adolescents knew that ticks feed on blood. The majority aged 10–26 years was convinced that ticks live on vegetation, and 23% supposed that ticks jump on humans from trees. On the other hand, 93.5% of youths knew that ticks transmit LB. The main sources of information about LD for students and pupils older than 10 years of age are television and radio (41%) and the press (38%). The frequency of contact of young people with ticks is high – 90% of children younger than 12 years, and 94% of youths from 10–26 years of age had at least once had an attached tick. Fifty-six per cent of youths older than 10 years use oil to remove an attached tick. Almost 24% remove ticks with bare hands although even children younger than 12 years of age knew that it was incorrect.

As infections by different *Borrelia* species may result in different clinical manifestations, determining the species present in a given area is essential to an understanding of the epidemiology of the infection. During the years 1996–2000, a total of 2398 *I. ricinus* were collected in three areas of southern Moravia and eastern Bohemia and examined for the spirochaetes. The prevalence of *B. burgdorferi* s. l. in *I. ricinus* rose from year to year. In 1996, prevalence was 6.8% and during 1997 and 1998, it increased to 8.4% and 12.3%, respectively. The lowest prevalence was in 1999 (3.6%), and in 2000 it increased to 4.0%. The mean rate of infection was 6.5%, and the proportions of infected ticks were 12.2% in 263 male ticks, 8.3% in 289 female ticks, 6.0% in 1621 nymphs and 1.3% in 225 larvae. From 156 highly infected ticks (>100 spirochaetes per sample), 13 isolates were obtained. Ten isolates were *B. afzelii*, and the other three were *B. garinii*. The results indicate the epidemiological importance of *B. afzelii* and *B. garinii* in central Europe (Janouskovcova *et al.*, 2004).

Demark

The WHO (1990) reported that the number of cases of LD in Denmark has been increasing; 9054 suspected cases were investigated in the laboratory in 1988, 462 of which were antibody-positive. The incidence of LD is estimated at 15–25 cases/1 million per year. A total of 2647 *I. ricinus* were collected from vegetation in 31 sites in eastern Jutland; 317 ticks (202 nymphs and 115 adults) from three sites were examined for *B. burgdorferi*; frequency of infection varied from 7–22% (Landbo & Flong, 1992). A survey in deer populations found antibodies in 52% of roe deer (*Capreolus capreolus*), 38% of fallow deer (*Dama dama*) and 27% of red deer (*Cervus elaphus*). The high antibody prevalence indicates that deer are exposed to tick-borne *B. burgdorferi* throughout Denmark (Webster & Frandsen, 1994). Six rodent species from six localities were examined for antibodies to *B. burgdorferi*. A total of 1097 specimens were tested. Wood mice (*Apodemus sylvaticus*) and yellow-necked mice (*A. flavicollis*) had high prevalences of *B. burgdorferi* antibodies of 42.1% and 27.9% respectively, but short-tailed voles (*Microtus agrestis*) also showed an exceptionally high prevalence of 32.7%. Bank voles (*Clethrionomys glareolus*) had a low average prevalence of 17.4%. The lowest and highest prevalences of rodents seropositive for *B. burgdorferi* were 6.5% and 100% for the harvest mouse (*Micromys minutus*) and the house mouse (*Mus musculus*) respectively (Frandsen *et al.*, 1995).

Lyme neuroborreliosis has become a clinical condition of some importance in Denmark and has become one of the most frequent causes of neuroinfections (Hansen, 1994).

Jensen & Frandsen (2000) found that the prevalence of *B. burgdorferi* s. l. in *I. ricinus* nymphs in Denmark was approximately 5 %. The mean abundance of infected nymphs varied from 0.3 to 4.4 per 100 m^2 according to site. The seasonal occurrence of infected nymphs in a beech forest coincided with seasonal distribution of neuroborreliosis cases. They suggested that high temperatures and low precipitation in the autumn is essential for transmission of *B. burgdorferi* s. l. to reservoir hosts or its development within ticks, ensuring high tick infectivity the following season.

France

The incidence of LD in France is not well known but is estimated by the Pasteur Institute in Paris to be in the order of 5000 to 10 000 cases per year. *Borrelia*-infected ticks are found throughout the country with the exception of a small area along the Mediterranean.

The first isolation of *B. burgdorferi* in France was reported in 1987 (Vieyres *et al.*, 1987). In a survey of 210 dogs in the Central Pyrenees region with a history of tick bites, 108 or 54.4% were found to be positive for borreliosis (Euzeby & Raffi, 1988).

Christian *et al.* (1996) estimated that there was a high incidence of LD in the region of Berry Sud and that the national incidence was 16.5/100 000. Farmers are mainly at risk of EM, which has been observed in 49% of the cases of LD seen and has a possible important economic consideration due to work loss.

A 3-year prospective study included all patients seen for suspected LD at the Strasbourg University Hospital; 132 patients, mean age 54 years, were determined to have LD. Within this study group, 77% of the patients were regularly exposed to tick bites and 64% could remember having been bitten by one. Erythema migrans (EM) occurred in 60% of the patients and was the only sign of LD in 40%. Nervous system involvement, the second most common clinical manifestation, was found in 40% of the patients and was the only sign of LD in 22%. Musculoskeletal involvement was present in 26% of the patients and was an isolated finding in 14%. The authors concluded that the clinical expression of LD in northeastern France is similar to other European countries but differs from that in North America (Lipsker *et al.*, 2001).

Ferté *et al.* (1994) reported the first isolation in France of *B. afzelii* from *I. ricinus* collected in the Marne Region. In the spring of 1994, Pichon *et al.* (1995) collected 249 unfed nymphs from vegetation in Rambouillet Park near Paris. Thirty of the nymphs were positive for *B. burgdorferi*, 19 nymphs were infected by a single species of *Borrelia*, 4 by *B. garinii*, 15 by *B. afzelii* and 6 by more than one species, 2 by both *B. burgdorferi* s. str. and *B. garinii*, 3 by *B. garinii* and *B. afzelii* and 1 by *B. burgdorferi* and *B. afzelii*. Co-infections could be due to a larval meal on a host mouse infected with more than one species of *Borrelia* or a successive interrupted meal or by mixed transovarial transmission. In a survey of adult and nymphs of *I. ricinus* in Rambouillet Park and the Fontainebleau forest, the infection rates of nymphs, males and females were 12.4% (314), 2.8% (35) and 2.9% (34) respectively. Isolates from the two forests were identified as *B. burgdorferi* in Rambouillet and *B. garinii* in Fontainebleau (Zhioua *et al.*, 1996). To determine the periods and areas of greatest risk for humans in Rambouillet Forest, a survey from September 1994 to October 1995 found significant variation in nymphal tick abundance between different zones in the Park according to the density of cervid populations. PCR was used to detect DNA of *B. burgdorferi* s. l. in 461 unfed nymphs. DNA was detected in 38 nymphs (8.2%). By genospecific PCR based on the OspA gene, three pathogenic spirochaetes were detected with occurrences of 10.3, 31.1 and 58.6 for *B. burgdorferi* s. str., *B. garinii* and *B. afzelii*, respectively, indicating that *B. afzelii* is probably the main *Borrelia* species in Rambouillet Forest. Finally, 11.5% of positive nymphs had a double infection. Nymphal infection rates were not significantly different between zones with a high density of deer (more than 100 animals per 100 ha) and zones with lower deer density (less than 20 animals per 100 ha). In addition to the role of deer as an amplifier of tick populations, these data indicate that zones with a high density of cervids are high-risk areas (Pichon *et al.*, 1999).

Gilot *et al.* (1996) examined 4673 *Ixodes ricinus* from several regions for *Borrelia burgdorferi* to obtain a comprehensive view of the spatial risk linked to the distribution of the species in France. Percentages of infection in the ticks were 4.95% in 3247 nymphs, 11.2% in 699 males and 12.5% in 727 females. Practically all the tested tick pools were positive. The percentage of tick samples absolutely free of *Borrelia*, wherever they came from, was very low (not exceeding 10% of the sampled forests). The study confirmed the widespread distribution of LB in France.

Fournier *et al.* (1998) studied relationships between the phyto-ecological characteristics of grazing pastures being infested by *I. ricinus* and, hence, the risk of transmission of *Borrelia*. Some 128 pastures in 20 dairy farms in western France were observed from April to July 1994. The average infestation rate was 40.2%. Types of pastures were significantly related to infestation rates, making it possible to provide a predictive value in risk assessment. Tick infestation rates were high (96% on average) in two types of pastures characterized by their proximity to woods, and low (13%) in two other types characterized by seeded grass species at some distance from woods, and intermediate (39% on average) in the last two types.

Many cases of LD have been reported in the region of Lyon. Quessada *et al.* (2003) investigated the identification and prevalence of *Borrelia burgdorferi* s. l. in *Ixodes ricinus* in the surroundings of Lyon between October 1994 and September 1995 and in June 1998. The overall prevalence of *B. burgdorferi* s. l. was 13.2% (91/688). No significant differences in prevalence were noted between different stages and sex of the ixodids or between collection areas. The majority of infections were simple infections (82.4%; 75/91), mostly due to *B. afzelii* (41.4%); co-infections (12.1%) were predominantly (54.5%) a combination of *B. valaisiana* and *B. garinii*. No tick was infected with more than two borrelia species, nor was *B. lusitaniae* identified. *Borrelia valaisiana* species was detected for the first time in France. Thus the surroundings of Lyon are risk areas for contracting LD; no particular clinical manifestations predominate due to the heterogeneous distribution of *Borrelia* genospecies.

While the incidence of LD in France is not accurately known, the estimate that there are between 5000 and 10 000 cases annually and the widespread presence of *Borrelia* in tick populations indicate that the infection is of public health importance in the country. Improved surveillance would provide better information of the areas of risk.

Germany

The annual number of cases of LD in Germany has been estimated as between 20 000 and 60 000 (Wagner, 1999). *Borrelia*-specific antibody is said to be detectable in some 7% of German residents. This extraordinarily high frequency of LD may be related to the large number of Germans living in forested areas or who frequent forests for recreational purposes. LD is not a notifiable disease in Germany, but six of Germany's 16 states – Berlin, Brandenburg, Mecklenburg-Vorpommern, Sachsen, Sachsen-Anhalt and Thüringen, have enhanced notification systems, which do include Lyme borreliosis.

The first record of a condition now associated with Lyme disease dates back to 1883 in Breslau, Germany, where a physician named Alfred Buchwald described a degenerative skin disorder now known as acrodermatitis chronica atrophicans (ACA). The meningopolyneuritis first described in 1922 by Garin and Bujadoux develops after a tick bite. In some of the cases, erythema migrans (EM) develops first. The first clinical description of LD and the importance of its association with tick bite was by Ackermann (1976). Ackermann *et al.* (1984) recovered 19 isolates of a spirochaete from *I. ricinus* ticks from endemic locations of EM in North Rhine-Westphalia; the infection rate was 16%. The sera of 90 patients with EM from this area showed antibody titres, which correlated with the clinical course. Similarly, antibodies were demonstrated in the sera of 21 patients with ACA. The results suggested to the authors an aetiologic role for the *I. ricinus* spirochaete in European EM disease.

It was apparent that the infectious agent causing EM was widely distributed in Germany. Schmidt *et al.* (1987) found that during a period of 19 months, serological and clinical investigation of 2955 patients rendered 1106 cases of infection with widespread incidence: of the 328 administration districts of Germany, 205 were affected. Antibodies against *B. burgdorferi* were demonstrated in an average of 15.7% of the rural population (2830 persons). Typical clinical signs were encountered in 817 of 1106 infected persons with a broad spectrum of symptoms.

After the realization that EM and other clinical symptoms were associated with tick-borne infection, increasing attention was given to determining the distribution and rates of infection with *B. burgdorferi* to ascertain the degree of risk of LD in different regions. In southern Germany, among 2403 *I. ricinus* tested in 1985 for *Borrelia*, 328 (13.6%) were infected (adults about 20%, nymphs about 10%, larvae about 1%). The highest prevalence of infected ticks was in the Isar region north of Munich (33.8%). Among 9383 persons whose serum was examined in 1985 and 1986, 1035 (11%) had raised *Borrelia* antibodies greater than, or equal to, 1:64. In 18.7% only IgM antibodies were demonstrated. Among 375 proven cases there were 78 with EM, 211 with neurological signs, 48 with Lyme arthritis and 36 with acrodermatitis (Wilske *et al.*, 1987).

Studies have been carried out in Germany to identify possible reservoir hosts for LB in the country and to determine if *I. ricinus* feeds frequently on certain species of rodents and if the abundance of these hosts corresponds to the seasonal feeding activity of the tick. The most abundant rodents in the study sites were yellow-necked field mice (*Apodemus flavicollis*), which predominated in a wooded site, and bank voles (*Clethrionomys glareolus*), in brush or grass-covered sites. Although *A. flavicollis* comprised only about a third of rodents collected, nearly 60% of all *I. ricinus* fed on this mouse. These ticks were more abundant on mice than voles in each of the study sites and throughout the year, and more larvae fed on these rodents than did nymphs. These observations suggest *A. flavicollis* as an important reservoir host for *I. ricinus*-borne infections (Matuschka *et al.*, 1990). In a survey on the presence of *B. burgdorferi* in domestic animals in Berlin, 189 dogs, 29 cats, 224 horses and 194 cows were investigated; 5.8% of the dogs and 24.5% of the cows investigated showed a positive reaction at titres of 1:128 or higher. Horses and cats gave negative results. With an enzyme-linked immunosorbent assay (ELISA), 10.1% of the dogs, 16.1% of the horses and 66% of the local cows showed positive reactions showing domestic animal contact with *B. burgdorferi* (Kasbohrer & Schonberg, 1990). Foxes are common in Berlin; 15% of 100 red foxes (*Vulpes vulpes*) examined in the city had antibodies to *B. burgdorferi* (Schoffel *et al.*, 1991). Ticks may be so frequent in the urban parks of European cities that all Norway rats (*Rattus norvegicus*) in the parks are infested by larvae and nymphs of *I. ricinus* and all the rats are infected by *B. burgdorferi* (Matuschka *et al.*, 1997). *Borrelia afzelii* is well adapted to these reservoir hosts and ticks detach in a manner that concentrates them in rodent burrows.

Birds are frequently positive for *B. burgdorferi*. Kaiser *et al.* (2002) determined the presence of *B. burgdorferi* s. l. DNA in tick larvae feeding on birds at study sites along the Rhine valley in SW Germany between August 1999 and March 2001. A total of 987 *I. ricinus* larvae were collected from 225 birds of 20 species. *Borrelia* DNA was analysed in the ticks and for the nightingale (*Luscinia megarhynchos*) blood samples were taken. *Borrelia* DNA was detected in 6 out of 9 larval ticks from the nightingale, in 1 out of 10 ticks from the dunnock, in 3 of 9 ticks from the chiffchaff (*Phylloscopus collybita*), and in 2 out of 21 larval ticks from reed warblers (*Acrocephalus scirpaceus*). Five out of nine ticks removed from robins (*Erithacus rubecula*) in winter were *Borrelia* positive. Blood samples of nightingales were positive in 71 of 138 birds (51%).

In the 1990s the presence in Germany of at least three genomic groups of *Borrelia burgdorferi* (*B. burgdorferi* s. str., *B. garinii* and *B. afzelii*) were recognized.

Borrelia burgdorferi s. l., the agent of LB has a very considerable degree of hetero-geneity in Europe.

Schaarschmidt *et al.* (2001) considered that the prevalence of different genospecies of *Borrelia burgdorferi* s. l. in infected ticks could be a determinant for the risk of acquiring LB. They collected 373 ticks and 2761 samples from LD patients in an area of southwest Germany, which were analysed to assess the frequency of the occurrence of LB-associated genospecies. Fifteen per cent of the tick samples and 19% of the human samples were positive for *B. burgdorferi* s. l. Further identification of 1106 *B. burgdorferi* s. l. positive tick samples revealed the occurrence of *B. afzelii*, *B. burgdorferi* s. str., *B. garinii* and *B. valaisiana*. Both single-species and mixed infections were noted and a similar distribution of the different genospecies was found in ticks and human samples. The distribution of the genospecies in ticks is the decisive factor for the occurrence of the different *Borrelia* genospecies in samples from LD patients. *Borrelia afzelii* was the predom-inant genospecies in all samples from the area. There seemed to be no asso-ciation of particular *Borrelia* genospecies with distinct clinical manifestations of LD.

In studies near Bonn, *I. ricinus* were collected in three areas and examined for infection with *B. burgdorferi* s. l. In 2001, 1754 ticks were collected; 374 ticks were analysed for *B. burgdorferi* s. l. by both an immunofluorescence assay IFA and at least two different PCR tests and 171 ticks were analysed by PCR only. By combining all assays, an average of 14% of the ticks tested positive for *B. burgdorferi* s. l.: 5.5%, 15.8% and 21.8% in the three collection areas. Of nymphs and adults examined, 12.9% and 21.1% respectively were infected. Genotyping of *B. burgdorferi* s. l. revealed high relative prevalences of *B. valaisiana* in 43.1% of infected ticks and *B. garinii* (32.3%), but *B. afzelii* (12.3%) and *B. burgdorferi* s. str. (1.5%) were relatively rare. *Borrelia burgdorferi* s. l. infection has increased since a survey 15 years earlier due to unknown ecological changes, perhaps related to climate change or wildlife management (Kampen *et al.*, 2004).

Knowledge of the distribution of tick-borne agents and the risks of contract-ing disease are essential to ensure an effective response and to decide the appro-priateness of vaccination should it become available (Holbach & Oehme, 2002). The prevalence of TBE-virus and *B. burgdorferi* in ticks was investigated and the respective risk of contracting disease from a bite assessed in the town of Lohr a. M. (Bavaria). A total of 1657 ticks obtained from five different biotopes around Lohr were examined for tick-borne encephalitis virus, and 408 ticks for *B. burgdorferi*. The risk of contracting illness was estimated on the basis of trans-mission and manifestation rates, together with epidemiological data from the region. The prevalence of TBE virus was 0.12% (95% CI: 0.05–0.44%) in the ticks investigated. These are comparable with rates in four other regions rated as

TBE-risk regions, but significantly lower than that in rated high-risk regions. *Borrelia burgdorferi* was detected in 14.9% (11.8–19.0%) of adult ticks, roughly twice the prevalence found in nymphs (7.2%, range 4.6–11.7%). The risk of contracting meningitis/encephalitis from a tick bite was calculated to be about 1:10 000, and the risk for LD to be about 1:100, the latter requiring that the tick remains attached for at least 2–3 days.

The prevalence of LB and LD in Germany is quite high and the incidence of human cases is increasing. The presence of several genotypes of *Borrelia* and the occurrence of multiple infections in ticks results in multiple infections in humans. The reasons for the increasing incidence are uncertain but they may be due to climate or ecological changes. Changes in the natural dynamics of European tick-borne zoonoses appear to have occurred towards the end of the twentieth century, largely brought about by human impact on the habitat and wildlife hosts of ticks. Purely climatic factors may have played some part. At the same time, raised awareness of ticks as vectors, and the intense interest in LB-In Germany as elsewhere, have undoubtedly stimulated surveillance and protective measures.

Italy

Erythema migrans (EM) was first described in Italy only in 1971 although Italian dermatologists had been familiar with the symptom for a long time. In 1983, the first case of LD was identified. The first isolation of *B. burgdorferi* was made from *I. ricinus* in the Trieste area (Cinco *et al.*, 1989). The endemic areas are mainly the Liguria, Friuli-Venezia and Giulia regions, the areas bordering the Alps and the region surrounding the city of Bologna. In Liguria, the incidence of Lyme disease is about 17/100 000 inhabitants per year. The infection is also found in Sicily at a low prevalence (Cimmino *et al.*, 1992). The number of cases in the country in the past is not known because Lyme disease was not a notifiable disease in Italy until 1990, but, on the basis of literature data, at least 1324 cases have been observed in the 14-year period 1983–1996; the infection is probably underestimated.

In the Italian Alpine regions from 1987–1991, antibodies to *B. burgdorferi* were found in 19% of the rangers and forestry workers, with lower values in farmers (10%) and hunters (8%); most infections were asymptomatic (Nuti *et al.*, 1993). Cinco *et al.* (1993) took serum samples from all the forestry workers of the Friuli Venezia Giulia region in the spring and autumn and tested for IgG antibodies of *B. burgdorferi*. Seroprevalence was significantly higher in this population when compared with the controls (18.5% vs. 13.3–14.5%) and increased in the 6-month interval giving an incidence of 6.56. Only 6 individuals among the seropositives in the first serum sampling had LD symptoms and 8 new subjects developed LD-related pathologies during the 6-month observation.

There is evidence that the incidence of infection is increasing in Italy. In 1996, Grazioli reported that in the province of Belluno, a mountainous area in the Veneto region of Italy, 284 cases of Lyme disease had been seen in 1977–1993; since that time 100 new cases have been reported.

The seroprevalence of *B. burgdorferi* in patients or at-risk subjects varies in northern Italy from the lowest incidence in Lombardia (3.2%) to the highest in Friuli (22.3%) and in Central Italy, from the lowest incidence in Emilia (Parma) (0.2%) to the highest in Toscana (18.3%). The range of antibodies to *B. burgdorferi* in blood donors or control subjects shows the lowest rates are in Lazio (1.5%), while the highest are in Sicilia (10.9%) (Santino *et al.*, 1997). It is often difficult to draw valid comparisons as different serological methods are used to detect *B. burgdorferi*. The overall findings in a survey in the Friuli Venezia Giulia (FVG) region indicated that *B. afzelii* and *B. garinii* are the prevalent genospecies in the area though strains of *B. burgdorferi* s. str. have been isolated (Cinco *et al.*, 1995). In forest workers the prevalence may be up to 27.2%.

A study on rates of infection of *I. ricinus* with *B. burgdorferi* s. l. was carried out in an endemic focus of LB in the Trieste area. A total of 227 ticks collected in 10 different stations were tested individually for the presence of the spirochaetes using PCR techniques able to identify both *B. burgdorferi* s. l. and the four genospecies (*B. burgdorferi* s. str., *B. garinii*, *B. afzelii* and group VS116 (*B. valaisiana*)). Multiple infections of individual ticks were found. Infection rates ranged from 0–70%. Infection of *I. ricinus* with *B. burgdorferi* group VS116 was found for the first time in Italy in endemic foci of LD (Cinco *et al.*, 1998).

No cases of LD had been reported in the Calabria region of southern Italy where a study was carried out by Santino *et al.* (1996) on 300 patients. The results showed a positivity rate of 7.3% with ELISA and 4.5% with Western Blot. Lyme borreliosis was found to a degree of positivity second to that of Campania (9.1%) and higher than Lombardy (3.2%) and Umbria (2.8%). Further surveys will no doubt show other foci in Italy.

The Netherlands

All four genotypes of *B. burgdorferi*, i.e. *B. burgdorferi* s. str., *B. afzelii*, *B. garinii* and *B. valaisiana*, have been found in the Netherlands. Santino *et al.* (1997) reported that the incidence of antibodies to *B. burgdorferi* in patients or in at-risk subjects in the Netherlands was 28%.

In 1989 a study was made of the rate of infection of *B. burgdorferi* in 1838 nymphs and adults of *I. ricinus* from 20 locations; of the pooled ticks, 63 isolates were identified as *B. burgdorferi*. In nymphs the minimal infection rate was 2.4%

and in adults it was 14.3%. In all locations examined *I. ricinus* was infected with *B. burgdorferi*. Thus, the infection is widespread in ticks in the country and the risk of acquiring LD by a tick bite is high (Nolmans *et al.*, 1990).

As in other countries, forestry workers are a high-risk group for LD in the Netherlands (Kuiper *et al.* 1991). Of 127 forestry workers tested by Western blot, seven (6%) were classified as being a case of LD. In only one person had a clinical diagnosis been made before the investigation. Five people had a history of EM, 1 of arthritis and 1 of persistent infection. The disease was often not diagnosed among this high-risk group.

In a survey in April 1995, all general practitioners (GPs) in the Netherlands were asked to complete a postal questionnaire on the number of tick bites and EM case-patients they had seen in 1994 and the size of their practice. The response rate was 79.9%. In 1994, GPs reported seeing approximately 33 000 patients with tick bites and 6500 with EM. The incidence rate of EM was estimated at 4.3/10 000 population. Ecological risk factors for both tick bites and EM were the proportion of the area covered by woods, sandy soil, dry uncultivated land, the number of tourist-nights per inhabitant and sheep population density (de Mik *et al.*, 1997).

In a second postal survey in April 2002, all GPs were asked to complete a short questionnaire on the number of cases of tick bites and EM seen in 2001 and the results were compared with the results of the 1995 study. The response of the GPs was 64.5% (4730/7330). Altogether, GPs reported seeing approximately 61 000 patients in 2001 with tick bites and 12 000 patients with EM. The incidence of EM was estimated at 73/100 000 inhabitants. There were obvious risk areas. At the municipal level, tick bites and EM were positively associated with the area covered by forest, sandy soil, the number of roe deer (*Capreolus capreolus*) and tourism. There was a negative association with the degree of urbanization. Increases in tourism in areas with many ticks, new forests in urban regions and an increased number of horses were positively associated with increase in tick bites and EM since 1994. The number of patients with tick bites and EM seen by GPs has doubled between 1994 and 2001. This increase may be due to changes in ecological risk factors or human behaviour (Boon *et al.*, 2004). Thus, LD is a growing problem in the Netherlands.

Poland

There has been a rapid increase in LD in Poland since 1992; the infection is widely endemic with high percentages of ticks infected with *Borrelia burgdorferi* in surveyed areas of the country. *Borrelia burgdorferi* was first isolated from *Ixodes ricinus* in 1994 in Olsztyn province (Dabrowski *et al.*, 1994). All four genotypes are present (Sinski and Rijpkema, 1997). In 1996–2000, 4933 of borreliosis cases were registered in Poland (Pancewicz *et al.*, 2001).

In the Gdansk region in 1993–1995, the incidence of diagnosed cases of LD has increased markedly; 1993 with 38 cases, 1994 with 50 cases and 96 cases in 1995 (Ellert Zygadlowska *et al.*, 1996). Infection is often related to occupation; in a study in the Lublin city region in the mid 1990s of 836 forestry workers and 56 farmers, there was a 26% positive response among the forestry workers and 11% among farmers; by ELISA 38.6% of the forestry workers and 28% of the farmers were positive (Chmielewska Badora, 1998). A study in Szczecin Province on antibody prevalence among foresters from four districts and a group of people sporadically exposed to tick bite showed that the infection was not restricted to forest areas but was present in some parts of Szczecin and in parks and gardens in its suburbs (Niscigorska, 1999).

A clinical review carried out in northeastern Poland of data on 262 421 case histories from 15 regional outpatient departments was retrospectively analysed and 1235 cases of EM were found. The annual number of cases increased from only a few, in the years 1969–1976, to over 436 in 1993 (Chodynicka & Flisiak, 1998). In Podlasie Province, LD incidence rose from 9.09% in 1996 to 32.2% in 2000, rising to 43.56% among forestry workers; morbidity increased with age (Pancewicz *et al.*, 2001).

In 1996–1998, 2285 *Ixodes ricinus* (1063 nymphs, 637 males, 585 females) were collected from 25 different localities in the eight provinces and examined for *Borellia burgdorferi* by PCR and IFA. The overall infection rate in ticks was 10.2%. The highest percentage of infected *I. ricinus* (37.5%) was in Katowice province and the lowest (4.1%) in Bialystok province, thus *B. burgdorferi* s. l. is present throughout the distributional areas of *I. ricinus* in Poland (Stanczak *et al.*, 1999). In the West Pomeranian province PCR showed the dominant genospecies to be *B. burgdorferi* s. str. (83.5% of ticks in 2000 and 87.7% in 2001), followed by *B. garinii* (12% and 10.2% in 2000 and 2001, respectively). *Borrelia afzelii* was detected in 4.4% of the ticks in 2000 and 2% in 2001. Co-infection with *B. burgdorferi* s. str. and *B. garinii* occurred in 3.4% in 2000 and 1.8% of ticks in 2001 (Bukowska, 2002).

Rodents are important reservoir hosts of LB in Poland as shown in a study of bank voles (*Clethrionomys glareolus*), yellow-necked mice (*Apodemus flavicollis*) and common voles (*Microtus arvalis*) from the Mazury Lakes district. *Clethrionomys glareolus* had an exceptionally high prevalence of *B. burgdorferi* antibodies (58%) and *A. flavicollis* and *M. arvalis* showed significant prevalence of 16.6% and 10.5%, respectively (Pawelczyk & Sinski, 2000). In the same area 1254 birds of 42 species were captured in a 3-year study period; overall, PCR positive results were obtained in 4.2% of all blood samples from passerine birds (Gryczynska *et al.*, 2004).

Lyme disease represents one of the most serious problems of infectious occupational diseases in Poland. It is also found among persons with only occasional contact with ticks such as in city parks where a single tick bite may lead to LD. The incidence of infections with LB is increasing in Poland.

Russia

There are at least 5000–7000 LD cases annually. Korenberg (1998) considered that with better diagnosis, the number of cases per year in Russia would be 10 000 to 12 000. In an unsigned article in *Med. Parazitol.* (2005 Apr–Jun;(2):40–42) reviewing Lyme disease in Russia, the statement is made that 'The proportion of Lyme borreliosis seropositive residents in Russia is 1–2 thousand times greater than that of officially notified patients'. Whatever the actual incidence, the disease is very widespread and of major public health importance in the country.

The vectors are *I. ricinus* in the west of Russia and *I. persulcatus* in the east. Infection rates are up to 30% in *I. ricinus* and up to 50–60% in *I. persulcatus*. Where both LD and TBE coexist in the same area, the incidence of LD is generally much higher. Cases of LD have been registered in 46 of 50 Russian regions inhabited by *I. persulcatus* and *I. ricinus* ticks.

The identity of EM with LD and the relation with *B. burgdorferi* were confirmed in Russia by Dekonenko *et al.* (1988); by 1992 LB was recognized as being present from the Baltic to the far east. The range of human LD in Russia coincides with that of tick-borne encephalitis (TBE). Although the vectors are identical, there are twice as many people who suffer from LD in Russia than TBE. Risk of infection in those having occupational contact with forest is greater than in other workers; 94% of the reported cases occur in the Russian Republic (Korenberg, 1994). Clinical symptoms of LD in Russia are similar to those in western Europe.

Borrelia burgdorferi s. str., the classical causative agent of LD was first isolated in Russia in 2000 from *Ixodes ricinus*. *Borrelia garinii, B. afzelii, B. valaisiana* and *B. lusitaniae* had been found earlier (Gorelova *et al.*, 2001).

Kovalevskii & Korenberg (1995) compared parasite loads in the two tick vectors in the area of Leningrad (St. Petersburg) where the species are sympatric. The average number of *Borrelia* in *I. persulcatus* and *I. ricinus* were 34.7 and 23.3 per 100 microscopic fields respectively. The maximal values each year were several hundred times the minimal values. Ticks carrying a relatively small number of *Borrelia* predominated. Parameters were generally significantly higher in areas with a predominance of *I. persulcatus* and therefore risk of infection with LB is greater in *I. persulcatus*-dominant areas than where *I. ricinus* predominates.

As has been stated above, it is evident that LD is widespread in Russia. As described earlier tick control programmes have ceased and little more can be done against the infections than make efforts to educate the public on the hazards of tick bites. The actual annual incidence of LD in Russia is uncertain but it is certainly an important public health problem.

Slovenia

Lyme disease has been notifiable in Slovenia since 1987. *Borrelia* infections have replaced tick-borne encephalitis as the most prevalent tick-borne illness. Three *Borrelia* species have been found in *I. ricinus* ticks with some regions showing >40% of examined ticks positive for spirochaetes. In 1993, 2267 new cases were registered and in 1994, 2733 cases. The highest incidence was in the centre of the country and erythema migrans (EM) was the most frequent manifestation (Strle & Stantic Pavlinic, 1996). The incidence has been increasing. In 1997 155/100 000 cases were recorded and in some regions the incidence was substantially higher. The disease affects both sexes (as a rule more often women than men) and all age groups. The incidence is the highest in persons 30–50 years of age, followed by children aged 6–15 years.

The first isolation of *B. burgdorferi* s. l. from Slovenian patients was made in 1988; the first isolation from ticks was in 1993. Thus far four *Borrelia* species have been found by isolation to cause disease in humans: *B. afzelii*, *B. garinii*, *B. burgdorferi* s. str. and *B. bissettii*. The majority of typed isolates belong to *B. afzelii* (Strle, 1999).

Ruzic Sabljic *et al.* (2002) assessed the genotypic and phenotypic diversity among clinical isolates of *B. burgdorferi* s. l. from patients in Slovenia. Among 706 *B. burgdorferi* s. l. human isolates, 599 (85%) were *B. afzelii*, 101 (14%) *B. garinii* and 6 (1%) *B. burgdorferi* s. str. Heterogeneity of *Borrelia* strains may play a significant role in the virulence and pathogenesis of the infection.

Zore *et al.* (2003) studied ticks in bird populations in Slovenia; *B. burgdorferi* s. l. was isolated from 13 ticks. *Borrelia valaisiana* was isolated from seven ticks from three black birds (*Turdus merula*). *Borrelia valaisiana* was also isolated from the skin of one of those blackbirds. *Borrelia afzelii* was isolated from the skin and from a tick of one hedge accentor or dunnock (*Prunella modularis*). *Borrelia garinii* was isolated from five ticks collected from various bird species.

Spain and Portugal

Lyme disease was not usually diagnosed in Spanish patients before 1987 and was first diagnosed in Portugal in 1989. The estimated yearly number of cases in Spain is about 500. The increasing number of LD cases in northern Spain represents a public health problem; in a study of hospitalized patients in Vizcaya, neurologic manifestations were most common (73%), followed by erythema migrans (EM) (62%), arthralgias (38%) and arthritis (15%). Fifty-eight per cent of the patients recalled a tick bite and rural professional or recreational activities were the main risk factors (Arteaga Perez & Garcia Monco, 1998).

Borrelia burgdorferi presence was determined in Ixodes ricinus collected in the Basque Country of Spain; nine isolates of B. burgdorferi were obtained, belonging to four genospecies (B. burgdorferi s. str., B. garinii, B. valaisiana and B. lusitaniae). Among 1535 ticks examined, B. burgdorferi DNA was detected in a mean of 9.3% and 1.5% of adults and nymphs, respectively and the authors concluded that the frequency of the agent in ticks in the Basque country is increasing (Barral et al., 2002). Borrelia burgdorferi s. l., B. afzelii, B. garinii, B. valaisiana and B. lusitaniae genotypes have also been isolated from questing ticks in the north of Spain (Escudero et al., 2000).

Borrelia lusitaniae is the only genospecies of B. burgdorferi s. l. isolated from Ixodes ricinus in Portugal and Tunisia; this genospecies appears to be restricted to the Mediterranean Basin. The first isolate of B. lusitaniae from a human patient and the first human isolate of a Borrelia in Portugal were reported in 2004 (Collares Pereira et al., 2004). Borrelia valaisiana has been isolated from a single human sample in Portugal. Baptista et al. (2004) genotyped spirochaete isolates from 719 I. ricinus from two different sites; five B. garinii, eight B. lusitaniae and one B. valaisiana strain were found. The authors believed that these findings indicate that B. burgdorferi s. l. agents are differentially maintained in nature in local tick populations in different geographic areas across Portugal and that the risk of acquiring LD in certain areas of Portugal is higher than previously assumed.

The presence of LB in ticks on subtropical Madeira Island was surveyed. Though spirochaetes infected <1% of the ticks sampled, the infections included all three genospecies implicated in human disease and spirochaetal diversity is, therefore, as great at the southern margin as it is in the centre of this pathogen's range (Matuschka et al., 1998).

Sweden

About 2000 cases of LD and 50–80 cases of TBE are estimated to occur in Sweden annually. The incidence of LD reported in 1997 was 69/100 000.

A population of 903 people in five LB- and TBE-endemic areas close to Stockholm was studied with regard to clinical manifestations and antibody prevalence. The study areas involved four groups of islands in the Baltic Sea and one island in Lake Malaren. A history of LB was reported by 1–21% of the participants and antibodies to B. burgdorferi were found in 7–29% of the individuals from the various areas. Also studied were 362 orienteers from the county of Stockholm; a past history of LB was reported by 6% of the individuals and 9% of them were seropositive. A past history of TBE was reported by 0.3% of the orienteers and 1% of the individuals were seropositive. A total of 3141 I. ricinus, 2740 adults and

401 nymphs, were collected from different localities in 23 of the 25 provinces in Sweden. The prevalence of *Borrelia*-infected ticks varied from 10–20% in the southern and central parts of Sweden to about 5% in the north (Norrland). Only *I. ricinus* was positive to *Borrelia* (Gustafson, 1994).

The first isolates of *B. burgdorferi* in Sweden were made from 13% of *I. ricinus* collected from an island near Stockholm where human borreliosis is endemic. *Borrelia burgdorferi* was cultivated from wild rodents, bank voles (*Clethrionomys glareolus*) and yellow-necked mice (*Apodemus flavicollis*) on the island (Hovmark *et al.*, 1988).

All the known human-pathogenic species (*B. garinii*, *B. afzelii* and *B. burgdorferi* s. str.) and *B. valaisiana* found elsewhere in Europe are present in the Swedish tick population; a *B. miyamotoi*-like *Borrelia* species seems to be present in *I. ricinus* ticks in Europe (Fraenkel *et al.*, 2002).

Borrelia burgdorferi s. l. is present throughout the distributional area of *I. ricinus* in Sweden. On average, 10% of nymphs and 15% of adult *I. ricinus* are positive for *Borrelia* (Gustafson *et al.*, 1995).

Climate change may be having an effect on the distribution of *I. ricinus* and the distribution and incidence of LD in Sweden. The findings of Lindgren *et al.*, (2000) indicating a northward shift of the northern distribution limit of the tick, have been referred to. With the extension of the range of the tick vector, the incidence of LD, as well as TBE, may increase.

Switzerland

As early as 1969, Aeschlimann *et al.* (1969) included erythema migrans among the possible tick-borne diseases in the country, observing that the condition was quite common. Caflisch *et al.* (1984) reported three paediatric cases of tick-borne meningoradiculitis noting that the disease is characterized by a sequence of symptoms including EM. Some 20% of *Ixodes ricinus* collected in the area of Lucerne were infected by spirochaetes, suggesting that the condition was spirochaetal and tick-borne. Five *Borrelia* species have been identified in the country: *B. garinii*, *B. burgdorferi* s. str., *B. afzelii*, *B. valaisiana* and *B. lusitaniae* (Jouda *et al.*, 2004).

The incidence of antibodies to LB among people at risk in Switzerland is about 26%, which is among the highest in Europe (Santino *et al.*, 1997). Risk occurs primarily from May through October in wooded, forested areas below 1500 metres elevation in an area extending from Lake Geneva in the west to Lake Bodensee in the northeast. Risk may be elevated on the northern Swiss plateau. Starting in 1988, LD has been monitored and the yearly incidence is between 1500 and 2000 cases. Asymptomatic infections are common.

Satz & Knoblauch (1989) estimated that 5–35% of *I. ricinus* in Switzerland carry *B. burgdorferi*, that there is a high risk of transmission of this infectious agent from any tick bite and 4–5% of affected subjects subsequently contract LB. Ticks on rodents are heavily infected. In one study, about 14% of larvae and 50% of nymphs collected on small mammals were infected; prevalence of infected rodents ranged from 20–44% (Humair *et al.*, 1993).

United Kingdom

Probably the first report of LD in the UK was that of a child in Hampshire (Williams *et al.*, 1986). Surveillance indicates that LD has increased in the UK from 0.06/100 000 during 1986–1992 to 0.32/100 000 since 1996. Case reports peak in the third quarter of each year. Lyme disease appears to be relatively rare in the UK although *B. burgdorferi* s. l. has been detected by PCR in many tick populations and in archived specimens collected over the last 100 years. Forty-five per cent of reports are of cases in three contiguous counties in southern England: Hampshire, Wiltshire and Dorset. This area includes foci of LB in and near the New Forest and Salisbury Plain. Other counties with a relatively high number of cases are Devon, Somerset and Norfolk. One hundred and eighteen (14.9%) of the 796 cases were reported to have been acquired abroad, mainly in the USA, France, Germany, Austria and Scandinavia. Most of these cases occurred in tourists. Forty-six (5.8%) cases were apparently acquired occupationally: 9 through deer hunting, 7 through forestry work, 3 through farm work, 1 through exposure during tick ecology studies, and another through work as a veterinary surgeon. For the others, occupational details were not reported. The proportion of LD patients with a history of tick bites has risen steadily, from 24% in 1986–1992 and 32% in 1993–1996 to 39% in 1997–1998 (Smith *et al.*, 2000).

A study was made of the genetic diversity of *Borellia burgdorferi* s. l. in southern England in questing *Ixodes ricinus*; three genospecies of *B. burgdorferi* s. l. were detected, with the highest prevalences found for *B. garinii* and *B. valaisiana*. *Borrelia burgdorferi* s. str. was rare (<1%) in all tick stages. *Borrelia afzelii* was not detected in any of the samples. More than 50% of engorged nymphs collected from pheasants (*Phasianus colchicus*) were infected with borreliae, mainly *B. garinii* and/or *B. valaisiana* (Kurtenbach *et al.*, 1998). *Ixodes ricinus* has been confirmed as the vector for Lyme disease in the UK (Muhlemann and Wright, 1997).

Conclusions on the public health importance of Lyme disease in Europe

Lyme disease is the most important vector-borne disease in Europe and a growing public health problem, possibly due to changing ecological conditions which result in greater contacts of human populations with ticks. In the absence of a usable vaccine, control can only be through avoidance of tick bites by personal protection measures.

Tick-borne rickettsial infections

The tick-borne rickettsial diseases in Europe include Boutonneuse spotted fever, species of ehrlichiosis, rickettsial pox and Q fever.

Boutonneuse fever or Mediterranean spotted fever

Boutonneuse fever is also known as tick typhus and Mediterranean spotted fever (MSF); the causative agent is *Rickettsia conorii* and the vector in Europe is the brown dog tick (*Rhipicephalus sanguineus*) although other tick species are occasionally infected. The infection is endemic through many parts of Africa and Asia and in the Mediterranean area of Europe; in Africa the infection caused by *R. conorii* has different clinical systems such as local adenopathy and multiple eschars, but it is sometimes given as a separate species – *R. africae*. Patients usually present with fever, malaise, generalized maculopapulous rash and a typical black spot or 'Tache Noir'. Serious forms of the disease, including encephalitis are infrequent but when they occur, may have a high mortality. Raoult *et al.* (1986) studied 199 serologically confirmed cases and reported a mortality rate of 2.5%. Brouqui *et al.* (1988) reviewed the status of boutonneuse fever in the Mediterranean area and observed that more malignant forms were being found, that dog populations were increasing and the incidence of the disease and its distribution were also increasing.

Rickettsia conorii is widely endemic in southern Europe and most of the countries bordering the Mediterranean including Bosnia–Herzegovina, Croatia, Greece, France, Portugal, Slovenia, Spain and Turkey, and the countries bordering on the Black Sea as well as in Algeria, Morocco, Tunisia, Egypt and Israel and has also been isolated from several species of bird ticks in Cyprus (Kaiser *et al.*, 1974).

While *Rhipicephalus sanguineus* is both the vector and reservoir host of *R. conorii*, the role of the dog in the epidemiology of the agent is important as it serves as the primary host for the tick. *Rhipicephalus sanguineus* will also feed on humans and, if infected, can transmit the agent. Dogs taken to southern Europe may return to northern countries infested with *R. sanguineus*; *R. conorii* has been detected in dogs returning to Germany after travel to Italy or Greece (Gothe, 1999).

In the late 1970s, the Lazio, Liguria, Sicily and Sardinia regions of Italy registered a marked increase of boutonneuse fever; previously there had been about 30 cases a year reported. In 1979, 864 cases were notified, and the real incidence of the disease was thought to be much greater. Such an increase was not seen elsewhere in the Mediterranean. Ecological changes affecting ticks may be the reason for the increase. Bacellar *et al.* (1991) estimated that the annual number of cases in Portugal was not far from a rather astonishing figure of 20 000 in some years.

In Russia, an outbreak of MSF due to *R. conorii* occurred in the Crimea from 1947 to 1957. Only sporadic cases of the disease were then reported until 1995, when the incidence of spotted fever increased in central Crimea, with 40 cases in 1996 and more than 70 in 1997. Most cases occurred in the summer, when the *Rhipicephalus sanguineus* nymphs were active. A survey showed that 8% of the *Rhipicephalus sanguineus* contained *R. conorii* DNA, providing further evidence that an outbreak of MSF had occurred in the region (Rydkina *et al.*, 1999). This report considerably extends the area in which the infection may be endemic.

The public health importance of MSF

It is estimated that the incidence of boutonneuse fever in the Mediterranean region is some 50 cases/100 000 inhabitants per year. Asymptomatic cases are common and are frequently detected during serological surveys. Within the endemic areas, the incidence of *R. conorii* or MSF may vary considerably from one area to another depending on the distribution and density of dog populations and, consequently, of the dog tick vectors. The importance of MSF in Europe is determined mainly through retrospective surveys of the prevalence of antibodies in man and dogs in various areas of the continent as well as by clinical reports of individual cases.

Inasmuch as the reporting of MSF cases is not obligatory in all countries of Europe, information on the overall occurrence and incidence of the infection is largely dependent on a review of scientific literature. Table 12.4 provides a partial review of the incidence of infections/antibodies found in surveys carried out in Europe.

The surveys listed in Table 12.4 present only examples of the many surveys which have been published. Nevertheless, from these results, it is apparent that the serological prevalence rates in humans and dogs are high (though this may be due to surveys being carried out in areas at high risk of transmission). The infection is widespread in southern Europe and is almost certainly under-diagnosed due to the lack of familiarity of physicians with the disease or the difficulties in carrying out serological surveys. As has been described above, *Rhipicephalus sanguineus* is very frequently introduced into northern Europe on dogs returning from southern Europe; if the species is established in northern areas, the distribution of *R. conorii* and hence the transmission of MSF may also be extended.

Table 12.4 *Surveys of the incidence of* Rickettsia conorri *infections/antibodies in Europe, human and canine populations*

Country		Results	Reference
Croatia	dogs	62.4–69.9%	Punda Polic *et al.*, 1995
France - Corsica	human	48/100 000	Raoult *et al.*, 1985
France	human	18% of 325	Raoult *et al.*, 1987
France – Marseille	human	24.2/100 000	Raoult *et al.*, 1993
Greece	human	18.1% of 337	Gourgouli *et al.*, 1992
Greece	human	7.9% of 1584	Daniel *et al.*, 2002
Italy	dogs	85% of 55	Mansueto *et al.*, 1984
Italy	horses	42%	Rosa *et al.*, 1987
Italy	human	10.4% of 241	Federico *et al.*, 1989
Italy	dogs	35.5% of 163	Masoero *et al.*, 1991
Italy	humans	1.6 per 100 000	Giammanco *et al.*, 2003
Portugal	dogs	85.6% of 104	Bacellar *et al.*, 1995
Portugal	human	1,000 cases/year	Oliveira & Corte Real, 1999
Slovenia	human	20% of 315	Novakovic *et al.*, 1991
Spain	human	73.5% to 82%	Herrero *et al.*, 1992
	dogs	93% overall	"
Spain	human	11.6% of 200	Espejo Arenas *et al.*, 1990
	dogs	36.8% of 103	"
Spain	human	73.5% of 400	Ruiz Beltran *et al.*, 1990
Spain	dogs	58.6% of 58	Herrero *et al.*, 1992
Spain	human	8% of 150	Segura Porta *et al.*, 1998
	dogs	26.1% of 138	"

Other spotted fever rickettsiae in Europe

Rickettsia aeschlimanni

Rickettsia aeschlimanni was first isolated from *Hyalomma marginatum marginatum* ticks collected in Morocco. Identical PCR-RFLP profiles have been found in *H. marginatum marginatum* from Portugal and *H. marginatum rufipes* from Zimbabwe, suggesting that the distribution of this rickettsia reaches from the Mediterranean to southern Africa (Beati *et al.*, 1997). Although a case has been imported into France from Morocco, no autochthonous clinical cases have been reported from Europe.

Rickettsia helvetica

Rickettsia helvetica was first isolated from *Ixodes ricinus* in Switzerland in 1979; it was recognized as a new and hitherto undescribed spotted fever agent

and called the 'Swiss agent' (Burgdorfer *et al.*, 1979). It was identified as a new member of the spotted fever group of rickettsiae and given the name *Rickettsia helvetica* (Beati *et al.*, 1993). The species has since been isolated from *I. ricinus* in Central France (Parola *et al.*, 1998), from Italy (Beninati *et al.*, 2002), Denmark (Nielsen *et al.*, 2004) and, as will be seen below, in Sweden.

A serological survey in Sweden examined 748 *I. ricinus* collected in the southern and central parts of the country; 13 (1.7%) of the ticks were positive for rickettsia; no species of the genus *Rickettsia* had previously been found in Scandinavian ticks, nor has any case of rickettsial infection in humans or animals been reported. Sequencing showed that the isolations were apparently identical to *R. helvetica* (Nilsson *et al.*, 1997). At first the species was not linked with human disease. An investigation was made of two young Swedish men who died of sudden cardiac failure and who showed signs of perimyocarditis similar to those described in rickettsial disease; rickettsia-like organisms predominantly located in the endothelium were found. As *R. helvetica* transmitted by *I. ricinus* is the only non-imported rickettsia found in Sweden, it may be an important pathogen in the aetiology of perimyocarditis causing sudden unexpected cardiac death in young people (Nilsson *et al.*, 1999b). The identification of the causative agent in the above cases was confirmed as *R. helvetica* in a DNA study of rickettsiae isolated from *I. ricinus* (Nilsson, 1999a). In 1997 a 37-year-old man in eastern France was found to have seroconverted to *R. helvetica*, 4 weeks after the onset of an unexplained febrile illness. Results of a serosurvey of forest workers from the area where the patient lived showed a 9.2% seroprevalence of *R. helvetica* among them (Fournier *et al.*, 2000). Beninati *et al.* (2002), examined 109 *I. ricinus* collected in north and central Italy; PCR identified nine ticks positive for rickettsia. Among them, three different spotted fever groups of rickettsiae were found including *R. helvetica*. It is possible that species other than *R. conorii* are involved in rickettsial diseases in Italy and ticks other than *Rhipicephalus sanguineus* may be vectors.

Rickettsia helvetica is now known to be widely found in Europe and may be the cause of more clinical disease and mortality than is currently recognized. In view of the serious illness that may be associated with this agent, its widespread distribution as an emerging disease is a matter of concern.

Rickettsia massiliae

In 1990, 17 adult *Rhipicephalus turanicus* were collected in the south of France. Two spotted fever group rickettsiae, Mtu1 and Mtu5, were isolated from two of these ticks (Beati *et al.*, 1992); these isolations were later recognized as a new species and were given the name *Rickettsia massiliae*; PCR analysis showed the strains to be distinct from all previously recognized spotted fever group rickettsiae (Beati & Raoult, 1993). *R. massiliae* was isolated from *Rhipicephalus sanguineus* in Greece (Babalis *et al.*, 1994), from the same species in Catalonia, Spain (Beati *et al.*, 1996) and Portugal (Bacellar *et al.*, 1995). Most recently it has been isolated from *Rhipicephalus sanguineus* in the Canton of Ticino in the South of Switzerland (Bernasconi *et al.*, 2002). While widely distributed in Europe, its pathogenicity and public health importance remain unknown.

Rickettsia slovaca

Rickettsia slovaca was first isolated from *Dermacentor marginatus* in Slovakia in 1968, and has been implicated in human febrile illness (Sekeyova *et al.*, 1998). *Rickettsia slovaca* is reported from Armenia, Austria, the Czech Republic, France, Germany, Hungary, Italy, Kazakhstan, Lithuania, Portugal, Russia, Slovakia, Spain and Switzerland. It was isolated in France from *D. marginatus* in the south of the country in 1991 (Beati *et al.*, 1993). From 51 partially engorged *D. marginatus* females collected in 1975 in southern Germany, four strains of rickettsiae were isolated and were found to be similar to and almost undistinguishable from *R. slovaca* (Rehacek *et al.*, 1977). A rickettsia identified as being identical to *R. slovaca* was isolated from *Argas persicus* and *D. marginatus* collected in the Armenian S.S.R. in 1974 (Rehacek *et al.*, 1977) and its presence in that country was later confirmed by PCR analysis (Balayeva *et al.*, 1994). Antibodies to *R. slovaca* have been found in *R. sanguineus* collected on the outskirts of Rome, Italy (Cacciapuoti *et al.*, 1985). In Portugal *R. slovaca* was isolated from 18 of 632 adult *D. marginatus* (Bacellar *et al.*, 1995b). In Hungary rickettsia isolated from 7.2% of *D. marginatus* and 4.7% of *D. reticulatus* ticks were probably this species and the isolation represented the first rickettsiae of the spotted fever group found in the country (Rehacek *et al.*, 1979). It has been detected in man and isolated from *D. marginatus* in Austria (Bazlikova *et al.*, 1977) and found in *I. ricinus* in Lithuania (Tarasevich *et al.*, 1981). The finding of *R. slovaca* in *Dermacentor* ticks in Russia and Kazakhstan has considerably extended the known distribution of this species eastwards (Shpynov *et al.*, 2001).

The pathogenic role of *R. slovaca* was first noted in 1997 in a patient with a single inoculation lesion of the scalp and enlarged cervical lymph nodes, who had been bitten by a *Dermacentor* tick. Subsequently, the authors (Raoult *et al.*, 2002) evaluated the occurrence of *R. slovaca* infections among patients living in France and Hungary with these symptoms. *Rickettsia slovaca* infections were confirmed in 17 of 67 patients examined. Infections were most likely to occur in patients aged <10 years who were bitten during the colder months of the year. Sequelae included persistent asthenia (3 cases) and localized alopecia (4 cases). Antibodies were detected in 50% of tested patients. *Rickettsia slovaca* was detected by PCR in three *Dermacentor* ticks obtained from patients.

Data were collected in Hungary from 1996 through 2000 on 86 patients with similar symptoms following a tick bite most of whom had enlarged regional lymph nodes and/or vesicular-ulcerative local reaction at the site of the bite, 96% of which were located on or near the scalp. A characteristic eschar was seen in 70 (82%) patients. The most frequent general symptoms were low-grade fever, fatigue, dizziness, headache, sweat, myalgia, arthralgia and loss of appetite. Without treatment, the symptoms persisted as long as 18 months. One of the patients reported symptoms suggestive of encephalitis. The infection occurred most commonly in young children (Lakos, 2002).

Rickettsia monacensis

Rickettsia monacensis was isolated from *I. ricinus* collected in the English Garden Park in Munich, Germany. Analysis demonstrated that the isolate was a spotted fever group rickettsia closely related to several yet-to-be-cultivated rickettsiae associated with *I. ricinus*. *Rickettsia monacensis* joins a growing list of such rickettsiae found in ticks but whose infectivity and pathogenicity for vertebrates are unknown (Simser *et al.*, 2002). There have been no other reports of this species to date.

Rickettsia mongolotimonae

Rickettsia mongolotimonae was first isolated from *Hyalomma asiaticum* in Inner Mongolia in 1991 and described and named in 1996. The first reported human case was in France and was at first thought to be imported (Raoult *et al.*, 1996). However, when a second case due to the same agent occurred in Marseille, it appeared that the disease may be common around the city, representing a new clinical entity with a broader geographic distribution than previously documented (Fournier *et al.*, 2000). The identity of the vector in southern France remains unknown.

An agent given the name of Israeli spotted fever belonging to the *R. conorii* complex and transmitted by *Rhipicephalus sanguineus* (Manor *et al.*, 1992), has been isolated from ticks and humans in Israel where it has caused mortality among children. Three cases of infection with this species have been reported in Portugal, and it has been isolated from *R. sanguineus* in Sicily (Giammanco *et al.*, 2003); these reports indicate that the geographic distribution of Israeli spotted fever is quite wide. Because initial signs and symptoms of the disease are particularly uncharacteristic of the spotted fevers and appropriate treatment may be delayed by a failure of timely diagnosis, this rickettsia can cause life-threatening disease; 2 of the 3 cases reported in Portugal died despite intensive treatment (Bacellar *et al.*, 1999). The clinical course of Israeli spotted fever is more severe than MSF, from which it differs in having a very low proportion of cases with eschar at the site of the tick-bite (Walker *et al.* 1995). In view of the deaths associated with this infection in Israel and in Portugal, the appearance of this species in Europe is a matter of concern.

The public health importance of the Spotted fevers

Spotted fever group infections in Europe are widespread, their incidence is increasing and new species are emerging. As with other tick-borne infections in Europe, the increased distribution and rise in incidence is due, in part, to ecological changes, which have encouraged greater man-tick contact. The growing dog population in Europe, both pet and feral, has resulted in higher densities of the dog-tick vectors and hence, increased transmission of MSF. The recognition of new species and new strains of spotted fevers further increases the public health importance of this group. Physicians must be aware of the possibility of illnesses being due to a member of this group so as to ensure a timely diagnosis and treatment.

Ehrlichiosis

Ehrlichioses are a group of emerging diseases caused by small, intracellular bacteria belonging to the family Rickettsiaceae, genera *Ehrlichia* and *Anaplasma*. (A recent reclassification has resulted in the transfer of several species previously known as *Ehrlichia* to the genus *Anaplasma*.) *Ehrlichia* and *Anaplasma* are transmitted largely through the bite of infected ticks. The first recognized human case occurred in the USA in 1986 in a man who had been exposed to ticks in a rural area of Arkansas. In 1990, the agent of human ehrlichiosis was isolated from the blood of an Army reservist at Fort Chaffee, Arkansas (Dawson *et al.*, 1991) and was named *Ehrlichia chaffeensis* (Anderson *et al.*, 1991). The two ehrlichial diseases recognized are human monocytic ehrlichiosis (HME) and human granulocytic ehrlichiosis (HGE). The causative agent of HGE in man is *Anaplasma equi*, first reported in the USA in 1994 and now reported elsewhere in North America, Europe and the Middle East.

Prior to the discovery of *E. chaffeensis, Ehrlichia sennetsu* was the only species known to infect humans. *Ehrlichia sennetsu* causes Sennetsu fever and occurs primarily in Japan; it is rare and usually benign, and no fatalities have been reported. The other *Ehrlichia* cause animal diseases and include *E. canis* (canine ehrlichiosis), *E. ewingii* (canine granulocytic ehrlichiosis), *E. risticii* (Potomac Horse Fever), *E. equi* (disease in horses), *E. phagocytophila* (disease in sheep and cattle), among others.

Symptoms of human ehrlichiosis resemble those of Rocky Mountain spotted fever (RMSF) ranging from a mild or asymptomatic illness to a severe, life-threatening condition. In patients with severe complications, acute renal or respiratory failure is common. There have been a number of fatalities in the USA and Europe (Hulinska *et al.*, 2002).

The first human case of ehrlichiosis in Europe was reported from Portugal (Morais *et al.*, 1991). Species of *Ehrlichia* and *Anaplasma* have now been reported from many countries in Europe as shown in Table 12.5.

The public health importance of ehrlichiosis

Ehrlichiosis is widespread in Europe and, as with Lyme disease and other tick-borne infections, its incidence appears to be increasing and spreading geographically. In several instances ticks have been found co-infected with both agents. In part the increased incidence of ehrlichial infections may be due to better recognition and diagnosis of the disease but changing ecological conditions favouring increased human-tick contact must be considered a major factor. Greater awareness of the potential severity of ehrlichiosis is needed by physicians to ensure that proper treatment is initiated early in the course of the disease.

Q Fever

Q Fever is a worldwide zoonosis caused by *Coxiella burnetii*, an obligate intracellular parasite. As with all rickettsiae, humans are dead-end hosts. The disease usually takes the form of an acute atypical lung pneumonia disease; subclinical or non-typical forms are known. Unlike other rickettsial diseases, no cutaneous exanthema is seen. Q fever is rarely fatal, with a mortality rate of about 1% in untreated patients and lower among treated patients.

The most important sources of infection are cattle, sheep and goats; transmission to man may occur by ticks but more often it is by aerosol, or by non-pasteurized milk and dairy products. A large proportion of the cases occur among farm and abattoir workers. Ticks may play a role in the silent maintenance of the rickettsiae, transmitting the agent from one animal to another (Walker & Fishbein, 1991). Many mammals and birds have been found infected and may have a role in the maintenance of the infection. The seroprevalence of *C. burnetii* in a population of brown rats, *Rattus norvegicus*, ranged from 7–53% in four Oxfordshire farmsteads and nine Somerset farmsteads in the UK (Webster *et al.*, 1995). On the other hand, in southern Bavaria in Germany, where 12% of 1095 cattle were seropositive for *C. burnetii*, neither rodents nor ticks in the same area were positive and the authors assumed that an independent natural cycle involving only cattle maintained the infective agent (Rehacek *et al.*, 1993).

Table 12.5 *Reports of the presence of ehrlichiosis in Europe* *

Country	Species	Host	Reference
Belgium	*Ehrlichia chaffeensis*	humans	Guillaume *et al.*, 2002
Bulgaria	*Ehrlichia* sp.	humans	Christova & Dumler, 1999
Bulgaria	*E. phagocytophila**	ticks	Christova *et al.*, 2001
Czech Rep.	*E. chaffensis*	humans	Hulinska *et al.*, 2001
	E. phagocytophila	humans	
Czech Rep.	*E. phagocytophila*	humans	Zeman *et al.*, 2002
Czech Rep.	*E. phagocytophila*	humans, deer, boar, birds,	Hulinska *et al.*, 2002
Denmark	*E. chafeensis*	humans	Lebech *et al.*, 1998
Denmark	*E. equi*	dogs	Ostergard, 2000
Denmark	*E. chaffeensis*	humans	Skarphedinsson *et al.* 2001
France	*E. equi*	*Ixodes ricinus*	Parola *et al.*, 1998a
	E. phagocytophila	*I. ricinus*	
Germany	*E. canis*	dogs	Reider & Gothe, 1993
Germany	*E. canis*	dogs	Gothe, 1999
Germany	*E. phagocytophila*	humans	Fingerle *et al.*, 1997, 1999
Germany	*E. phagocytophila*	*I. ricnus*	Baumgarten *et al.*, 1999
Germany	*Ehrlichia* sp.	humans	Hunfeld & Brade, 1999
Germany	*A. phagocytophila**	*I. ricinus*	Hildebrandt *et al.*, 2003
Greece	*E. chaffeensis*	humans	Daniel *et al.*, 2002
Ireland	*E. phagocytophila*	cow	Purnell *et al.*, 1978
Italy	*E. canis*	dogs	Buonavoglia *et al.*, 1995
Italy	*E. phagocytophila*	humans	Nuti *et al.*, 1998
Italy	*E. chaffeensis*	humans	Santino *et al.*, 1998
Italy	*E. phagocytophila*	*I. ricinus*	Cinco *et al.*, 1998
Netherlands	*Ehrlichia* sp.	*I. ricinus*	Schouls *et al.*, 1999
Netherlands	*E. phagocytophila*	cows	Siebinga & Jongejan 2000
Norway	*E. phagocytophila*	moose *I. ricinus*	Jenkins *et al.*, 1998
Norway	*Ehrlichia* spp.	humans	Bakken *et al.*, 1996
Norway	*Anaplasma phagocytophila**	sheep	Stuen *et al.*, 2002
Poland	*Ehrlichia* sp.	*Microtus arvalis*	Bajer *et al.*, 1999
Poland	*E. phagocytophila*	*I. ricinus*	Grzeszczuk *et al.*, 2002
Portugal	*E. canis*	humans	Morais *et al.*, 1991
Portugal	*E. canis*	dogs	Bacellar *et al.*, 1995

Table 12.5 (*cont.*)

Country	Species	Host	Reference
Russia	E. phagocytophila	I. ricinus	Alekseev et al., 2001
Russia	E. phagocytophila	voles	Telford et al., 2002
		Clethrionomys glareolus	
Slovenia	E. equi	humans	Lotric-Furlan et al., 1998
Slovenia	Ehrlichia sp.	humans	Cizman et al., 2000
Slovenia	E. phagocytophila	humans	Lotric-Furlan et al., 2001
Spain	E. phagocytophila	I. ricinus	Oteo et al., 2000
Spain	E. phagocytophila	humans	Oteo et al., 2001
Sweden	E. chaffeensis	humans	Bjoersdorff et al., 1999a
Sweden	E. equi	humans	Bjoersdorff et al., 1999b
Sweden	E. equi	I. ricinus	Gustafson & Artursson, 1999
Sweden	E. equi	horses	Artursson et al., 1999
Switzerland	E. phagocytophila	cows	Pusterla et al., 1997
Switzerland	E. canis	dogs	Pusterla et al., 1998a
	E. phagocytophila	dogs	
Switzerland	E. phagocytophila	humans	Pusterla et al., 1998b
Switzerland	E. equi	I. ricinus	Liz et al., 2000
	E. phagocytophila	rodents	
UK	E. phagocytophila	cow	Purnell & Brocklesby, 1978
UK	E. phagocytophila	humans	Sumption et al., 1995
UK	E. phagocytophila	I. ricinus	Ogden et al., 1998
UK	E. phagocytophila	roe deer (Capeolus capreolus)	Alberdi et al., 2000
		I. trianguliceps	

*Species names are as reported in articles; *Ehrlichia phagocytophila* has now been renamed *Anaplasma phagocytophila* or as cited by certain authors, *Anaplasma phagocytophilum.*

Tick vectors of Q fever

In Slovakia, strains of *Coxiella burnetii* were recovered from *Ixodes ricinus, Dermacentor reticulatus, D. marginatus, Haemaphysalis concinna, H. inermis* and *H. punctata*; the agent was thus isolated from all important ticks in the country (Rehacek *et al.*, 1991). In Bulgaria, antibodies to *C. burnetii* were found in *I. ricinus, D. marginatus, Rhipicephalus bursa* and *Hyalomma plumbeum* (Aleksandrov *et al.*, 1994). In Cyprus *C. burnetii* was found in *R. sanguineus* and *Hyalomma* species (Spyridaki *et al.*, 2002). In Ticino, Switzerland, Bernasconi *et al.* (2002) identified a *Coxiella* species (probably *C. burnetii*) in *R. sanguineus* and *R. turanicus*. *Rhipicephalus sanguineus* were also infected by *Rickettsia massiliae*.

The public health importance of Q fever

Overall, the public health importance of Q fever in Europe is no longer great. The incidence is low in most countries and mortality is quite uncommon. Improved working conditions on farms have reduced exposure. Yet, the infectious agent and the putative vectors of *C. burnetii* remain widely endemic in Europe. Raoult *et al.* (2000) carried out examinations of both Q fever patients and sera (1383 patients hospitalized in France for acute or chronic Q fever), and an analysis on 74 702 sera in their reference centre; they concluded that Q fever is grossly underestimated. Cases of Q fever are also reported from Austria, Bosnia–Herzegovina, Croatia, Czech Republic, Greece, Italy, Lithuania, Netherlands, Poland, Russia, Spain, Sweden, Portugal and the Ukraine but in small numbers. In the UK and Ireland, between 100 and 200 cases of human Q fever are encountered annually. Most cases are sporadic but occasionally large outbreaks occur. In Germany, 40 documented outbreaks have been identified since 1947; in 24 of these sheep were implicated as the source of transmission (Hellenbrand *et al.*, 2001).

Babesiosis

Human babesiosis is an emerging tick-borne zoonotic disease occurring in significant numbers in Europe. It is an intra-erythrocytic parasitic infection caused by protozoa of the genus *Babesia* and transmitted via the bite of *Ixodes* ticks. The disease most severely affects elderly patients, those who are immunocompromised, or have undergone splenectomy. Babesiosis is usually asymptomatic in healthy individuals. Babesial parasites (and those of the closely related genus *Theileria*) are some of the most widespread blood parasites in the world, second only to the trypanosomes and malaria, and have a considerable worldwide economic, medical and veterinary impact (Homer *et al.*, 2000).

The first human infection caused by *Babesia* was described in the former Yugoslavia in 1957 from an asplenic farmer. Many species of *Babesia* have now been reported from Europe. There have been hundreds of cases in the USA, mostly caused by *Babesia microti*. Human babesiosis in Europe is caused by *B. divergens*, and it frequently causes a more serious disease particularly in asplenic persons. Human babesiosis is relatively rare in Europe, but cases have been reported from the former Yugoslavia, Ireland, France, Portugal, Russia, Switzerland and the UK. Most human cases have occurred in France, Ireland and the UK. In Europe, it is symptomatic in a large proportion of the infections and may often be fatal (Rowin *et al.*, 1984). Splenectomy is the main factor of risk found in 86% of the patients. Unfortunately, the effect of treatment is poor in splenectomized patients especially if delayed (Hohenschild, 1999). All of the cases which have involved *B. divergens* have occurred in individuals who were splenectomized.

Tick vectors of babesiosis

The tick vectors of babesiosis in Europe are species of *Ixodes*, *Dermacentor*, *Rhipicephalus* and *Haemaphysalis*. In western Europe, the most frequent vector species are *I. ricinus* and *D. marginatus*. In Russia *I. trianguliceps* has been found infected with *B. microti* (Telford *et al.*, 2002). In Germany and France, *D. reticulatus* and *D. marginatus* are vectors of *Babesia canis* (Zahler *et al.*, 1996). In Germany, France, Ireland and the UK, *I. ricinus* is the primary vector. Sixl *et al.* (1989) provided evidence in Austria for the presence of *Coxilla burnetii* in hedgehogs (*Erinaceus europaeus*) and the role of *I. hexagonus* and *I. ricinus* as vectors of the causative agent. In Switzerland, *B. microti* DNA has been detected in *I. ricinus* in the eastern part of the country; 1.5% of 396 human residents in the region had antibodies to *B. microti*. This constitutes the first report of *B. microti* in a human-biting vector, associated with evidence of human exposure to this agent, in a European site (Foppa *et al.*, 2002). However, in Poland, a high infection rate of DNA of *B. divergens* and *B. microti* has also been found in *I. ricinus* (Skotarczak & Cichocka, 2001). Experimental studies have shown that strains of *B. microti* from the USA infect and are readily transmitted by *I. ricinus* from one gerbil to another. This suggests that European *B. microti* strains are probably infective for *I. ricinus*, supporting the view that infection of humans with European *B. microti* may be a regular occurrence (Gray *et al.*, 2002). Studies in Slovenia examined questing *I. ricinus* adult and nymphs collected in various parts of the country for the presence of babesial parasites; 13 of 135 ticks contained babesial DNA identical with *B. microti* and with a high degree of homology with *B. odocoilei* and *B. divergens*. The authors considered that this represents the first genetic evidence of *B. microti* and *B. divergens*-like parasites in *I. ricinus* in Europe (Duh *et al.*, 2001).

The public health importance of babesiosis

The importance of babesiosis in Europe lies not so much in the relatively small number of clinical cases so far diagnosed and reported but in the severity of the disease which may develop in infected persons. The surveys cited above show that emerging babesial infections in both man and animals are widespread on the continent. Spielman (1994), in considering the emergence and spread of Lyme disease, and babesiosis in the USA and in Europe, believed that their emergence derives from the recent proliferation of deer populations, and the abundance of deer derives from the process of reforestation now taking place throughout the North Temperate Zone of the world. Residential development seems to favour small tree-enclosed meadows interspersed with strips of wood-land, a 'patchiness' prized by deer, mice and humans. As a result, large numbers of people live where risk of Lyme disease and babesiosis is intense. The agents of these infections that were once transmitted enzootically by an exclusively rodent-feeding vector have become zoonotic. It is thus possible that the inci-dence of babesiosis in Europe may increase as has been the case with other tick-borne diseases.

Tularaemia

Tularaemia is caused by the gram-negative bacterium, *Francisella tularen-sis*. *Francisella tularensis* occurs worldwide in more than 100 species of wild ani-mals, birds and arthropods. Infection with this agent produces an acute febrile illness in humans. The route of transmission and factors relating to the host and the organism influences the presentation. Humans can become infected through diverse environmental exposures: bites by infected arthropods; handling infec-tious animal tissues or fluids; direct contact with or ingestion of contaminated food, water or soil; and inhalation of infective aerosols. Untreated, tularaemia has a mortality rate of 5–15% or even higher with the typhoidal form. Appropri-ate antibiotics lower this rate to about 1%.

In Europe, a large outbreak of tularaemia occurred in Kosovo in 1999–2000. A total of 327 serologically confirmed cases were identified in 21 of 29 Kosovo municipalities. Matched analysis of 46 case households and 76 con-trol households suggested that infection was transmitted through contamin-ated food or water and that the source of infection was rodents. Environmen-tal circumstances in war-torn Kosovo led to epizootic rodent tularaemia and its spread to resettled rural populations living under circumstances of sub-standard housing, hygiene and sanitation (Reintjes *et al.*, 2002). By 5 February 2002, the Institute of Public Health in Pristina had reported 715 cases with no deaths.

According to the statistics of the international animal disease office, OIE, in the year 2000 tularaemia was diagnosed in 917 people in Finland and 468 people in Sweden. The infection is endemic in both of these countries particularly in hares (*Lepus europaeus*) and other small mammals. An extensive epidemic of tularaemia with 529 cases, 400 of which were confirmed by laboratory tests, occurred in the northern part of central Sweden during the summer of 1981. The outbreak was of short duration and was restricted to certain communities within a narrow geographical area. During the 2 years preceding the outbreak only 3 and 7 cases were reported in Sweden. The infection was reported as being mainly transmitted by mosquitoes. The later cases in September and October were infected by contact with hares or rodents. All age-groups were affected, with a slight predominance of women and the 30–60 year age-groups (Christenson, 1984).

In 2003, 6 cases of tularaemia were described from Oppland and Hedmark Counties, Norway. All the cases described lived in areas where rodent activity is high and where tularaemia has been diagnosed previously. There were no known associations between the patients. None of the patients reported drinking water from unchlorinated sources. Physician awareness of tularaemia in Norway is fairly low, and some of the cases described were not recognized at initial presentation. Tularaemia has been a notifiable disease in Norway since 1975. Between 1975–2004, the annual incidence has varied from 0 to 47 cases (incidence in 1985). In Sweden, the incidence has been much higher in recent years – up to many hundreds of cases annually (Hagen *et al.*, 2005).

An increasing number of cases of tularaemia infection are being observed in eastern Europe, above all in Bulgaria, Hungary, Russia and, to some extent, Yugoslavia and Slovakia. However, isolated cases are reportedly sporadically from other European countries.

Tick vectors of tularaemia in Europe

In central and western Europe, *I. ricinus* is probably the most important vector while *Dermacentor nuttalli* may be the vector in Russia. In the Czech Republic, 26 478 ticks were examined in a survey; three strains of *Francisella tularensis* were isolated, one each from *I. ricinus* and *D. reticulatus* males in southern Moravia, and one from *D. marginatus*-engorged females collected from sheep in eastern Slovakia. A survey from 1984–1993 in the region of Bratislava collected 6033 ticks, mostly adults of *D. reticulatus* (4994) and *I. ricinus* (1004), and 35 *Haemaphysalis concinna* nymphs. Out of 4542 starving ticks, 34 *F. tularensis* strains were isolated from *D. reticulatus* (30), from *I. ricinus* (3) and *H. concinna* (1). Natural infection with *F. tularensis* was further shown in 27 of 1491 adult *D. reticulatus* (Gurycova *et al.*, 1995).

Other arthropods are also involved in the transmission of tularaemia organisms, although less so than ticks. In the original study of tularaemia in the USA, transmission by the deer fly (*Chrysops discalis*) was shown to occur. Mosquitoes may be involved in transmission, at least in Scandinavia according to reports from Sweden (Christenson, 1984; Eliasson *et al.*, 2002).

The public health importance of tularaemia in Europe

The annual number of cases of tularaemia is fairly large and the infection is widespread. It frequently appears as epidemic outbreaks though these are usually not due to tick transmission. Due to the virulence of the infection and its broad geographical distribution, physician awareness of the disease and a good surveillance system are necessary. Though not of broad public health importance at this time, it may periodically be of considerable focal importance.

Tick paralysis

Tick paralysis occurs worldwide. It is not an infectious disease but is caused by the introduction of a toxin from tick salivary glands during blood-feeding. It is a relatively uncommon neuromuscular disease with a higher prevalence among young girls, although older men who are exposed to tick bites may also be affected. It typically presents as an acute ascending paralysis occurring a few days after tick attachment and may result in respiratory failure and death. Generally, this condition is associated with ticks attached around the head area, particularly at the base of the skull. Symptoms begin a day or two after initial attachment. The victim loses coordination and sensation in the extremities. The paralysis progresses in severity, the legs and arms becoming useless; the face may lose sensation; and speech becomes slurred. If the breathing centre of the brain is affected, the victim may die. If the tick or ticks are found and removed, recovery begins immediately, and the effects disappear within a day.

A particularly severe form of this condition is seen in Australia and many cases have been reported from the USA and Canada where the causative ticks are *Dermacentor* species. In Europe *Ixodes ricinus* is the species known to cause tick paralysis.

Only sporadic cases of tick paralysis have been reported from Europe. Lefebvre *et al.* (1976) described a series of cases in France and reviewed the literature while Lefebvre (1981) cautioned that seasonal inflammatory neurological disorders following tick-bites have been reported more and more often during the last few years especially during the summertime and especially in high-risk subjects (campers, agricultural workers, troops during field exercise). Schmidt & Ackermann (1985) described the clinical aspects of a series of cases in Germany.

Though the number of cases of tick paralysis in Europe is small, death can result from an infestation, which can be readily and easily prevented by removal of the tick. Persons showing neurological symptoms described above and who may have been at risk by having been exposed to tick bites, should be carefully examined to see if they have a tick attached to them, particularly on the head, and, if found, the tick should be very quickly removed.

13

Mite-borne infections and infestations

Scabies

Scabies is an intensely pruritic and highly contagious infestation of the skin caused by the mite, *Sarcoptes scabiei*. It lives its entire life on the human host. A variant is canine scabies, in which humans become infected from pets, mainly dogs. Canine scabies (i.e., mange) causes patchy loss of hair and itching in affected pets. A highly contagious form of scabies, known as Norwegian or crusted scabies, is being increasingly found in individuals who are immuno-compromised, aged, physically debilitated or mentally impaired. In this type of infestation, extensive widespread crusted lesions appear with thick hyper-keratotic scales over the elbows, knees, palms and soles. Extensive proliferation occurs in immunocompromised patients. It is readily spread by close contact within families, in institutions and by sexual contact and it has been estimated that the global prevalence of scabies is 300 million cases annually (Walker & Johnstone, 2000).

Although the *Sarcoptes* mite does not transmit a disease, they are the cause of a diseased condition; scabies spreads in households and neighbourhoods in which there is a high frequency of intimate personal contact or sharing of inanimate objects; and fomite transmission is a major factor in household and nosoco-mial passage of scabies (Burkhart *et al.*, 2000). Scabies is often a serious problem in institutions such as hospitals and other health-care institutions, especially homes for the aged; van Vliet *et al.* (1998) considered that there were six impor-tant factors that contribute to the transmission and return of scabies in health-care institutions: (a) among residents in health-care institutions a considerable number are, once infested, at risk of developing crusted scabies, (b) many peo-ple are exposed through close contact, (c) the generally long diagnostic delay,

Learning Resources
Centre

(d) insufficient survey of the epidemiological problem, (e) treatment failures and (f) incomplete post-intervention monitoring.

Particular attention must be given to the prevention of scabies and especially Norwegian scabies in institutions such as hospices for AIDs patients. Patients with Norwegian scabies require strict barrier nursing if nosocomial transmission is to be avoided. Virtually all reviews of scabies in Europe note that the incidence is increasing both of uncomplicated scabies and especially of Norwegian scabies; the reasons for such an increase is variously ascribed to greater mobility of human populations, greater promiscuity and, in the case of Norwegian scabies, to the much greater susceptibility of immunosuppressed individuals. The systemic drug ivermectin usually provides effective control in individuals but the pyrethroid insecticide permethrin may also provide safe and effective control (Walker & Johnstone, 2000).

Scabies infestations may have a direct public health impact; as an example, from 1966–1986 there were 1 675 213 cases of scabies reported in Poland. The highest incidence (580.5) per 100 000 inhabitants was noted in 1968, the lowest (41.7) in 1986. During this period, the losses brought about by this disease among people 20–64 years of age amounted to 1 836 234 days of sick leave, while the average absence in 1980 was 82 500 days (Zukowski, 1989). There are reports from several countries of an overall increase in the incidence of scabies. This is particularly evident in institutional outbreaks and among persons living under poor socioeconomic conditions.

Reports of increased incidence of scabies infections have been received from Denmark, Germany, Italy, Russia, Sweden and the UK. In a recent review of cases of infectious disease reported by occupational and specialist physicians to the UK-based schemes collecting data on occupational ill-health in 2000–2003, no less than 11.1% of the 5606 reports received were due to scabies (Turner *et al.*, 2005), presumably representative of the industrial sectors associated with occupational disease. Scabies is a persistent and serious public health problem in Europe which must be dealt with by rapid diagnosis and effective control, especially in institutional settings.

Effective treatment and control will be enhanced by rapid diagnosis. Scabies control by the application of topical pesticides has become increasingly difficult as toxicological concerns have reduced the number of compounds available. Effective treatment of scabies can be achieved with a 5% permethrin cream. Fortunately, ivermectin, a potent macro-cyclic lactone of the avermectin family is now widely available and approved because of its high degree of safety and effectiveness of its control of the mites. Used as a systemic, it is well tolerated. Ivermectin is particularly effective in large institutional outbreaks where control as been hitherto difficult (Buffet & Dupin, 2003).

Rickettsial pox

The causative agent of rickettsial pox is *Rickettsia akari*, a member of the spotted fever group of rickettsia. This rickettsial agent is transmitted by the house mouse mite (*Liponyssoides sanguineus*) from rodents, particularly, the house mouse (*Mus musculus*), to man. Rickettsial pox is a mild, self-limited, zoonotic febrile illness characterized by a papulovesicular skin rash at the site of the mite bite. It was first recognized clinically in New York City in 1946 (Huebner *et al.*, 1946).

The infection was detected in southern Europe for the first time with the isolation of *R. akari* from the blood of a sick patient in Croatia (Radulovic *et al.*, 1996). The authors considered that rickettsial pox would probably be found more often on the continent if physicians gave greater consideration to the possibility of its presence and if laboratory diagnostic methods were better able to distinguish between the different spotted fever group rickettsiosis. In view of the scarcity of cases and its lack of pathogenicity, rickettsial pox is not an infection of public health importance in Europe.

Mites and allergies

Although not an infectious disease, allergy due to house-dust mites, (*Dermatophagoides pteronissinus* and *D. farinae*) is an important causative factor for allergic asthma or rhinitis in children throughout the world and may, in part, be a factor in 50–80% of asthmatics. There is a clear relationship between the degree of allergen exposure and the subsequent development of asthma or the risk of sensitization. In fact, allergens produced by these mites are probably the single most important allergens associated with asthma worldwide (Tovey, 1992). Along with respiratory symptoms, high levels of dust mite allergens have also been correlated with atopic dermatitis, characterized by itchy, irritated skin. In general, these studies suggest that those persons susceptible to mites (i.e. those likely to form IgE antibodies) are also likely to develop skin sensitization if exposed to high concentrations of mite allergens. ('Atopic' refers to a group of diseases where there is often an inherited tendency to develop other allergic conditions, such as asthma and hay fever.)

There is a general agreement that house dust mites in the home feed on shed skin of man. The average individual sheds 0.5 to 1.0 g of skin daily. Mites have a life cycle of about 2–3.5 months. The size of the mite is up to 0.3 mm. They live in house dust and thrive in warmth and high humidity. Mites' faeces seem to be the major source of allergenic exposure. They are about the size of a pollen grain and could therefore very easily become airborne and penetrate the lung alveolus. Among farm or warehouse workers, allergies may also be caused by storage mites, which are especially common under warm, humid conditions.

The presence of mites in house dust was known as early as 1928. In 1964 the significance of this finding was enlarged upon when *D. pteronissinus* and *D. farinae* were identified in house dust samples from all over the world (Voorhorst *et al.*, 1964). *Dermatophagoides pteronissinus* dominates most of Europe and the UK. *Dermatophagoides farinae* is an important allergen in North America and Japan, and also is common in parts of Italy and Turkey. In the Far East both species are equally prevalent. Altitude plays a role as low damp areas with high humidity are preferable for the mites. In total 13 species have been found in house dust and recorded from locations throughout the world, including the USA, Canada, Europe, Asia, the Middle East and parts of Australia, South America and Africa (Colloff, 1998). However, three species *D. pteronissinus*, *D. farinae* and *Euroglyphus maynei* (Em) are most common, comprising up to 90% of the mite fauna in houses (Blythe *et al.*, 1974). *Euroglyphus maynei* is common in homes in the UK, Europe, the southern USA and other parts of the world. It is often present in densities greater than 100 mites per gram of dust. *Euroglyphus maynei* usually occurs with *D. pteronissinus* and *D. farinae* and it is frequently more abundant than *Dermatophagoides* species (Arlian *et al.*, 1993). Sensitization to storage mites in Germany is frequently associated with sensitivity to *D. pteronissinus*. Overall, skin test sensitivity to storage mites in grain and other foodstuff was greater in rural than in city dwellers (Musken *et al.*, 2002). Perennial conjunctivitis due to house dust mites is the most frequent form of allergic conjunctivitis in urban environments.

Currently, several studies in France have shown that sensitivity to mites is frequently found in the general population and in people suffering from respiratory allergic disorders. From 5–30% of the general population have positive skin tests for mite allergens whilst such sensitivity is found in 45–85% of asthmatic subjects (Vervloet *et al.*, 1991). There is apparently a relationship between altitude and allergy to house-dust mites in France; Charpin *et al.* (1988) found that subjects living in high altitude where house-dust mites are known to be uncommon exhibit a lower prevalence of asthma and allergy to house-dust mites.

The successful treatment of asthma can be achieved in good part through the control of house dust mites although this is often difficult and may require vinyl covers for mattresses, hot washing of bedding, removal of carpets from bedrooms and the use of acaracides (WHO, 1988). The use of house dust mite-impermeable mattress covers has been shown to provide a measure of relief but Rijssenbeek *et al.* (2002) found that while the use of anti-allergic mattress covers results in significant reductions in *D. pteronissinus* concentrations in carpet-free bedrooms, patients with moderate to severe asthma, airways hyper-responsiveness and clinical parameters are not affected by this otherwise effective allergen avoidance. Custovic *et al.* (1996) investigated the rate of increase in mite allergen levels in new, ordinary mattresses. They found that new mattresses can become a significant source of exposure to mite allergens after a short period of time (< 4 months); they considered that there was little justification for advising mite-sensitive patients to replace their mattresses as part of an avoidance regime. The most effective method of prevention of house dust mite allergies is by immunotherapy.

The public health importance of mite allergies

The number of people affected by allergies due to mite allergens is rising throughout Europe (and indeed worldwide), impairing the health and quality of life of a substantial proportion of children, and many adults, and placing a significant burden on health services. The problem of asthma resulting from exposure to house-dust mites is particularly severe in the UK and the house dust mite is the most important environmental allergen implicated in the aetiology of childhood asthma in the UK. Mites can be found in great numbers in bedrooms and in bedding. Studies in other countries of Europe have also demonstrated the importance of mites as a cause of asthma, particularly among children; there seems little doubt that house dust mites will be shown to be of similar importance wherever they are studied, particularly in areas of high humidity. Their presence has little to do with the degree of cleanliness of a home. Allergies caused by house dust mite remain a serious, if often unrealized, public health problem throughout most of the countries of Europe.

14

Cockroaches and allergies

Cockroaches are known to contain powerful allergens; many asthmatics are allergic to cockroaches and in some areas cockroach sensitization is a very important indoor allergen, second only to that of house dust mites. A number of studies in the USA, which will be discussed in the section dealing with that country, have shown that cockroaches are an important risk factor for allergic asthma. However, in Europe, the prevalence of cockroach allergy in atopic or asthmatic populations seems to be much lower than in North America, ranging between 4.2% in Germany (Hirsch *et al.*, 2000) and 27.8% in France (Dubus *et al.*, 2001). Another study in Marseille 'determined that the presence of *Blatella germanica* hypersensitivity among patients with asthma was 35% and was the most sensitive allergy after mite, grass pollens and cat allergens' (Birnbaum *et al.*, 1995). Studies in Croatia suggested that there is no relevant exposure to cockroach allergen in house dust samples from inland areas of Croatia (Macan *et al.*, 2003). Observations in Switzerland suggested that allergy to cockroaches is uncommon in the country but can sometimes be detected (Mosimann *et al.*, 1992). On the other hand, among Polish children, cockroach allergen is a very important factor of sensitivity after dust mites; concentrations of cockroach antigens in homes studied were higher than previously reported in other European countries. Children with hypersensitivity to cockroach allergens have severe asthma more often than children with other allergies (Stelmach *et al.*, 2002).

In Madrid, Spain, a study on the degree of importance of three species of cockroach, *Periplaneta americana*, *Blatella germanica* and *Blatta orientalis* in producing allergens that resulted in rhinitis and asthma concluded that cockroach sensitization is the most important indoor allergen in the area, and *B. orientalis* accounts for most cockroach sensitization (Sastre *et al.*, 1996).

The clinical significance of allergy to cockroaches and its relation to cockroach allergen exposure is not, as yet, clearly understood in Europe; further studies should be done to resolve the question of the relative importance of cockroaches as causes of allergies and on the measures needed to reduce cockroach infestations in homes.

15

Vector-borne disease problems associated with introduced vectors in Europe

Increased travel and increased exchange of goods and the manner in which they are transported, particularly by cargo containers, has resulted in the introduction and establishment of vector species not previously found in Europe. The most notable example of an undesirable introduction is that of the mosquito *Aedes albopictus*. This mosquito is of Asian origin, and has spread to the American and African continents in the last two decades: it is now established in North and South America, Africa, Oceania and Europe. Its first detection in Europe was in Albania, in 1979 (Adhami & Murati, 1987). The species was probably introduced into that country from China in the mid-1970s. The initial infestation was likely to have been at a rubber factory adjacent to the port of Durres (Durazzo), from where the eggs of the mosquito were inadvertently shipped in tyres to recapping (retreading) plants in other parts of the country. This was the first recorded infestation of *Ae. albopictus* outside Oriental and Australasian regions (Adhami & Reiter, 1998). *Aedes albopictus* was thereafter found in Genoa, Italy in 1990 (Sabatini *et al.*, 1990) as well as in Padua. The Padua introduction may have resulted from tyre imports from the USA. Eighty-five per cent of the imported tyres came from a single source in Atlanta, Georgia; the remaining 15% of the tyres came from the Netherlands. By late 2001, *Ae. albopictus* has rapidly become the most important pest mosquito species in areas of northern Italy and is now present in 9 of Italy's 21 political regions, i.e. Veneto, Lombardy, Emilia Romagna, Liguria, Tuscany, Lazio, Piedmont, Campania and Sardinia. It has become a serious pest mosquito in Rome (Di Luca *et al.*, 2001). Because of its cold tolerance it is an active biter through much of the year even in the colder regions of northern Italy. The species appeared in France in 1999 when it was found in villages in the Basse-Normandie and Poiou-Charentes departments. Larvae, pupae and adults were collected in the used tyre stock of a tyre recycling company, importing tyres especially from the USA, Italy and Japan (Schaffner *et al.*, 2001). The international shipping trade of used tyres provides *Ae. albopictus* with an ideal mechanism of dissemination. *Ae. albopictus* has also been introduced into Belgium (Schaffner *et al.*, 2004), Greece (Samanidou Voyadjoglou *et al.*, 2005), Yugoslavia and Serbia (Petric *et al.*, 2001), Spain (Aranda *et al.*, 2004) and Switzerland (Flacio *et al.*, 2004), amongst others.

In 1996, another introduced mosquito species, *Aedes atropalpus*, was discovered in the Ventao region of northern Italy, also found breeding in tyres stored at a recapping (retreading) company importing used tyres from Eastern Europe and North America (Romi *et al.*, 1997); the species now appears to have been established in Italy. Hitherto, *Ae. atropalpus* was known only from North America; unlike *Ae. albopictus*, its distribution in Italy remains limited to the area in which it was first found (Romi *et al.*, 1999). Most recently *Aedes japonicus*, an Asian mosquito, has been reported from Normandy in France where it was found breeding in tyres (Schaffner *et al.*, 2003). This same species has also been introduced into the USA in 1998 and has spread very widely in that country; West Nile virus has been isolated from *Ae. japonicus* in the USA.

The establishment of exotic mosquito species in Europe is a cause for serious concern; the newly introduced species may be vectors or potential vectors of disease and, possibly, even more efficient vectors than indigenous species of mosquitoes. Introduced species may spread rapidly, as has occurred with *Ae. albopictus* and *Ae. japonicus* in the USA. Both in Europe and the Americas, *Ae. albopictus* has become an important pest and is a potential vector of disease. As has been discussed, infected *Anopheles* vectors may be introduced by aircraft and if infected with malaria parasites, they may transmit infections such as has occurred with airport malaria.

As has been noted, the tick *Rhipicephalus sanguineus*, the vector and reservoir host of MSF, is a southern European species; however, it is frequently found on dogs which have returned from southern Europe where they had been taken during holidays and they carried the tick back to their homes. There are several instances in which introduced ticks have transmitted *Rickettsia conorii* in areas outside of their normal range of distribution. There have been many findings of tick species as an occasional invader in countries in which it is not normally present and its distribution now appears to be moving north; as an example, *Rhipicephalus sanguineus* has been reported in Switzerland since 1940 as an introduced species but it has now become established in the Canton of Ticino in southern Switzerland where two different tick species co-exist, i.e. *R. sanguineus* s. str. and *Rhipicephalus turanicus*. In this area it has been found infected with *Rickettsia* and *Coxiella* (Bernasconi *et al.*, 2002).

Jaenson *et al.* (1994) reported the frequent introductions of several tick species into Sweden, among them *Ixodes persulcatus, Hyalomma marginatum* and

R. sanguineus as well as the first European record of the American dog tick, *Dermacentor variabilis* and remarked that there is a risk of the potential introduction to Sweden of exotic pathogens with infected ticks e.g. *I. persulcatus* and *H. marginatum* on birds or *Dermacentor* species and *R. sanguineus* on mammals. Imported ticks have been found in most countries of Europe although they have only infrequently become established. *Rhipicephalus sanguineus* and *Hyalomma detritum* were found on seven occasions in Brittany between 1975 and 1985; they were found free in houses and on dogs, and *H. detritum* was also found on a man (Couatrmanac'h *et al.*, 1989). *Argas reflexus*, a pigeon tick which may cause severe allergic reactions in humans exposed to it, has been found in many houses in Berlin and is probably an introduced species (Dautel *et al.*, 1991). Migrating birds are a common source of introduction of ticks; Bjoersdorff *et al.* (2001) estimated that during the spring of 1996, a calculated 581 395 *Ehrlichia*-infected ticks were imported into Sweden by migrating birds. The gene sequences of the *Ehrlichia* in the birds were similar to the granulocytic ehrlichiosis found in domestic animals and humans in Sweden thus demonstrating that birds have a role in the dispersal of both ticks and disease.

Many species of arthropods are found as occasional introductions in countries in which they do not become established. Cases of the invasion of dipteran larvae in humans i.e. myiasis, are often found on travellers returning from tropical countries; among the species recorded are *Cordylobia anthropophaga* (the 'tumbu-fly'), *Cordylobia rodhaini*, *Dermatobia hominis*, *Hypoderma lineatum*, *Oestrus ovis* and *Wohlfahrtia magnifica*. *Dermatobia hominis* is generally acquired in the American tropics and *Cordylobia anthropophaga* almost exclusively in Africa though the species has been recorded in Saudi Arabia.

The sand flea (*Tunga penetrans*) is occasionally found infesting travellers returning from Africa or South America but rarely constitutes a serious medical problem.

Cockroaches are ubiquitous pests aboard ships and aircraft and exotic species are quite often transferred from one country to another; some species may become established as has *Pyenoscelus surinamensis*, a southeast Asian species which has been found in a heated greenhouse in Sweden (Hagstrom & Ljungberg, 1999). *Periplaneta australasiae* has been recorded as introduced into the Czech and Slovak Republics and *Periplaneta brunnea* and *Supella longipalpa* into the Czech Republic (Kocarek *et al.*, 1999); *Periplaneta brunnea* has been found infesting an airport in the UK; the species was probably imported from the southern USA on an aircraft and seems to have established itself in the heated rooms at the airport (Bills, 1965).

16

Factors augmenting the incidence, prevalence and distribution of vector-borne diseases in Europe

After decades of decline following World War II, there is now a serious recrudescence of several vector-borne diseases in the European region; in addition the densities of some vector and potential vector populations of mosquitoes and ticks are increasing in many areas. There has been an emergence of new diseases and disease syndromes such as the co-infection of leishmaniasis/HIV. It is important to understand the reasons for the recrudescence of these diseases as a basis for their control or prevention.

Ecological changes

As elsewhere, there have been massive ecological changes over the continent of Europe in the last 50 years. These include increases in human populations and increased population densities as a result of urbanization. After years of urban growth there has been a trend towards movement of populations to suburban areas in more affluent countries resulting in greater exposure to vectors and to animal reservoir hosts of infections. Changed leisure habits have also increased exposure to vectors, especially ticks.

Unfortunately, there have been major displacements of human populations from some countries within the region because of conflicts; the search for economic betterment has also been the cause of population migrations. Substantial immigration has occurred from outside Europe, often from countries endemic for vector-borne diseases. This has resulted in the frequent introduction of exotic infections and occasional secondary transmission of both tick- and mosquito-borne diseases; there is a risk that these newly introduced infections may become established as has occurred with West Nile virus in North America. Greatly increased tourism has brought with it the introduction of tropical diseases, particularly malaria, among persons returning from travel to disease-endemic areas. As physicians and health-care workers are often not familiar with the symptoms of these tropical diseases, diagnosis and treatment may be delayed. The frequent appearance of Mediterranean spotted fever (*Rickettsia conorii*) in dogs brought back from family travel to southern Europe has been discussed.

Changes in agricultural practices, especially increases in the extent of irrigated areas and the use of pesticides, have been the cause of important changes in vector and potential vector population densities in rural and peri-urban areas. The spread of rice cultivation in Italy has resulted in the reappearance of *Anopheles labranchiae* in areas from which it was previously eliminated (Bettini *et al.*, 1978). Reforestation has increased and caused important changes in the fauna and flora of newly wooded areas. As has been noted above, the extension of forested areas has increased the density of deer populations and hence the density of deer ticks and the incidence of Lyme disease (*Borrelia burgdorferi* s. l.) and babesiosis.

Documented increases in the incidence of tick-borne encephalitis over the last decades have been reported from Belarus, the Czech Republic, France, Germany, Latvia, Lithuania, Poland, Russia, the Slovak Republic, Sweden and Switzerland. It is generally agreed that the rising trend in tick-borne encephalitis (TBE) incidence in central and northern Europe has resulted from increased densities of the vector tick, *Ixodes ricinus*. In France, where there has been a notable increase in TBE incidence in the region of Lorraine since the mid 1990s, the proliferation of ticks is a major factor in the increased incidence of cases, and is mainly due to a modification of the ecosystem (George & Chastel, 2002). In the area of St. Petersburg, Russia the increase in incidence of TBE is ascribed to changing recreational habits which bring people into greater contact with ticks (Antykova & Kurchanov, 2001).

Virtually all of these changes have had an impact on the distribution and incidence of the vector-borne diseases; some of the ecological changes have resulted in a decline in vector-densities and of the vector-borne diseases as was the case with malaria at the end of the nineteenth century. Other ecological changes have resulted in increased vector densities and greater exposure of human populations to them. These changes will continue and will require continued surveillance of vector and potential vector population densities and of the possible appearance of vector-borne diseases.

17

The potential effect of climate change on vector-borne diseases in Europe

The International Council of Scientific Unions and the Intergovernmental Panel on Climate Change established by the World Meteorological Organization and the United Nations Environmental Programme, have estimated that by the year 2100, average global temperatures will have risen by between 1.0 and 3.5 °C. Important ecological changes may come about in the future due to global warming. Globally, 1998 was the warmest year and the 1990s the warmest decade on record. The distribution and seasonality of diseases that are transmitted by insects or ticks, are very likely to be affected by climate change. Such increases in temperatures in Europe might allow the establishment of tropical and semitropical vector species, permitting transmission of diseases in areas where low temperatures have hitherto prevented their over-wintering. Kovats *et al.* (1999) have pointed out that for the last few decades, Europe has experienced significant warming and this is likely to continue. A change in the distribution of important vector species may be among the first signs of the effect of global climate change on human health. Indeed, there is evidence that the distribution of tick vectors in Sweden has expanded to the north between 1980 and 1994 as reported by Lindgren *et al.* (2000), and that this is consistent with observed changes in climate. The Swedish study indicated that the reported northern shift in the distribution limit of ticks was related to fewer days during the winter seasons with low minimum temperatures below − 12 °C. At high latitudes, low winter temperatures had the clearest impact on tick distribution; in the south of the country, the increased temperatures resulted in increased tick population densities.

Another consequence of the warmer weather in Sweden has been an increase in the incidence of tick-borne encephalitis; the number of cases has substantially increased since the mid 1980s. Lindgren & Gustafson (2001) believe that this is associated with increased tick abundance, a longer life-span of the tick vectors (*Ixodes ricinus*), more persons visiting endemic areas and increases in host animal abundance. Randolph (2001) believes that increased incidence in tick-borne encephalitis in northern Europe is related to climate change, either directly by increasing tick population densities or indirectly, in that warmer weather in spring and autumn may permit longer activity seasons for both ticks and humans, which is likely to be the most important factor in northern regions such as Scandinavia, where prolonged low temperatures are limiting factors for tick development and activity.

Global warming may have a more immediate effect on mosquito populations and only later on mosquito-borne diseases. With increased average temperatures, the length of the breeding seasons for mosquito populations could be extended and population densities would increase. Warmer northern climates could cause an expansion of distribution of species now found only in warmer climates. Should these changes occur, it seems likely that the distribution of mosquito-borne diseases would also subsequently expand.

There is concern about the possible recrudescence of malaria transmission with rise of temperatures in Europe. Such a recrudescence has already been seen in eastern Europe. In light of the constant introduction of malaria cases from endemic countries, local transmission could reoccur in western Europe if there were substantial increases in the densities of vector populations. Increased temperatures would also favour the development of *Plasmodium falciparum* in regions where temperatures have been too low to permit completion of its development cycle. Although the well-established surveillance programmes throughout Western Europe should provide adequate warning of any marked increase in malaria incidence and allow for the organization of effective control operations, even a modest re-establishment of transmission would prove costly in terms of medical care and control operations. In eastern Europe where recrudescence of malaria transmission has occurred, the health services of some countries have had only limited success in containing malaria transmission and there is a need for donor support of surveillance and control operations.

18

The rodent-borne diseases of Europe

Rodents are implicated in the epidemiology of many of the vector-borne diseases in Europe; this chapter will consider only those infections that are directly transmitted from rodents to humans. This group of infections is of considerable importance to human health.

Few other animal groups live in closer contact with man than rodent populations and particularly the commensal rodents. Rodents are the largest family of mammals with some 1500 species divided into 30 families. Rodents are found native on all continents except Antarctica. Globally, rodents are of immense importance both because of their depredations on crops, the structural damage they may cause and as carriers of disease.

The hantaviruses

Hantaviruses, the causative agents for haemorrhagic fever with renal syndrome (HFRS) and hantavirus pulmonary syndrome (HPS), are viruses in the genus *Hantavirus* of the family Bunyaviridae with a worldwide distribution; there are some 25 distinct viruses in the group. Hantaviruses are not known to be transmitted by an arthropod vector. The natural hosts of these viruses are rodents. Hantavirus virions are excreted from infected rodents via saliva, urine and faeces, and humans may become infected through inhalation of aerosols of dried excreta, inoculation through the conjunctiva, or entry through broken skin or rodent bites.

Haemorrhagic fever with renal syndrome is a group of clinically similar diseases that occur throughout Eurasia. It is mainly seen in Europe and Asia; one causative agent, the Seoul virus, is found worldwide and has been associated with cases of HFRS in the USA. The epidemiology and clinical manifestations of these diseases are linked to the ecology of their associated rodent hosts. Murine rodents (Old World mice and rats) are associated with the severe forms of HFRS in Asia and the Balkans due to hantaan, Dobrava and Seoul viruses. Arvicoline rodents (voles) are associated with a mild form of HFRS in Europe called nephropathia epidemica due to Puumala virus (Khan & Khan, 2003). HFRS is also caused by the hantaan virus in China, Russia and Korea, the Puumala virus in Europe, Russia and Scandinavia, and the Dobrava virus in the Balkans. The very severe hantavirus pulmonary syndrome (HPS) has been reported only from the Americas and will be described below.

During the 1950s, hantavirus carried by the striped field mouse (*Apodemus agrarius*) caused approximately 3000 cases of Korean haemorrhagic fever among United Nations troops participating in the Korean conflict. The aetiologic agent, defined as hantaan virus, was first isolated in 1977 from the rodent reservoir host and was named after the Hantaan River which runs along the 38th parallel which divides North Korea from South Korea. A similar disease was described in 1934 in Scandinavian countries (epidemic nephrosonephritis, nephropathia epidemica or nephropathy). Since 1982, similar clinical manifestations of the disease, differently named in various parts of the world, have been assigned one unique name – haemorrhagic fever with renal syndrome, as recommended by the World Health Organization; over 20 other hantaviruses from rodent species in Asia, Europe and the Americas have been serologically or genetically characterized. Currently, the worldwide infection rate of hantaviral disease is estimated to exceed 200 000 cases annually. The majority of these cases are HFRS which occurs in eastern Europe and central and eastern Asia. Each species within the genus *Hantavirus is* primarily associated with a single rodent species, although accidental infections have been reported in other mammals. Four primary reservoir hosts for hantaviruses are found in Europe: the brown rat (*Rattus norvegicus*), Seoul virus; the bank vole (*Clethrionomys glareolus*), Puumala virus; the yellow-necked mouse (*Apodemus flavicollis*), Dobrava virus; and the common vole (*Microtus arvalis*), Tula virus (Scharninghausen *et al.*, 1999).

Hantaviruses are widely endemic in Europe; several strain types with some human pathogenicity, i.e. Puumala, Dobrava and Saaremaa, have been reported; hantavirus is suspected to be the prevailing cause of renal failure associated with infectious diseases in Central Europe (Faulde *et al.*, 2000).

Puumala virus

The following description of the importance of Puumala virus is taken from Clement *et al.* (1997). The Puumala serotype, carried by the bank vole (*Clethrionomys glareolus*), is the most important western and central European serotype, with at least 1000 serologically confirmed nephropathia epidemica cases per year in Finland and hundreds per year in Sweden. The number of documented cases in other European countries was more than 1000 in the former Yugoslavia, 531 in France by the end of 1994, approximately 250 in Belgium by the end of 1996, some 200 in Germany by the end of 1995, 138 in Greece by the end of 1993, and 39 in the Netherlands by the end of 1994. In Germany, the seroprevalence from Puumala and Dobrava viruses in the normal German population is about 1%. In professionally exposed risk groups, e.g. forest workers, a seroprevalence higher than that in the normal population was observed. Endemic regions for hantavirus infections are located mainly in Baden-Wurttemberg. In the years 2001–2003 an annual number of about 200 clinically apparent hantavirus infections were reported in Germany (Ulrich *et al.*, 2004) and the number of cases annually has shown a rising trend.

In the former Soviet Union, hantavirus disease has been recognized since 1934 and officially registered since 1978. Seroprevalence studies carried out by immunofluorescence assay or direct blocking radioimmunoassay involving 115 765 persons showed an overall seropositivity rate of 3.3%, ranging from 3.5% in the European part to 0.9% in the far eastern part. A total of 68 612 cases were registered between 1988 and 1992 (65 906 from the European part and 2706 from the far eastern part), with morbidity rates of 1.2 (1982) to 8.0 (1985) per 100 000 inhabitants. The peak year was 1985 with 11 413 registered cases. In the European part of Russia, most cases were due to milder infection with Puumala-related viruses, with mortality rates of 1% to 2% (Clement *et al.*, 1997).

In Europe, the parallel spread of Puumala and hantaan (HTN, or HTN-like) viruses has been noted in Belgium and the Netherlands, Germany and European Russia. A partial explanation could be that the HTN-like infection is due to serologic cross-reaction with Seoul virus. The brown rat (*Rattus norvegicus*) is the only hantavirus reservoir host with a worldwide distribution, including Europe, and Seoul infections are probably underestimated. The first documented hantavirus disease in Portugal was an HTN-like infection with acute renal failure and icterus. In Portugal *Clethrionomys glareolus* is not common but hantavirus-seropositive wild rats have been documented. Sixteen cases of acute disease, mostly with acute renal failure and reacting almost exclusively against a SEO strain have been described in Northern Ireland, another country where *C. glareolus* is not prevalent, but the most important hantavirus vector seems to be *R. norvegicus*. In France, three Seoul-induced cases of acute renal failure have been reported south of the Puumala-endemic region, from rural areas where the rat is an agricultural pest. Between December 1991 and February 1992, 14 Seoul-like cases were detected in the Tula region (300 km south of Moscow) and confirmed by plaque reduction neutralization tests and positive virus isolation in three of the cases; these cases are awaiting further confirmation.

An exceptionally large increase in hantavirus infections has been detected simultaneously in Belgium, Germany and France since spring 2005. The following information regarding this increase is taken from Mailles *et al.* (2005).

From 1 January to 15 June 2005, 120 cases were reported in Belgium and 115 cases in France. In Germany, 258 laboratory-confirmed hantavirus cases were reported between 1 January and 30 June, and in contrast to previous annual trends, the increase in cases occurred earlier in the year. The total number of cases for 2003 and 2004 was respectively 122 and 47 in Belgium and 128 and 55 in France. In Germany the number of reported cases for 2003 and 2004 from 1 January to 30 June was 72 and 64 respectively. In 2005, most cases were in men; the male: female sex ratio was 3.0 in Belgium, 3.4 in France and 2.5 in Germany. Their mean age was around 41 years in all three countries: 42.9 (range: 11–82 years) in Belgium, 42.8 (range: 16–83 years) in France and 40.9 (range: 5–75 years) in Germany. In Germany, most infections were caused by the hantavirus species Puumala (n = 212, 82%); 5 infections (2%) were caused by Dobrava and for 41 cases the causative virus was not specified.

In Belgium, the area most affected in this outbreak was Luxembourg province, where there were 34 cases, an incidence of 2.0/100 000 inhabitants, followed by Liège province (27 cases, 2.6/100 000) and the Namur province (25 cases, 5.5/100 000). A large number of cases in Liège province was observed in 2003. In Belgium, hantavirus epidemics are characterized by a minor spring peak and a major summer peak, and so a further increase in the registered number of cases was anticipated for 2005. According to local health professionals, part of this increase is due to a greater awareness among health professionals and a higher recourse to hantavirus testing.

Since 1980, over 1200 cases have been diagnosed in Belgium, and outbreaks of hantavirus infections in humans have been described in 1985, 1990, 1991, 1993, 1996, 1999, 2001 and 2003, with most cases having onset in summer. Known endemic areas are the provinces of Hainaut, Namur and Luxembourg.

Most of the 2005 cases in France are in the Ardennes (40 cases, 13.8/100 000) and the Aisne administrative départements (18 cases, 6.9/100 000) in northern France, bordering on Belgium. The département of Jura bordering on Switzerland is also an epidemic area, with 13 cases to 15 June 2005 (5.2/100 000). Since 1980, over 1000 cases of hantavirus infection have been diagnosed in France. Hantavirus outbreaks have been described in 1985, 1990, 1991, 1993, 1996, 1999 and 2003, with most cases having onset in summer. The known endemic areas in previous years have been northeastern France, along the Belgian and German borders (in the administrative départements of Ardennes,

Aisne, Nord and the administrative regions of Lorraine, Picardie and Franche-Comté). Clusters of hantavirus infections had rarely been reported in the Jura before 2005.

The increase of hantavirus infections in Germany between 1 January and 30 June 2005 was also seen in federal states such as Nordrhein-Westfalen, Niedersachsen and Hessen, that had not had high hantavirus prevalence in previous years. The federal states most affected were Nordrhein-Westfalen (92 cases, 0.5/100 000), Niedersachsen (51 cases, 0.6/100 000), Baden-Württemberg (46 cases, 0.4/100 000), Hessen (23 cases, 0.4/100 000) and Bayern (23 cases, 0.2/100 000).

In Germany, the average incidence for hantavirus infections over the time period 2001–2004 was 0.25/100 000 persons. An increase in hantavirus infections was observed in 2002 and 2004. In both years the increase was due to outbreaks in a known endemic area of Baden-Wurttemberg, and an outbreak of 38 cases also occurred in Niederbayern in 2004. In Germany, the known regions with higher prevalence of human hantavirus infections are Schwäbische Alb in Baden-Wurttemberg and parts of Unterfranken in Bavaria. Hantavirus infection became a notifiable disease in Germany with the introduction of the Infektionsschutzgesetz (the Protection against Infection Act) in 2001.

In many forest and agricultural regions of Belgium, France and Germany, a significant increase in the population density of rodents, especially voles, has been observed since the autumn of 2004, with no sign of this increase abating.

In the Balkans, and particularly in the former Yugoslavia, outbreaks of hantavirus disease have been recorded since the early 1950s, often with a mortality rate of 5–10% or even higher. The elevated rates of illness and death in early reports suggests the spread of one (or several) hantaviral strains, in addition to the mild Puumala serotype prevalent in the rest of Europe. These HTN-like viruses were later called Plitvice and Fojnica. In 1987, an HTN-like virus (Porogia) was isolated from the urine of a Greek soldier who became ill after a military exercise near the border in northern Greece and had both acute renal failure and severe pulmonary oedema. Dobrava virus is probably identical to Belgrade virus reported in the Balkans.

Dobrava virus

Dobrava virus is probably the most virulent of the European hantaviruses. The severity of Dobrava-associated HFRS can reach fatality rates of up to 12%, as has been reported from southeast Europe in Croatia, Greece and Slovenia. An increasing number of cases of Dobrava virus are being identified in Germany and elsewhere in central Europe where their virulence appears to be less serious than in the Balkans. They may be a different strain of the virus, since the central European Dobrava virus infections have a different, less severe clinical outcome. However, Schutt *et al.* (2004) reported on a female patient from northern Germany, who suffered primarily from severe acute respiratory distress syndrome-like pulmonary failure due to Dobrava hantavirus infection that was complicated by acute renal insufficiency. The reservoir hosts of Dobrava virus are the yellow-necked mouse (*Apodemus flavicollis*) and the striped field mouse (*A. agrarius*) and probably the wood mouse (*A. sylvaticus*).

Saaremaa virus

There are few available data about the clinical picture of Saaremaa infections, but epidemiological evidence suggests that it is less pathogenic than Dobrava virus, and more similar to nephropathia epidemica caused by Puumala. Along with its rodent host the bank vole (*Clethrionomys glareolus*), Puumala is reported throughout most of Europe (excluding the Mediterranean region), whereas Dobrava, carried by the yellow-necked mouse (*Apodemus flavicollis*), and Saaremaa, carried by the striped field mouse (*A. agrarius*), are reported mainly in eastern and central Europe. Whereas Puumala is distinct, Dobrava and Saaremaa are genetically and antigenically very closely related and were previously thought to be variants of the same virus (Vapalahti *et al.*, 2003) but are now considered as unique serotypes.

Lymphocytic choriomeningitis virus

Lymphocytic choriomeningitis is a rodent-borne arenavirus infectious disease that may cause aseptic meningitis (inflammation of the membrane, or meninges, that surrounds the brain and spinal cord), encephalitis (inflammation of the brain), or meningoencephalitis (inflammation of both the brain and meninges). Asymptomatic infection or mild febrile illnesses are common clinical manifestations. Its causative agent is the lymphocytic choriomeningitis virus, a member of the family Arenaviridae, first isolated in 1934 in the USA.

Lymphocytic choriomeningitis infections have been reported in Europe, the Americas, Australia and Japan, and may occur wherever infected rodent hosts of the virus are found. The disease has historically been under-reported, often making it difficult to determine incidence rates or estimates of prevalence by geographic region. Several serologic studies conducted in urban areas have shown that the prevalence of lymphocytic choriomeningitis infection among humans ranges from 2% to 10%. Cases have been reported from France, Germany, Hungary, the Netherlands, Russia, the Slovak Republic, Spain, Switzerland, the

UK and the Ukraine; the number of cases reported from each country is generally small.

Lymphocytic choriomeningitis is naturally spread by the common house mouse, *Mus musculus*. Once infected, mice can become chronically infected by maintaining virus in their blood and/or persistently shedding virus in their urine, a common characteristic of arenavirus infections in rodents. Chronically infected female mice may transmit infection to their offspring, which in turn become chronically infected. Humans become infected by inhaling infectious aerosolized particles of rodent urine, faeces or saliva, by ingesting food contaminated with virus, by contamination of mucous membranes with infected body fluids, or by directly exposing cuts or open wounds to virus-infected blood. Person-to-person transmission has not been reported, excepting vertical transmission from an infected mother to foetus. Often human infections may be associated with laboratory mouse colonies or with pet mice or hamsters procured from pet shops.

Bacterial diseases

Rodent-borne leptospirosis

Leptospirosis, a spirochaetal infection, causes a wide spectrum of disease ranging from asymptomatic infection, or influenza-like symptoms, to severe jaundice and renal failure. Humans become infected through skin or mucous membrane contact with infected animal urine or urine-contaminated water or soil. The most common source of human infection worldwide is rats, although other animals may also be sources. Pathogenic leptospires live in the kidneys of the rodent host and are excreted in the urine. Five to ten per cent of those infected will have severe leptospirosis with jaundice, known as Weil's disease. Leptospirosis is probably the most widespread zoonotic infection in the world and the WHO has estimated that there are about 5 million human cases and some 1000 deaths worldwide annually, due to infection by *Leptospira* (23 serogroups and more than 200 serovars). It remains under-diagnosed largely due to the broad spectrum of symptoms. The reported incidence of leptospirosis reflects the availability of laboratory diagnosis and the clinical index of suspicion as much as the incidence of the disease. Symptoms in humans can range from benign to severe febrile disorder, to kidney and/or liver failure, to internal haemorrhaging and death (5–20%, but fatalities of 50% have been reported).

Clear descriptions of leptospiral jaundice can be recognized as having appeared earlier in the nineteenth century. It has been suggested that *Leptospira interrogans* serovar *icterohaemorrhagiae* was introduced to western Europe in the eighteenth century by westward extension of the range of the brown rat (*Rattus norvegicus*) from Eurasia (Alston & Broom, 1958). In Europe, serovars copenhageni and icterohaemorrhagiae, carried by rats, are usually responsible for infections.

Occupational risk groups include sewage workers, miners and persons engaged in livestock farming. Persons involved in water sports may be at risk if water courses have been contaminated by infected urine from rodents or livestock.

The infection is present in most countries of Europe but the overall incidence is generally not high. However, there are some areas with concentrations of cases and where a high mortality has been reported; in the period 1994–1996, 222 cases of leptospirosis were reported in Italy with an overall fatality rate of 22.6% (Ciceroni *et al.*, 2000). Five out of eleven brown rats (*R. norvegicus*) trapped in the most ancient urban area of Rome along the Tiber were found positive for *Leptospira* (Pezzella *et al.*, 2004). Among the 118 confirmed cases in Denmark from 1970–1996, the fatality rate was 7% (Holk *et al.*, 2000). The rate of infection of rodents in endemic foci may be as high as 30%. In Germany, after a steady decrease of leptospirosis incidence from 1962 to 1997, surveillance data indicate an increase in disease incidence to 0.06/100 000 (1998–2003). Of the cases, 30% were related to occupational exposure whereas residential exposure accounted for 37%. Direct contact with animals, mostly rats and dogs, was observed in 31% of the cases. The recent changes in transmission patterns of leptospirosis appear to be partially caused by an expanding rat population and the resurgence of canine leptospirosis (Jansen *et al.*, 2005),

Rat bite fever

The causative agent of this disease in Europe is *Streptobacillus moniliformis*, a gram-negative bacterium found in various laboratory and wild animal species; the human disease is also known as Haverhill fever. Contact with the saliva or faeces of rats can lead to infection, and fever accompanied by headache, nausea and myalgia develops within 10 days. Complications can be fatal. The infectious agent can also be transmitted from pet rats and species of gerbils. Less commonly transmission may occur through the ingestion of raw milk contaminated by rodent faeces or urine; such an epidemic occurred among 130 school children in the UK in 1983 (Shanson *et al.*, 1983).

Although the infection is rare, it has been reported from several countries in Europe including Belgium, Denmark, France, Germany, Greece, Netherlands, Norway, Spain, Sweden, Switzerland, Spain and the UK. Engel (1949) reported that at the time of the study, two-thirds of the laboratory and wild rats in Sweden carried the causative agent in their intestines. Physicians should be aware of the possibility of this infection among persons at occupational risk of exposure to rodents such as farmers or sanitation workers or among persons who have been exposed to laboratory or pet rodents.

Rodent-borne salmonellosis

A large number of serotypes of the genus *Salmonella* are pathogenic for both humans and animals; *S. typhimurium* is the most commonly reported serotype. Many different species of animals serve as reservoir host of *Salmonella* and while rodents are not the main reservoir host, there is ample evidence to show that their importance is considerable both in transmitting the organism to man by contamination of food or water or transmitting the infectious agents to other animal species which in turn may transmit to man (Gratz, 1988). While rodents may not be the main reservoir hosts of *Salmonella*, rats and other rodents are frequently found infected and may contaminate human foodstuffs or water and thus transmit the infection. In 1982, only a low proportion of 91 *R. norvegicus* caught in the sewers of Lyon, France were carriers of *S. typhimurium* (6%) (Seguin *et al.*, 1986). In the UK only rarely were more than 10% of the rodents found to carry *Salmonella*. The organisms were mostly cultured from spleen or liver and less than 3% of the animals were found to carry these organisms in the intestine. Ironically, the extensive use of the so-called 'virus' rodenticides (consisting of *Salmonella* cultures) in the first half of the twentieth century increased the prevalence of *Salmonella* carriage in rodents and was a hazard to human health. Although rodents certainly acted as sources of human *Salmonella* infection in the UK in the past, their role has probably never been great (Healing, 1991). A more recent study by Hilton *et al.* (2002) in the West Midlands of the UK found *Salmonella* in 10% of the rectal swabs taken from *R. norvegicus*.

Salmonella enterica serotypes Typhimurium and Enteritidis have been used as rodenticides since the late nineteenth century. Researchers soon realized that the strains of *S. typhimurium* used as rodenticides were identical to strains causing 'meat poisoning' and might cause disease among humans. Use of *S. typhimurium* rodenticides was discontinued early in the twentieth century, but *S. enteritidis* continued to be used as a rodenticide in the UK and Denmark until the early 1960s. In 1954 and again in 1967 the WHO recommended that *Salmonella*-based rodenticides not be used because they posed a hazard to human health stating: '. . . that salmonellas should under no circumstances be used as rodenticides. Rodents rapidly develop resistance to *Salmonella* serotypes; thus, this method has little practical value. Moreover, it has been shown in different countries that such practices are a public health hazard because the serotypes used are dangerous to man' (WHO/FAO, 1967).

In spite of these recommendations, *Salmonella*-based rodenticides are still produced and used in Central and South America and in Asia. 'Biorat' (Labiofam, Cuba), one *Salmonella*-based rodenticide currently used in several countries, is made by coating rice grains with a combination of *S. enteritidis* and warfarin. Currently, the Biorat product label offers no warning regarding the risk for human salmonellosis. Indeed, product information indicates that this product contains a strain of *Salmonella* that is pathogenic to animals but not to humans (Painter *et al.*, 2004).

Protozoal diseases

Toxoplasmosis

Toxoplasma gondii is an intracellular coccidian parasite and one of the most common parasitic diseases of animals and man. The definitive hosts for the parasite are members of the Felidae family, mainly cats. *Toxoplasma gondii* is of primary importance since the range of intermediate hosts which can become infected encompasses virtually all warm-blooded animals, including humans. Although infection with *T. gondii* is extremely common, it is rarely a cause of significant disease in any species. It has been estimated that 30–50% of the world's human population has been infected with *Toxoplasma* and harbours the clinically unapparent cyst form. Intrauterine infection is the most serious; infection of the foetus may take place when the mother acquires a primary infection in the second trimester of pregnancy. Approximately 10–20% of pregnant women infected with *T. gondii* show clinical signs.

Cats may become infected by eating raw meat, birds, mice and other small rodents including rats containing oocysts or by ingesting oocysts from faecal contamination. *Toxoplasma gondii*-infected rats are important in the epidemiology of toxoplasmosis because they can serve as reservoirs of infection for pigs, dogs and cats (Dubey & Frenkel, 1998); a high prevalence of *T. gondii* was found recently in *Rattus norvegicus* on farms in England, suggesting that *Toxoplasma* infections can be perpetuated in wild rodent populations without the presence of cats (Webster, 1994); *Toxoplasma gondii* infection may enhance the likelihood of infected rats being predated by cats. The very high incidence of human infection with toxoplasmosis in France and other countries in Europe is thought to result from the consumption of undercooked and raw meats rather than through contact with cats.

Thus while the prevalence of toxoplasmosis in Europe is very high, most infections are subclinical; the public health importance of the infection lies in the risk to pregnant mothers.

Rodent-borne cestode infections

Hymenolepiasis

Hymenolepiasis is caused by two cestode (tapeworm) species, *Hymenolepis nana* (the dwarf tapeworm, adults measuring 15–40 mm in length) and *H. diminuta* (rat tapeworm, adults measuring 20–60 cm in length). *Hymenolepis diminuta* is a rodent cestode infrequently seen in humans but frequently found in rodents; *H. diminuta* occurs throughout the world. In rare instances, it can infect humans, when by accidental ingestion of infected arthropods, cysticercoids find their way to the small intestine. Rats and other rodents are usually *H. diminuta's* definitive targets and natural hosts. Coprophilic arthropods (e.g. fleas acquired when childen kiss or fondle animals), act as obligatory intermediate hosts. When the infected arthropod is eaten by the definitive host, cysticercoids present in its body cavity develop into an adult worm, and its eggs are eliminated in faeces. *Hymenolepis diminuta* human infection is rather uncommon. Surveys of different populations have reported parasitization rates ranging between 0.001% to 5.5% (Tena *et al.*, 1998). Cases of *H. diminuta* have been reported from Italy, Spain and Russia in Europe but are rare and the tapeworm is of no public health importance.

Eggs of *H. nana* are immediately infective when passed with the stool and cannot survive more than 10 days in the external environment. When eggs are ingested by an arthropod intermediate host (various species of beetles and fleas may serve as intermediate hosts and may be ingested by children especially when playing with pets), they develop into cysticercoids, which can infect humans or rodents upon ingestion and develop into adults in the small intestine. *Hymenolepsis nana* is the most common cause of all cestode infections, and is encountered worldwide. In temperate areas its incidence is higher in children and institutionalized groups.

Echinococcus (hydatid disease)

Two species of *Echinococcus* are known to occur in central Europe, *E. multilocularis* and *E. granulosus*, causing the alveolar and the cystic form of echinococcosis in humans, respectively. Infestation by the larvae of *E. multilocularis* is the most dangerous life-threatening parasitic disease of man in central Europe. The mortality of the disease exceeds 95% in untreated or inadequately treated patients. This is due to its location in the liver but still more so because of the proliferative and infiltrating growth of the larval tissue. Recent studies have shown that in central Europe *E. multilocularis* occurs further north, south and east than previously believed. This parasite is endemic in Austria, Belgium, the Czech Republic, Finland, France, Germany, Greece, Liechtenstein, Luxembourg, the Netherlands, Switzerland and Poland. The prevalence rates of *E. multilocularis* in foxes (*Vulpes vulpes*), the principal definitive host, are alarmingly high in some areas with average rates > 40%. Infection rates in dogs and cats are much lower. In recent years accidental infections with the metacestode stage of *E. multilocularis* have been observed in various animal species (dogs, domestic pigs, wild boar, nutria or coypu (*Myocaster coypus*), and monkeys) and in humans. The intermediate hosts are microtine rodents and occasionally house mice (*Mus musculus*). The mean annual incidence rates of alveolar echinococcosis in humans are low varying between 0.02 and 1.4/100 000 inhabitants in several European countries and regions.

Concern is growing in Europe about alveolar echinococcosis resulting from the increase in grassland rodent and red fox (*Vulpes vulpes*) populations, the intermediate and definitive hosts of the agent, respectively. A study by Viel *et al.* (1999) confirmed that human alveolar echinococcosis is strongly influenced by the densities of arvicolid species. Foxes, which feed almost exclusively on grassland rodents, could mediate this relation when the latter expand.

In France infestations of *E. multilocularis* have been found in the water vole *Arvicola terrestris* in the Auvergne (Deblock & Petavy, 1983) and in the bank vole (*Clethrionomys glareolus*) and in the cuates vole, *Arvicola terrestris* in the Lorraine and in the common vole (*Microtus arvalis*) in the Jura region (Bonnin *et al.*, 1986). In Switzerland, 388 foxes (*Vulpes vulpes*) from Zurich were examined for intestinal infections with *E. multilocularis* and other helminths. The prevalence of *E. multilocularis* in foxes sampled during winter increased significantly from 47% in the urban to 67% in the adjacent recreational area. Ten of these foxes (8%) were infected with more than 10 000 specimens and carried 72% of the total biomass of *E. multilocularis* (398 653 worms). In voles (*A. terrestris*) trapped in a city park of Zurich, *E. multilocularis* metacestodes were identified by morphological examination and by PCR. The prevalence was 20% among 60 rodents in 1997 and 9% among 75 rodents in 1998 (Hofer *et al.*, 2000).

An extensive review of the data on alveolar echinococcosis in Europe was made by Kern *et al.* (2003). The study confirmed that infection with this parasite is still dangerous. A low annual incidence persists in the previously known foci. However, case reports from regions remote from the core area indicate that the disease is spreading. All regions with a proven occurrence of *E. multilocularis* in red foxes (*Vulpes vulpes*) indicate a 'potential risk area', irrespective of the magnitude of prevalence rates. This view is the basis for the current concept of a continuous distribution of the parasite in Europe from central France to Poland.

Rodent-borne nematode infections

Trichinosis

Trichinella spiralis is a nematode worm that is passed between carnivores as a result of eating larvae inside muscle tissues. The parasite normally cycles in rodents (rats and mice). Infection in cats and pigs will result if they eat infected mice or other rodents. Pigs are the most commonly consumed reservoir hosts. If humans then eat infected pig meat, the larvae mature into worms in the small intestine and release eggs and larvae into the blood. The larvae travel from the blood into muscle tissues. This causes the symptoms of trichinosis, fever and muscle pain. Animals can become a source of infection for other animals when larvae inside the muscle are eaten.

The infection in Europe is endemic in Spain, France, Italy, Latvia, Lithuania, Poland, Russia and the countries of the former Yugoslavia. The incidence in Europe has been low because of mandatory inspection of pork for *Trichinella* species. Infection with *T. spiralis* is rare in those countries that have adopted laws limiting the feeding of raw garbage to commercially raised pigs. Infections in western Europe usually now result from consumption of pork products originating in eastern Europe where swine herds currently have a 50% prevalence of trichinosis, and thousands of human cases have been documented (WHO, 2001b).

The re-emergence of a domestic cycle has been due to an increased prevalence of *T. spiralis*, which has been primarily related to a breakdown of government veterinary services and state farms (e.g. in countries of the former USSR, Bulgaria, Romania), economic problems and war in countries of the former Yugoslavia; this has resulted in a sharp increase in the occurrence of this infection in swine herds in the 1990s, with a prevalence of up to 50% in villages in Byelorussia, Croatia, Latvia, Lithuania, Romania, Russia, Serbia and the Ukraine, among other countries (Pozio, 2001). In the past 7 years (1998–2004), 247 cases have been reported in Latvia. Annual case numbers peaked in 2000 with 91 cases, which included four outbreaks involving a total of 77 cases. In the period 2001–2004, the number of cases reported annually has remained steady (range: 20–24), with an incidence of between 0.7 and 1/100 000 inhabitants. In the last 5 years, cases of trichinellosis have been identified in all the age groups above 1 year old, (Perevoscikovs *et al.*, 2005).

Studies in Croatia (Stojcevic *et al.*, 2004) found that *T. spiralis*-infected *Rattus norvegicus* was present only on farms with *T. spiralis*-positive pigs and low sanitation or formerly with low sanitation, yet no infected rat was detected on farms with *T. spiralis*-negative pigs. The finding that no infected rat was found on farms with *T. spiralis*-negative pigs suggests that, in the investigated area, the brown rat is not a reservoir host but only a victim of improper pig slaughtering.

Conclusions

Rodent-borne diseases, and those in which rodents are implicated in the epidemiological cycle of infection, are of considerable importance in Europe. The incidence and severity of haemorrhagic fever with renal syndrome in Europe is growing. Taking into consideration the importance of rodents in the epidemiological cycle of Lyme disease and of a group of other newly emerging infections such as ehrlichiosis, the incidence of rodent-associated infections in Europe is already considerable and their importance is growing.

19

The economic impact and burden of vector- and rodent-borne diseases in Europe

Obtaining economic data about the incidence and prevalence of disease are essential for estimating the costs and benefits of strengthening and maintaining prevention and control programmes, improving existing surveillance systems, and introducing other proposed interventions, such as vaccines. To ascertain the financial costs to countries and individuals of vector and rodent-borne diseases many different aspects must be taken into account; one would have to determine the medical costs to the affected person which would include medical care, the costs of hospitalization, if necessary, and the cost of losses in productivity caused by illness and premature death. The costs to local or national governments might include all aspects of medical care depending on the health system of the given country or community, the cost of epidemiological investigations and the costs involved in the control of the arthropod vectors or rodent reservoir hosts of disease. The number of mosquito control organizations in Europe is rapidly growing while other groups are responsible for the control of tick vectors. An attempt to determine such costs would be a major undertaking and beyond the scope of this book. Nevertheless such costs are real and are a severe 'burden' on the individual and on the community. Some examples will be given below of instances in which the costs of vector- or rodent-borne disease outbreaks have been calculated.

Costs associated with arboviruses

Information has been given earlier on the general increase in tick-borne encephalitis (TBE) incidence in the Czech Republic in the 1990s; people at risk are encouraged to have a vaccination against the infection. Vaccines against TBE in the Czech Republic cost about US $ 10 a dose in 1991 and about

250 000 doses are sold yearly. Tick-borne encephalitis vaccination is not a part of the immunization programme in any of the countries in the region, but in Latvia children in high endemic areas have been offered free TBE vaccines. Normally, the cost of vaccines has to be covered by people themselves or by their employers.

In Austria and in Sweden, a defined post-encephalitic TBE syndrome is present, causing long-lasting morbidity that often affects the quality of life and sometimes also forces the individual to a change in life-style. The sequelae render high costs for individual patients and society.

The epidemic of West Nile virus in Romania in 1966 was certainly very costly to the country in terms of medical costs to the state and to individuals and their families but no information is available on the actual expenditures made during the severe outbreaks nor the costs to the affected individuals or their families.

Costs associated with malaria resurgence in Europe

The resurgence of malaria in a number of countries in eastern Europe has been described above. The dramatic resurgence of malaria in certain countries (mostly eastern ones), is due to political and economic instability, massive population movements and the impact of large-scale irrigation projects. Most of the affected countries have encountered great difficulties in adequately funding malaria control measures. Several project proposals for emergency aid at country and sub-regional level were prepared and the governments of Italy, Japan and Norway provided funding for surveillance and control. In 1996–1997, Japan provided financial support for a large malaria control project in Tajikistan, and Norway supported activities carried out in 1997 to deal with a malaria outbreak in Armenia. In 1997–1998, Italy supported malaria prevention activities in Kazakhstan, Kyrgyzstan and Uzbekistan and some of the malaria prevention activities carried out in Tajikistan.

The disruption of traditional links among the former republics of the USSR has resulted in difficult economic conditions, human migrations, and a sudden reduction in the quality of health care. The shortage of essential equipment and supplies for malaria prevention and control, particularly those that used to be purchased from abroad by the Ministry of Health of the former USSR (such as antimalarial drugs and insecticides) has weakened malaria prevention activities. Lack of knowledge and experience in malaria prevention and control among health-service staff who have not seen malaria for 30 years, is another obstacle in the planning and effective implementation of these measures.

No accurate quantification can be given as to the magnitude of the costs due to the resurgence of malaria in those countries of eastern Europe and western Asia which have been listed above though it must be assumed that they were great. The costs were borne both by the affected countries and by donors who payed for insecticides, equipment and training. The cases of autochthonous malaria in the region of Moscow, Russia have involved funding for medical care and the carrying out of large-scale vector control operations.

Costs associated with imported malaria

As has been noted above, the number of autochthonous cases of malaria in western Europe is relatively small. However, the number of cases of imported malaria is so large as to constitute both a public health and economic burden on the countries into which the cases of malaria are imported. The cost of treatment of malaria and the economic costs of deaths from imported malaria are considerable. Legros *et al.* (1998) estimated that in France the overall cost of an uncomplicated case of malaria (medical expenses and an average sick leave of 2 weeks) has been estimated at 6400 Euros for inpatients and 1400 Euros for outpatients. Thus for the more than 8000 cases of malaria imported into France in 2000, the total cost to the country must have been between a minimum of 10 to 20 million Euros. Pugliese *et al.* (1997), reported that in 1995, 33 patients were hospitalized in Nice hospital from malaria. In 32 of these cases, malaria infection was due to the traveller having taken no or poor prophylaxis. The cost of such poor prophylaxis was high in terms of the consequent human suffering and financial costs. Four patients had to be hospitalized in the intensive care unit and one died during hospitalization. The cumulative cost for these 33 cases was evaluated at that time at 660 000 French Francs. In 1996, 2117 cases of imported malaria were registered in the 22 regional subdivisions of continental France (overseas departments excluded). During 2001–2002, 1255 cases of imported malaria were reported at 36 hospitals in metropolitan France. *Plasmodium falciparum* was identified in 1147 cases; many additional cases were not hospitalized.

Schlagenhauf *et al.* (1995) observed that in Switzerland, the cost of the treatment of single cases of malaria could be as high as 44 000 Swiss Francs or the equivalent of US $32 000 at the time of publication of the article.

Harling *et al.* (2004) studied the cost of imported infections which had been treated in infectious disease units (IDUs) in England and Wales in 1998–1999. The IDUs reported a total of 421 travel-related admissions during the 2-year period. The most common diagnosis was malaria. The average cost per bed day on the IDUs was around £100, and on this basis, the total cost of treating imported infections, most of which were malaria, on the four IDUs in 2 years was around £289,000. The relative risk of acquiring an imported infection requiring treatment on an IDU was greatest for travellers to Africa. From December 1997 through January 1998 a single hospital in the UK treated five patients for severe malaria and gave advice on a further 20 patients with malaria who had been admitted to intensive care units throughout England. The cost to the National Health Service for intensive care for these patients exceeded UK £160 000 (Reid *et al.*, 1998).

Costs associated with mosquito control

In Europe, large-scale, intensive mosquito control operations have been conducted from the beginning of the twentieth century. A large proportion of the control efforts were directed against *Anopheles* vectors of malaria. Mosquito abatement activity in Europe has regained momentum in wake of recent trends of mosquito-borne infections. The establishment of the European Mosquito Control Association (EMCA) in 2000, and initiation of the Roll Back Malaria (RBM) programme by the World Health Organization in the late 1990s, demonstrated the need for developing regional mosquito abatement cooperation. Although the phenomenon of re-emerging vector-borne diseases in Europe represents a serious problem, the main problem being addressed by the countries in the EMCA is the seasonal outbreaks of nuisance mosquito populations, which plague ecologically sensitive tourist and urban areas and cause significant economic damage. At the end of 2004, the EMCA had 168 members in 23 countries, reflecting the growing interest (and expenditures) for mosquito control in Europe.

There are a growing number of mosquito control organizations in Europe, particularly in France, Germany, Italy and Spain. Some of these programmes are very large; as an example, the mosquito control organization in the Piedmont, Italy in 2004 had a budget of well over a million Euros per year and the Rhine Valley mosquito control organization in Germany has a similar budget. The largest number of control organizations is in Italy. There are also various programmes for tick and sandfly control in Europe. No study has been made of the total costs of mosquito control in Europe but the EMCA estimates that these costs are substantial and growing. A few instances are known; the Rome city council spends more than 2 million Euros per year on controlling *Aedes albopictus*, which has become a major pest in the city. In response to a questionnaire sent out in 1985 to 482 health officers and local authorities in the UK, 47 local authorities stated that they had carried out mosquito control in the previous 25 years and 22 had carried out measures in 1985; five of the returns reported that the annual cost of mosquito control ranged from £10 – 4000 with an average of £790 (Snow, 1987). Concerned by the possibility of West Nile virus transmission occurring in the UK, expenditures for vector surveillance have been substantially increased.

Overall, mosquito control activities in Europe are mainly pest or nuisance control but control of the mosquito vectors of West Nile virus has been undertaken in the south of France and, to some extent, in Romania. The existing control organizations would be capable of undertaking disease vector control with greater cooperation with public health authorities. The costs of individual protection measures including household insecticide sprays and repellents can be very considerable both among persons living in areas with high mosquito populations and tourists to these areas.

Costs associated with lice and scabies control

Surprisingly large sums are spent for the purchase of insecticides for the control of head lice; a report by the UK Government estimated that UK £25 million is spent each year in the UK on head louse treatments. In Denmark in 1996, 138 163 individual preparations were sold for head lice control. In France in 1989, 4 656 000 such preparations were sold and by 1994–1995 the value of the sales of individual lice control formulations was almost 100 000 000 French Francs; sales in other countries of Europe were similarly high (Gratz, 1988).

The costs associated with scabies infestations are often difficult to determine as many of the patients are, in any event, institutionalized. Hospital or nursing home outbreaks are often quite serious when they occur, are not easily controlled and require extra work and treatment expenditures beyond normal levels. In Poland, the severe losses in the 20–64-year-old age group from scabies infestations involved 1 836 234 days of sick leave during a 20-year period; there was an average absence in 1980 of 82 500 days of sick-leave as has already been described (Zukowski, 1989).

Costs associated with Lyme disease

While the incidence of Lyme disease (LD) in the UK may not be high, the cost of the infections that do occur is substantial. Joss *et al.* (2003) analysed LD costs incurred by patients over an 18-month period of 2110 samples in Scotland taking into account the costs of consultation, laboratory tests, antibiotic treatment and management of any sequelae, as well as costs of the loss of healthy time through illness. Of the sample, 295 patients had evidence of early LD and 31 had late LD. Based on these figures, the total annual cost for LD, when projected to the whole of Scotland, was estimated to be significant at UK £331 000 (range UK £47 000–615 000. In addition, some late LD sequelae required management for more than 1 year. The authors observed that these data raise the question of whether there is sufficient focus on prevention and management of this disease. One may suppose that costs were similar for other endemic areas of the UK.

No studies have been made on the costs of LD in Germany but it can be readily assumed that all types of medical care costs and lost time from work that result from the estimated 60 000 cases each year must be very high indeed.

Conclusions

Although relatively little data are available, it is evident that the costs for medical care alone for the autochthonous and imported cases of vector-and rodent-borne diseases must be staggering. Several of the diseases have long-term, chronic sequelae that may be the cause of expensive medical care for years. The magnitude of the expenditures for the control of arthropod vectors and rodent reservoir hosts in Europe has not been studied; in light of the magnitude of the number of infections, it would surely be cost-effective to undertake control operations for those diseases that could be controlled in this manner or prevented by vaccines. The growing number of mosquito control organizations in Europe serve to enhance health and quality of life through the suppression of mosquitoes and other vectors of public health importance.

PART II THE VECTOR- AND RODENT-BORNE DISEASES OF NORTH AMERICA

20

Vector- and rodent-borne diseases in the history of the USA and Canada

Infectious diseases took a heavy toll in illness and death from among the first European settlers in North America. While a number of vector-and rodent-borne diseases were certainly present when foreign settlement began, little is known of their history among the North American Indians. A considerable body of literature is available on the history of such diseases.

The arboviruses

Yellow fever

It is hypothesized that Dutch slave traders brought yellow fever to the Americas from Africa during the mid seventeenth century (Bryan *et al.*, 2004) while *Aedes aegypti* was probably imported in the drinking water casks on the decks of slave ships arriving from Africa. However, as jungle yellow fever circulates in South America, the disease may have been circulating on that continent for a long time. Whatever the origin of the disease in the Americas, yellow fever epidemics swept through the settlements, towns and cities along the eastern seaboard, the southern states and up the Mississippi valley in the eighteenth and nineteenth centuries. Thousands of people died and many more fled from the towns and cities when the epidemics struck. By the middle of the nineteenth century it was routine for people to leave the towns and cities in the summer months to avoid the fever. In 1793 the largest yellow fever epidemic in the history of the USA struck Philadelphia. Some 10% of the population or 4000 people died in Philadelphia. In 1821 an epidemic caused heavy mortality in St. Augustine, Florida. In 1855 Norfolk, Virginia, experienced an epidemic. Later in the century, in 1878, Memphis was one of the major southern towns to be struck. In that year 20 000 died from yellow fever in the south. Other cities, like Charleston, South Carolina, suffered more than 20 epidemics in as many summers during the eighteenth century. In the city of New Orleans, the epidemic which developed in the summer of 1853 caused more than 7000 deaths. Later, in 1878 which was the major epidemic year, yellow fever invaded 132 towns in the USA, causing the loss of 15 932 lives out of a total number of cases which were estimated to have reached to more than 125 000: New Orleans alone suffered a mortality of 4600 at that time and the state of Mississippi nearly 15 000 cases (Harden *et al.*, 1967).

Yellow fever was first suspected to be associated with *Aedes aegypti* by the Cuban scientist, Carlos Finlay, and its transmission cycle was established by the US Army Major, Walter Reed. After the development of yellow fever vaccine by Theiler and Smith, yellow fever outbreaks ceased in North America in 1905 and the last urban case in the Americas occurred in South America in 1942, although small outbreaks of jungle yellow fever continue to occur in the endemic countries of Bolivia, Brazil, Colombia, Ecuador, French Guiana, Peru and Venezuela. Many of these countries have dense urban populations of *Ae. aegypti* and *Ae. albopictus* and the threat of urban outbreaks of the disease persists, especially where rates of immunization are low.

Dengue

In the Americas, the first documented dengue fever epidemic occurred in 1779–1780 in the city of Philadelphia; it was clinically characterized as 'break-bone fever' (Halstead, 1980). Epidemics in the Americas were reported in the Caribbean region, affecting several countries, as well as in southern USA during the eighteenth and nineteenth centuries and through the first half of the twentieth century. In 1922, the disease struck many major cities in the southern states, including an estimated 500 000 cases in Texas. Another widespread outbreak occurred in 1947–1948 (Ehrenkranz *et al.*, 1971). Hundreds of thousands of cases of dengue occurred in the USA in the early part of the twentieth century and persisted until changes in life style reduced the densities of the vector *Aedes aegypti*; these changes included screened houses, air-conditioned buildings and, probably the most important change, the installation of reliable, potable water supplies in homes. The last outbreak of dengue in the USA was in 1945 in Louisiana; small numbers of cases now occur in Texas in the continental USA though it remains a severe public health problem in Puerto Rico as will be discussed below.

Other arboviruses

There are four main virus agents of encephalitis in the USA and Canada: eastern equine encephalitis (EEE), western equine encephalitis (WEE), St. Louis encephalitis (SLE) and La Crosse (LAC) encephalitis, all of which are transmitted by mosquitoes. Eastern equine encephalitis is a rare but deadly disease with severe sequelae for survivors. The virus was first documented in 1931 and first isolated in the USA in 1933. Western equine encephalitis was first isolated in California in 1930. In 1941 there was an epidemic in the USA and Canada, in the northern Plains states of the USA and in Canada's Manitoba, Alberta and Saskatchewan provinces. The epidemic involved 300 000 horses and 3336 humans. Until the introduction of West Nile virus into the USA and Canada, SLE was the most important arbovirus disease in North America. The virus was first recognized in 1932 (in Paris, Illinois) and was first isolated in 1933 during the St. Louis epidemic from a human brain. Detailed information on these and the other endemic arboviruses of the USA and Canada will be provided below.

Epidemic or louse-borne typhus

Until the early part of the twentieth century body louse infestations were common, especially in the crowded city slums. Clinically, no differentiation was made between typhoid fever and louse-borne typhus fever until 1837. In that year, Gerhard, in Philadelphia, noted differences between the two, and was the first to call attention to the presence of typhus in the New World. Outbreaks were known to have occurred several times during the nineteenth century in cities of the eastern USA but no data are available on their severity.

Plague

Plague cases originating in China were documented in Honolulu in 1899. Several months later plague arrived in San Francisco. On 6 March 1900, the body of a working man was found in a basement in San Francisco's Chinese quarter, apparently dead from plague. The San Francisco Board of Health took prompt action, quarantining the entire Chinatown area. A house-to-house search, led by uniformed police officers, was made for other victims and for various insanitary conditions. Within hours, the Chinatown community was alarmed and the sick and dead were hidden, while fears were voiced that the entire quarter would be torn down as had happened in Honolulu's Chinatown. Only a few cases were subsequently identified and there was no epidemic but a second outbreak occurred in January 1901; by 1905 there were 121 cases and 113 deaths in San Francisco. The disease struck again in 1907 as a result of the disruptions which resulted from the great earthquake of 1906; plague was not fully eradicated from the city until 1908. However, as a result of this introduction, plague spread throughout much of the western USA and a part of Canada as will be seen below.

Rocky Mountain spotted fever

Rocky Mountain spotted fever is the most severe and most frequently reported rickettsial illness in the USA. The disease is caused by *Rickettsia rickettsii*. Rocky Mountain spotted fever was first recognized in 1896 in the Snake River Valley of Idaho. It was a dreaded and frequently fatal disease that affected hundreds of people in this area. By the early 1900s, the recognized geographic distribution of this disease grew to encompass parts of the USA as far north as Washington and Montana and as far south as California, Arizona and New Mexico. Beginning in the 1930s, it became clear that this disease occurred in many areas of the USA other than the Rocky Mountain region. It is now recognized that this disease is broadly distributed throughout the continental USA as well as southern Canada, Mexico, Central America and parts of South America. Between 1981 and 1996, this disease was reported from every US state except Hawaii, Vermont, Maine and Alaska (CDC website: http://www.cdc.gov/ncidod/dvrd/rmsf/).

Malaria

Although once widespread in North America, the last major outbreak of malaria in North America occurred in the 1880s, although the disease continued to occur through parts of the southern USA until the late 1940s.

There are no references to malaria in the 'medical books' of the Mayans or Aztecs. It is likely that European settlers and slavery brought malaria to the New World and the awaiting anophelines within the last 500 years. Malaria existed in parts of the USA from colonial times to the 1940s. One of the first military expenditures of the Continental Congress, around 1775, was for $300 to buy quinine to protect General Washington's troops.

During the American Civil War (1861–1865), 50% of the white troops and 80% of the black soldiers of the Union Army got malaria annually. The disease reached its peak in about 1875 and then began to decline in the northern regions but still remained one of the most important diseases in the south. More than an estimated 600 000 cases of malaria occurred in the USA in 1914, according to information from the Centers for Disease Control and Prevention in Atlanta, Georgia (Desowitz, 1991).

During World War II and in the years following the war, malaria control was greatly intensified and in 1947 an eradication programme was begun, based to a great extent on the residual applications of DDT in houses and by the mid-1960s local transmission virtually ceased.

Malaria was once common in parts of Canada; it was an important cause of illness and death in the nineteenth century in Upper and Lower Canada and out into the Prairies. In the summer of 1828 'swamp fever' broke out in the settlement of Bytown (Ottawa) and along the construction route of the Rideau Canal. According to some accounts, 'malaria' was not native to North America but had been introduced by infected British soldiers who had returned from India. Numerous deaths had occurred by the time the epidemic subsided in September when the mosquitoes disappeared.

During the period 1826–1832, malaria epidemics halted the construction of the Rideau Canal between Ottawa and Kingston, Ontario during several consecutive summers, with infection rates of up to 60% and death rates of 4% among the labourers. Malaria also appears to have had an important effect on the health of the Northwest Mounted Police in the Prairies. When the Montreal General Hospital opened, in 1823, 3% of the first 3665 patients admitted were ill with malaria, and 3% died in hospital as a consequence. Indigenous malaria gradually disappeared early in the twentieth century for a variety of reasons, including decreasing malaria in Europe and less frequent importation of infections, destruction of *Anopheles* breeding sites, use of window screens and more rapid treatment of febrile cases before the malaria parasite reached the mosquito-infecting, gametocyte stage (Maclean & Ward, 1999).

21

The mosquito-borne arboviruses

Of more than 500 arboviruses recognized worldwide, five were first isolated in Canada and 58 were first isolated in the USA. Six of these viruses are human pathogens: western equine encephalitis (WEE) and eastern equine encephalitis (EEE) viruses (family Togaviridae, genus *Alphavirus*), St. Louis encephalitis (SLE) and Powassan (POW) viruses (Flaviviridae, *Flavivirus*), LaCrosse (LAC) virus (Bunyaviridae, *Bunyavirus*) and Colorado tick fever (CTF) virus (Reoviridae, *Coltivirus*). Their scientific histories, geographic distributions, virology, epidemiology, vectors, vertebrate hosts, transmission, pathogenesis, clinical and differential diagnoses, control, treatment and laboratory diagnosis are reviewed. In addition, mention is made of the Venezuelan equine encephalitis (VEE) complex viruses (family Togaviridae, genus *Alphavirus*), which periodically cause human and equine disease in North America.

Western equine encephalitis, eastern equine encephalitis and St Louis encephalitis viruses are transmitted by mosquitoes between birds; Powassan and Colorado tick-borne fever viruses, between wild mammals by ticks; Lacrosse virus, between small mammals by mosquitoes; and Venezuelan equine encephalitis, between small or large mammals by mosquitoes. Human infections are tangential to the natural cycle. Such infections range from rare to focal but are relatively frequent where they occur. Epidemics of WEE, EEE, VEE and SLE viruses have been recorded at periodic intervals, but prevalence of infections with LAC and CTF viruses typically are constant, related to the degree of exposure to infected vectors. Infections with POW virus appear to be rare (Calisher, 1994). To this list should be added occasional autochthonous outbreaks of dengue in the southern continental USA and the rapid spread of West Nile virus in the USA and Canada following the introduction of the virus in 1999. The public health importance of a number of viruses which have been isolated only from arthropods or animals or rarely from humans and whose human or animal health importance is uncertain or unknown, will not be included.

Togoviridae

Western equine encephalitis (WEE)

The virus was first recognized in a horse in California in 1930. In 1938 it was recovered from the brain of a child who had died of encephalitis. During the summer of 1930, clinical cases of encephalitis occurred in horses in the San Joaquin Valley of California. This initial epizootic of WEE in the San Joaquin Valley in 1930 affected approximately 6000 horses with a case fatality rate of 50%. Several further outbreaks occurred in western states from 1931 to 1934. In 1938, more than 300 000 horses and mules were affected in the USA. In 1941, there were 1094 human cases of WEE reported in Canada and 2242 human cases in the USA. An outbreak in California's Central Valley in 1952 resulted in 813 cases of encephalitis in humans with an attack rate in Kern County of 50/100 000 humans and 1120/100 000 horses. Outbreaks of WEE in the 1930s and 1940s caused serious losses in horses and mules used as draught animals, adversely affecting agricultural production. According to CDC data, there have been 639 confirmed cases in the USA since 1964. In 1941, more than 3400 cases among humans occurred in the northern Plains states and in Canada's Manitoba, Alberta and Saskatchewan provinces. The attack rate reached 167/100 000.

In North America, the present general distribution of WEE virus is west of the Mississippi River in agricultural landscapes. This distribution parallels that of the most important mosquito vector of WEE, *Culex tarsalis*. Outbreaks have often affected wide areas of the western USA and Canada. Western equine encephalitis virus has been reported across south-central Canada from approximately Lake Superior to the Rocky Mountains, but it also has occurred in British Columbia. Disease in humans and horses has occurred most often in Saskatchewan and Manitoba. In western Canada, *Culex tarsalis* is the mosquito host of greatest importance, but the virus has been found in a variety of mosquito species of five genera: *Aedes, Anopheles, Coquillettidia, Culex* and *Culiseta*.

Passerine birds, primarily house sparrows (*Passer domesticus*) and house finches (*Carpodacus mexicanus*) are highly preferred hosts of *Cx. tarsalis* and are an important source of vector infections. Infections in house sparrows, particularly nestlings, amplify WEE virus that is present.

Most people infected with WEE suffer only a mild disease or have no symptoms at all. Children are more likely to suffer clinical disease than are adults;

approximately half of children under 1 year of age will become ill if infected while slightly less than 1 in 1000 people aged 14 years or older will become ill if infected. The mortality rate from WEE is reported to range from 8–15% in people who develop clinical disease after infection. While an effective vaccine is available for equines, none is used for humans. It appears that in California sociobiological changes resulting from the use of television and air conditioning have decreased exposure of California residents to vector attack. The prime time when *Cx. tarsalis*, the primary vector of encephalitis, bites people is around sundown which also is primetime for television watching (Reeves, 2001). Similar prevention has probably occurred in most endemic areas of WEE.

Eastern equine encephalitis

Eastern equine encephalitis is normally the least common of the mosquito-borne arboviral infections in the USA. It is a member of the Togaviridae, genus *Alphavirus*, and occurs only in the Americas. It was first recognized in the USA in 1931 but is thought to have been a cause of encephalitis in North American horses since 1831. The initial isolation of EEE virus was made from equine brain tissue in the eastern USA in 1933. The first human cases were confirmed in Massachusetts during 1938. From 1964–2003, 200 human infections with EEE were recorded in the USA. The enzootic transmission cycle is most common to coastal areas and freshwater swamps but human cases occur relatively infrequently, principally because the primary transmission cycle occurs in swampy areas where the presence of human populations is limited. In 1972 there was an outbreak in Quebec, Canada which occurred at the same time as an outbreak in Connecticut, USA; Sellers (1989) thought it possible that the virus could have been brought to Lac Brome in Canada by infected *Culiseta melanura* mosquitoes carried on surface winds from Meriden, Connecticut 400 km distant.

Although infection with this virus may cause serious illness in humans and horses, it normally cycles between wild birds and the enzootic vector mosquito, *Culiseta melanura*. Eastern equine virus has been isolated from, or antibodies have been found in, a multitude of wild and domestic birds, particularly passerines but also owls, whooping cranes (*Grus americana*) and shore birds. People and horses become infected by the bite of infectious 'bridge vector' mosquitoes that were infected by feeding on viraemic birds. The bridge vectors bringing the disease to humans and horses are *Aedes solicitans* (the salt marsh mosquito), *Culex nigripalpus*, *Coquillettidia perturbans*, *Aedes vexans*, and *Ae. atlanticus* among other species.

Most people infected with EEE have no clinical symptoms. During an outbreak in New Jersey in 1959, a study showed that the ratio of inapparent:apparent infections was 23:1. For people who develop an infection of the central nervous system, a sudden fever and severe headache can be followed quickly by seizures and coma. About half these patients die from the disease. Of those who survive, many suffer permanent brain damage and require lifetime institutional care. Symptoms usually appear 4–10 days after infection. Children <15 years old and adults >55 years old constitute 70–90% of cases during outbreaks. Death typically occurs 2–10 days after onset of symptoms. The equine mortality rate due to EEE ranges from 75–90%. There is a vaccine for horses and a killed vaccine for humans who work with the virus. The use of EEE immunization has successfully protected whooping cranes (*Grus americana*) from death. Large equine outbreaks of EEE have occurred in Florida in 1982, 2003 and 2005 almost entirely among unvaccinated horses.

Venezuelan equine encephalitis

In 1936, Venezuelan equine encephalitis (VEE) was first recognized as a disease in Venezuela following a major outbreak of equine encephalomyelitis. From 1936 to 1968, equines in several South American countries suffered devastating outbreaks of this disease. In 1969, the disease moved north throughout Central America, finally reaching Mexico and Texas in 1971. It has been responsible for numerous outbreaks of febrile illnesses and encephalitis involving thousands of equine cases and some human cases. Venezuelan equine encephalitis viruses are transmitted among equines and rodents by a variety of mosquito species, and have zoonotic reservoir hosts in bats, birds, rodents, equines (horses, donkeys, mules) and certain tropical jungle mammals. Rodents and other small animals are the most important amplifiers in endemic preservation of the virus in tropical forests, swamps and marshlands. Horses are the most important amplifier hosts in large epidemic outbreaks.

In humans the overall mortality rate from epidemics is 0.5–1%. In patients who develop encephalitis, the mortality rate is in the range of 20%. Encephalitis is clinically diagnosed in 2–4% of adults and 3–5% of children. In natural human epidemics, severe and often fatal encephalitis in Equidae always precedes disease in humans.

Venezuelan equine encephalitis is now rare in the USA but it is endemic in the southeast of the country. Serologic evidence of VEE virus was first documented in 1960 during a survey of Seminole Indians in Florida (Work, 1964). Follow-up investigations conducted by the CDC in 1963–1964 resulted in the isolation of VEE virus from *Culex melanoconian*, cotton mice (*Peromyscus gossypinus*) and cotton rats (*Sigmodon hispidus*). Data from outbreaks in South and Central America suggest that many more infections than the number reported have occurred as subclinical or mild forms. In the 1971 outbreak in Texas, there were more than 1500 equine deaths and 110 human cases but no human deaths were reported. *Psorophora confinnis* was considered the vector (Sudia *et al.*, 1975). *Psorophora columbiae, Aedes sollicitans, Ae. taeniorhynchus* and *Ae. albopictus* have been shown to transmit the virus experimentally (Turell & Beaman, 1992). Based on past experience, VEE is an arbovirus with an important human and veterinary potential. An effective equine vaccine is available.

Everglades encephalitis virus

Virulence studies showing distinct differences between the endemic Florida strain (Fe-37c), and the more virulent epizootic strains of VEE found in Central and South America led to the classification of the Florida strain of VEE as a separate Group A arbovirus which is now known as Everglades encephalitis virus (EVE). In the late 1960s the CDC and State Board of Health monitored SLE virus and EVE activity throughout south Florida. In 1968 the first documented human case of EVE ever reported in the USA was reported from the Florida City area in southern Dade County. To date, only 3 human cases have been reported in the state. Everglades encephalitis virus only causes a very mild flu-like illness in most people. Clinical disease in the reported cases, which ranged in age from 51–75 years, began with a gradual onset and prolonged prodrome with respiratory symptoms, progressing to profound obtundation followed by a gradual recovery with few residua. The probable vector is the black saltmarsh mosquito, *Aedes taeniorhynchus*.

No new cases of disease due to this virus have been reported in over 25 years and thus this VEE-related virus is, at present, of no public health importance in the USA.

Highlands J virus

In North America, Highlands J (HJ) virus appears to be an antigenic link with an intermediate virulence between epizootic WEE virus and the enzootic Fort Morgan virus which has been isolated from cliff swallows (*Petrochelidon pyrrhonota*) and house sparrows (*Passer domesticus*) and from cimicid bugs (*Oeciacus vicarius*) in eastern Colorado (Calisher *et al.*, 1980). Highlands J virus is primarily a veterinary pathogen causing disease in domestic birds including turkeys, chickens and partridges (*Perdix perdix*). It has an enzootic cycle similar to EEE. It is often used as an indicator of the presence of EEE in a surveyed area. Highlands J virus has been associated with rare cases of sporadic equine and human disease. It has a low pathogenicity in mammals and is rarely seen in humans or horses. During the 1990–91 St. Louis encephalitis (SLE) outbreak in Florida, four patients were reported to be dually infected with SLE and HJ; however, exposure to HJ virus has not been directly associated with human illness and it is not considered of public health importance. It has been attributed as the cause of death of a horse in Florida (Karabatsos *et al.*, 1988).

Highlands J virus is transmitted from *Culiseta melanura* mosquitoes to songbirds in freshwater swamps; it has been isolated from *Aedes vexans* in Rhode Island, *Cs. melanura* in Massachusetts and *Cs. melanura*, *Cs. morsitans*, *Ae. canadensis*, *Ae. vexans*, *Ae. stimulans*, *Ae. triseriatus*, *Anopheles punctipennis*, *Culex pipiens* and *Cx. restuans* in Connecticut.

Flaviviridae

Dengue

Dengue viruses are members of the Flaviviridae and transmitted principally in a cycle involving humans and mosquito vectors. The principal vector worldwide is *Aedes aegypti*. Infections are caused by any of the four virus serotypes (DEN-1, DEN-2, DEN-3 and DEN-4). The serious dengue haemorrhagic fever (DHF) syndrome and dengue shock syndrome (DSS) were first recognized in the 1950s during dengue epidemics in the Philippines and Thailand; dengue affects many countries in Asia and the Americas and has become a leading cause of hospitalization and death among children in several Asian countries. In its most severe form infections as DHF and DSS can threaten the patient's life, primarily through increased vascular permeability and shock, in which bleeding and sometimes shock occurs and may lead to death. The case fatality in patients with dengue shock syndrome can be as high as 44% but with good physiologic fluid replacement therapy, mortality rates should be no higher than about 3.5%.

In an effort to prevent epidemics of urban yellow fever transmitted by *Aedes aegypti*, the Pan American Health Organization organized a campaign that succeeded in eradicating *Ae. aegypti* from most of the Central and South American countries in the 1950s and 1960s. As a result, epidemic dengue continued to occur only sporadically in some Caribbean islands during this period. Records show no evidence of epidemic dengue in the Americas from 1946 through 1963, reflecting in part the benefits from the eradication programme. In all areas where *Ae. aegypti* had been eliminated, transmission of dengue virus was interrupted. The *Ae. aegypti* eradication programme was officially discontinued in the USA in 1970, and gradually eroded elsewhere, and the species began to reinfest countries from which it had been previously eradicated. By 1995, the geographic distribution of *Ae. aegypti* was similar to that of its distribution before the eradication programme.

Dengue was endemic in the USA through the nineteenth century and up until approximately 1945. The first indigenous transmission of dengue in the

USA since the 1940s occurred in Texas in 1980. Small outbreaks of dengue have subsequently been detected in south Texas in 1980 (63 cases), 1986 (9 cases) and 1995 (29 cases). An outbreak of DEN-3 that occurred in 1999 involved 55 cases and one death due to DSS. Usually these outbreaks have been associated with concurrent dengue epidemics in northern Mexico. Cases of dengue are imported into the USA and Canada every year by persons returning from travel to endemic areas; reporting of the infection is passive and a significant proportion of the introduced cases go unrecognized. There is no threat of autochthonous transmission in Canada for the present as neither *Ae. aegypti* nor *Ae. albopictus* are found in that country.

In 1966, when *Aedes aegypti* had been eradicated from all mainland countries of the Western hemisphere except the USA, Venezuela, Colombia and French Guiana, infestation rates of up to 80% were being found in Puerto Rico. The species was found breeding in tyres, auto bodies, tins cans, bottles etc. even in wooded rural areas up to half a mile from human habitations and tree hole breeding was found in most areas. Epidemics of dengue fever occurred in Puerto Rico in 1963–1964 and 1969 and were caused by dengue-2 and dengue-3. Dengue 3 disappeared and then was detected in Puerto Rico in January of 1998 after a 20-year absence. DEN-1, DEN-2 and DEN-4 have circulated on the island from 1985–1997 (Rigau Perez *et al.*, 2002). Expanding autochthonous transmission of DEN-3 was documented in 1998, an epidemic year where two other serotypes predominated (DEN-4, 45% and DEN-1, 35%).

Puerto Rico has a large urban population with high mosquito vector densities and has experienced nearly 20 years of dengue epidemics with a continuous upwards trend. Although dengue fever was recorded in Puerto Rico as early as 1915 (Dietz *et al.*, 1996), continuous transmission of all four serotypes has only occurred since the 1980s. Puerto Rico had five dengue epidemics in the first 75 years of the twentieth century, which was then followed by six epidemics in 11 years, with an estimated cost of over $150 million. The first epidemic in Puerto Rico, consisting primarily of DEN-4, was reported in 1981/1982, followed by another DEN-4-dominated outbreak in 1986, this one marked by high incidences of DHF/DSS (Dietz *et al.*, 1996). DHF/ DSS cases have occurred periodically since the 1980s, reaching record levels in the DEN-4 epidemic in Puerto Rico in 1998.

Dengue typically occurs in a seasonal pattern in Puerto Rico, with minimal occurrence from March to June and a transmission peak from September to November. The number of reported cases during the last 5 non-epidemic years (1992, 1993, 1995, 1996 and 1997) ranged from 4645–11 078 (an average rate of 2.0 cases per 1000 population). In 1994, 23 693 cases were reported (6.7 per 1000 population). From 1995–1997 dengue was reported in Puerto Rico at an average annual rate of 1.75/1000 population, compared to 6.73 in 1994, an epidemic year. From June 1994 through March 1995, Puerto Rico experienced the most severe dengue epidemic on record for the island, with DEN-2 as the predominant serotype. Another epidemic, with DEN-4 and DEN-1 predominating, occurred in 1998.

Dengue was first introduced into Honolulu, Hawaii in 1902, reportedly from Hong Kong by a steamship with 12 dengue cases on board. The ship had sailed about 25 days earlier. About 50 000 cases then occurred in the islands and a small number of cases persisted in Hawaii until 1912 (Usinger, 1944). Dengue was again reported in Honolulu on 24 July 1943, for the first time in over 30 years; the virus had been introduced by an infected Army Air Force pilot who, during his own incubation period, had flown a plane from the Fiji islands where a dengue outbreak was in progress. The outbreak reached its peak in October 1943. No cases were reported during the last 2 months of 1944 and the first 4 months of 1945. The total number of civilian cases was 1506; of military personnel, 56.

No further autochthonous dengue infections were reported in Hawaii after 1944 until September 2001, when the Hawaii Department of Health was notified of an unusual febrile illness in a resident with no travel history; dengue fever was confirmed. During the investigation, 1644 persons with locally acquired dengue-like illness were evaluated, and 122 (7%) laboratory-positive dengue infections were identified; dengue virus serotype 1 was isolated from 15 patients. No cases of DHF or shock syndrome were reported. Phylogenetic analyses showed the Hawaiian isolates were closely associated with contemporaneous isolates from Tahiti. *Aedes albopictus* was present in all communities surveyed on Oahu, Maui, Molokai and Kauai; no other *Aedes* species were found (Effler *et al.*, 2005).

Dense populations of *Ae. aegypti* and *Ae. albopictus* are present in many areas of the USA. The risk of further outbreaks of dengue is substantial especially as a result of the frequent introductions of the virus by travellers returning from endemic areas of the Americas.

St. Louis encephalitis

This virus is named after the city of St. Louis, Missouri in which it was first recognized and isolated in the late summer of 1933 when over 1000 cases and 200 deaths from the disease were diagnosed (Muckenfuss *et al.*, 1933; Monath, 1980); there had apparently been an outbreak a year earlier in Paris, Illinois. Prior to the introduction of West Nile virus in 1999, SLE was the most important mosquito-borne arbovirus in North America. A large proportion of the infections by SLE are subclinical; asymptomatic human infections are about 200 times more common than symptomatic infections but in those persons who develop an illness, mortality rates may be about 5% for persons under 50 or from 7%–24% or higher in patients over 50 years of age. Humans are the only vertebrates known to suffer clinical disease after infection with SLE.

The virus is widely distributed in the USA, and infections occur as periodic focal outbreaks of encephalitis in midwestern, western, and southwestern USA, followed by years of sporadic cases. It has caused large urban epidemics of encephalitis. Outbreaks occur from August through October. During the last 5 decades, about 10 000 cases were reported. St. Louis encephalitis is mostly sporadic. Annual incidence is 0.003–0.752 cases per 100 000 population. Median occurrence is 35 cases per year. In the USA, the virus exhibits an endemic pattern primarily along the western coast and sporadic infections occur in the east.

St. Louis encephalitis was first isolated in Saskatchewan in 1971 but disease in Canada due to this virus was not documented until 1975 when an outbreak of 66 cases occurred in southern Ontario; it was the first outbreak of arbovirus encephalitis in the province of Ontario. One case each occurred in Manitoba and Quebec in that year (Artsob, 1990). St. Louis encephalitis has caused a small number of cases of human disease in Ontario, Quebec and Manitoba, and was found once in a mosquito in Saskatchewan. The prevalence of antibodies to SLE was high in some wild mammals in southern Ontario sampled in 1975–1976.

The virus has been isolated from several different species of mosquitoes in different areas of the USA; in the area of the Gulf Coast, Ohio and Mississippi Valley, *Culex pipiens* and *Cx. quinquefasciatus* are the primary vectors, in Florida *Cx. nigripalpus*, while in the western States *Cx. tarsalis* is the most important vector species. In Canada, the most common vector, *Cx. pipiens*, is present everywhere in southern Ontario and southern Quebec. St. Louis encephalitis virus has been detected in *Cx. quinquefasciatus* and *Cx. stigmatosoma* in California. The virus has also been isolated from a tick although the epidemiological significance of this isolation remains unknown.

Birds are the primary hosts for SLE virus in most of North America; in endemic areas the prevalence of antibodies in reservoir host birds, especially the house sparrow (*Passer domesticus*) and the robin (*Turdus migratorius*) may be very high. House finches (*Carpodacus mexicanus*), blue jays (*Cyanocitta cristata*) and rock doves (*Columbia livia*) have also been found infected with the virus. Reisen *et al.* (2003) cautioned that birds with elevated field antibody prevalence rates may not be the most competent hosts for encephalitis viruses and that relatively few birds developed chronic infections that could be important in virus persistence and dispersal.

No vaccine is currently available though chimera viruses are thought to have the potential for use as diagnostic reagents and vaccines against SLE and WNV.

West Nile virus in North America

West Nile virus is a flavivirus in the same family as SLE – a member of the Japanese encephalitis complex. A description of its origin and its clinical aspects has already been provided in the section on arbovirus diseases in Europe.

The 1999 outbreak of WNV in the northeastern USA was the first known occurrence of this flavivirus in the western hemisphere. Sixty-two cases of severe disease, including seven deaths, occurred in the New York City area in late August 1999. Although initially attributed to SLE, examination of the virus based on positive serologic findings in cerebrospinal fluid and serum samples using a virus-specific IgM-capture enzyme-linked immunosorbent assay (ELISA) established that the cause of the outbreak was West Nile virus (CDC, 1999). Identification of virus was made in human, avian and mosquito samples. Laboratory confirmation was also received on the presence of WNV in horses. A total of 61 human cases of the fever were reported during the New York City outbreak, which resulted in 7 deaths and, in addition, one case involved a Canadian visitor to New York, who contracted the virus in late August and died on 5 September in Canada. None of the patients involved had travelled to Africa during the incubation period, and many of them had never ventured far from New York City.

The infection caused severe mortality among birds, especially crows (corvids). Before and concurrently with the human cases, health department officials had begun to investigate an increase in the number of dead birds in New York City. Crows were particularly affected, and officials at the Bronx Zoo had noted the deaths of a cormorant, two captive flamingoes and a pheasant. All were subsequently confirmed to have died of infection with WNV. The virus was subsequently found in many bird species in the area. Most of the birds were crows (*Corvus brachyrhynchos*), but the virus has also been isolated in the ring-billed gull (*Larus delawarensis*), yellow-billed cuckoo (*Coccyzus americanus*), rock dove (*Columba livia*), sandhill crane (*Grus canadensis*), fish crow (*Corvus ossifragus*), blue jay (*Cyanocitta cristata*), bald eagle (*Haliaeetus leucophalus*), laughing gull (*Larus atricilla*), black-crowned night heron (*Nycticorax nycticarax*), mallard duck (*Anas platyrhynchos*), American robin (*Turdus migratorius*), red-tailed hawk (*Buteo jamaicensis*) and broad-winged hawk (*Buteo platypterus*). Horses were also affected and had a high death rate.

The virus is thought to have been introduced to New York by an infected human traveller or through the importation of infected birds (Hoey, 2000). The possibility also existed that the virus may have been introduced by a migratory bird or by infected mosquitoes arriving aboard aircraft from an endemic area. Late in 1999, WNV was isolated from overwintering *Culex* mosquitoes in the New York City area (CDC, 2000a).

Since the introduction of WNV in 1999, the virus has spread to most of the USA, reaching California in 2003 (where there were a total of 826 West Nile virus infections and 26 deaths reported from 23 counties in California in 2004). West Nile virus is now endemic in Canada and Mexico, and has become a significant public health problem in both the USA and Canada. In Canada, by the end of 2003, disease activity had spread to a total of nine provinces and territories. In 2002, WNV-neutralizing antibodies were detected in samples from birds captured in Jamaica, the Dominican Republic and Guadeloupe, and horses in Mexico, suggesting establishment of the virus in the Neotropics. By January 2005 the presence of WNV had been recorded as either human or mosquito infections in all 48 continental states, seven Canadian provinces and throughout Mexico. In addition, WNV activity has been detected in Puerto Rico, the Dominican Republic, Jamaica, Guadeloupe and El Salvador.

The incidence of West Nile virus in the USA and Canada, 1999–2004

In 2000, 18 human cases and 65 equine cases were reported from 12 states. In 2001, there were only 66 cases but the virus had spread to 27 states. Although the virus had spread westward, only a modest disease activity was seen until 2002, when the number of cases increased dramatically; in 2002, 39 states and the District of Columbia reported 4156 human WNV cases with 284 deaths. Of these, 2942 (71%) were neuroinvasive illnesses (i.e., meningitis, encephalitis or meningoencephalitis). The number of cases of WNV has also continued to rise; in 2003, 9862 cases and 264 deaths were reported to the CDC. In 2004, 2470 human cases and 88 deaths were reported.

In Canada, human cases were first recorded in both Ontario and Quebec in 2002 with approximately 800–1000 probable, confirmed and suspected cases detected. By the end of 2002 viral activity had been documented in Nova Scotia, Quebec, Ontario, Manitoba and Saskatchewan. In the same year, cases of human infection were reported in Ontario and Quebec. In 2003 there was a further increase in the number of cases to 1388 reported and 1220 confirmed cases with 10 deaths reported from nine provinces or territories. In 2004, there was a considerable fall in the number of cases reported in Canada to only 25 cases and no deaths. (Canadian data from Health Canada website: http://www.phac-aspc.gc.ca/wnv-vwn/, accessed 27 June 2005.)

West Nile virus in avian populations in North America

Birds serve as the reservoir hosts for WNV and the transmission of the virus occurs predominantly via mosquitoes. Ornithophilic mosquitoes transmit the virus amongst the bird population while those mosquito species that feed on both birds and mammals can transmit the pathogen to humans. The differences so far observed between the epidemics in North America and in Europe might be explained by the following considerations: Europe has long had contact with WNV-endemic areas in Africa via migratory birds, whereas the pathogen was first imported into the USA in 1999 where it has infected a 'naïve' bird population. It is possible that the strain presently spreading through the USA is highly pathogenic, whereas strains of varying pathogenic potential circulate in Africa. This could result in a natural immunization of the bird population through contact with strains of low pathogenicity. The possibility of a natural resistance in European birds is also being considered since these animals have been confronted with the pathogen for long periods of time (Pauli, 2004).

Following the first outbreak of WNV in New York City, large numbers of dead crows (Corvidae) have characterized areas in which outbreaks of WNV have occurred. West Nile virus has been isolated from 198 bird species in North America (Komar, 2003), and mortality may approach 100% in some species. Passerine birds, including crows, house sparrows (*Passer domesticus*) and blue jays (*Cyanocitta cristata*), serve as the primary amplifying hosts of the virus, and develop a high-level viraemia that lasts for several days. A study of WNV transmission in 25 species of birds found cloacal shedding of virus in 17 of 24 species and oral shedding in 12 of 14 species. In addition, contact transmission was identified in four species, and oral transmission in five species (Komar *et al.*, 2003). Although the precise contribution of direct transmission to disease activity among birds has not been quantified, it has potentially significant implications for disease epidemiology, because even with effective vector control virus may be amplified and transmitted. Direct transmission may be aggravated in commercial settings because of cannibalism and feather-picking of sick birds (Banet-Noach *et al.*, 2003). As WNV has been marked by a very high level of mortality among crow populations, observations of crow mortality have become an important component of surveillance for WNV.

The mosquito vectors of West Nile virus in North America

The mosquito vectors are essentially species which may feed on both birds and humans. Humans are 'dead-end' hosts as the level of viraemia in humans is not high enough to be infective to mosquitoes. The species found to be the vector in the area of the original outbreak of WNV in New York in 1999 was *Culex pipiens*. By the end of the second year of transmission of WNV, the virus was isolated from, or WNV gene sequences were detected in, 470 mosquito pools in 38 counties in five states (352 pools in New York, 54 in New Jersey, 46 in Pennsylvania, 14 in Connecticut and 4 in Massachusetts). Of the

470 reported WNV-infected pools, *Culex* species accounted for 418, including 222
Cx. pipiens/restuans, 126 *Cx. pipiens*, 35 *Cx. salinarius*, 11 *Cx. restuans* and 24 unspecified *Culex* pools. *Aedes* species accounted for 29 positive pools, including 9 *Ae.
japonicus*, 9 *Ae. triseriatus*, 8 *Ae. trivittatus*, and 1 each of 3 other *Aedes* species. In
addition, WNV was detected in 3 pools of *Culiseta melanura*, 1 pool of *Psorophora
ferox* and 1 pool of *Anopheles punctipennis* (MMWR 2000b; Marfin *et al.* 2001). *Culex
nigripalpus* was detected as a vector in Florida in 2001 (Rutledge *et al.*, 2003). West
Nile virus has also been isolated from *Aedes taeniorhynchus, Deinocerites cancer* and
Anopheles atropos in Florida (Hribar *et al.*, 2003). The primary vector for WNV in
the western USA is *Cx. tarsalis* (CDC, 2000d).

Conclusions on the public health importance of West Nile virus in North America

The introduction of WNV into North America in 1999 has been followed
by a remarkably rapid spread of the disease to almost all parts of the USA and
many provinces of Canada; it is associated with high morbidity and significant
mortality. The infection has clearly become endemic in virtually all the areas
to which it has spread and one must expect further periodic outbreaks as have
occurred in Europe and Israel as well as sporadic individual cases in humans
and animals. In the absence of a useable vaccine, efficient mosquito programmes
represent the only preventive measure against outbreaks of WNV.

Bunyaviridae

Cache Valley virus

Cache Valley virus (CV), first isolated in Utah in 1956, has been recovered
mainly from mosquitoes and occasionally from vertebrates and has the widest
apparent distribution among this serogroup of viruses in the USA and Canada.
Antibodies against CV and other viruses of the Bunyamwera serogroup are prevalent in livestock, large wild mammals and humans from Alaska to Argentina
(Sexton *et al.*, 1997). Cache Valley virus is a bunyavirus that causes foetal death,
still-birth and congenital malformations such as arthrogryposis and anencephaly
in sheep. It has been suggested that CV is the cause of neural tube defects in
humans, although a study in 1997 could not serologically link infection with
anencephaly and other defects (Edwards & Hendricks, 1997). Cache Valley virus
was isolated from a deer hunter in North Carolina who suffered life-threatening,
multi-organ failure, and recent serologic evidence suggests that deer may be a
reservoir host in nature. However, the true zoonotic impact of CV remains to be
determined, and people are most likely infected via mosquito bites rather than
by direct contact with infected animals.

The virus has been isolated from *Anopheles quadrimaculatus*, *Coquillettidia pertur-bans*, *Culiseta inornata*, *Aedes sollicitans*, *Psorophora columbiae*, *Anopheles punctipennis*, *Ae. vexans* and *Ae. trivittatus* (Calisher, 1994). It has also been isolated from a single pool of *Ae. dorsalis* in New Mexico (Clark *et al.*, 1986).

While CV may be the cause of congenital abnormalities in sheep, its terato-genic potential for humans is uncertain. However, it has been associated with human mortality and, rarely, severe illness and close surveillance for the infec-tion is warranted. Cache Valley virus has been isolated from *Ae. albopictus* (Moore & Mitchell, 1997) but the vectorial importance of this species as a vector of CV (and other arboviruses in the USA) remains unknown.

Jamestown Canyon virus

Jamestown Canyon virus (JC) is a member of the California serogroup. This group of related viruses is one of 16 serogroups within the genus *Bunyavirus*, family Bunyaviridae. Several other human pathogens (e.g. La Crosse virus) also belong to the California serogroup. Few instances of human disease were associ-ated with these viruses until 1960 but it is now realized that clinical infections are very common.

Jamestown Canyon virus has a wide geographic distribution throughout much of temperate North America. It was first isolated in 1962 from a mosquito in Colorado and was first recognized as causing disease in humans in 1980 in an 8-year-old girl living in rural southwest Michigan. The child was released after 27 days of hospitalization, and 15 months after discharge she had resumed all her previous activities with no evidence of neurological sequelae (Grimstead *et al.*, 1982). Jamestown Canyon virus generally causes a mild febrile illness and, rarely, aseptic meningitis or primary encephalitis. Many persons may be infected but show no signs of illness.

Subsequent human infections with JC virus have been recognized from Alaska, California, Connecticut, Indiana, Illinois, Iowa, Michigan, Minnesota, New York, North Carolina, Ohio, Wisconsin and Canada. The first recognized case in Canada occurred in 1981 in Ontario and the virus is found in New-foundland, Quebec, Ontario, Manitoba and Saskatchewan, and the Northwest Territories in Canada.

Jamestown Canyon virus persists in cycles of infection among wild ungu-lates, especially the white-tailed deer (*Odocoileus virginianus*). Neither small mam-mals nor domestic animals are likely to be important in disease transmis-sion. Transovarial transmission has been demonstrated in *Aedes stimulans* and *Ae. triseriatus*. In the western USA, Jamestown Canyon virus has been isolated almost exclusively from *Culiseta inornata*. The most important vectors in the Midwest may be *Aedes stimulans* and *Ae. communis*, while *Anopheles punctipennis* and

An. quadrimaculatus are potential late season vectors. Over-wintering maintenance of the virus is probably in mosquito eggs and possibly in hibernating adult female *Anopheles*.

Jamestown Canyon virus is very widespread, particularly in the northern USA, and extends to California and Canada; in some areas the prevalence of antibodies may be quite high in local human and animal populations, e.g. 55% of 141 horses in California (Campbell *et al.*, 1990), 17% of 780 Michigan residents (Grimstad *et al.*, 1986). Serious illness is rare and the disease is not a public health problem in either USA or Canada.

La Crosse virus

La Crosse encephalitis (LAC) is a mosquito-borne disease that can be mistaken for herpes simplex encephalitis. It was first identified in LaCrosse, Wisconsin in 1963. It has been reported in 28 states but may be under-recognized in others and is generally thought to be under-diagnosed. The number of cases reported annually in the USA ranges from 30–180. It has been diagnosed in areas of the eastern portion of Canada. It has been estimated that there may be as many as 300 000 LAC infections annually in the Midwestern USA alone (Calisher, 1994). However, the vast majority of infections are not clinically apparent or are associated with mild symptoms, suggesting that humans have a powerful defence against LAC infections.

McJunkin *et al.* (2001) studied the clinical aspects of 127 LAC patients in the USA; their symptoms included headache, fever and vomiting (each in 70% or more of the patients), seizures (in 46%) and disorientation (in 42%). Thirteen per cent had aseptic meningitis. Hyponatremia developed in 21%, and there were signs of increased intracranial pressure in 13%. Follow-up assessments, performed in 28 children, suggested an increase in cognitive and behavioural deficits 10 –18 months after the episode of encephalitis. Infections from this virus are endemic to the midwestern USA and primarily affects children between 5–10 years of age. Death occurs in less than 1% of the cases.

La Crosse virus cycles in woodland habitats between *Aedes triseriatus* and vertebrate hosts, e.g. chipmunks (*Tamias striatus*) and the eastern grey squirrel (*Sciurus carolinensis*). The virus survives the winter in mosquitoes. During the summer the virus is amplified horizontally among small mammals. Studies in Wisconsin found natural infections of LAC in sentinel foxes and in raccoons (*Proyon lotor*) (Amundson & Yuill, 1981) and antibodies have been found in deer in Wisconsin and North Carolina. In most endemic areas of the mid-west, *Ae. triseriatus* is the principal, if not only vector; *Ae. canadensis* was shown to be a vector of LAC in Ohio through isolation of the virus from field-collected specimens. Frequent identification of *Ae. canadensis* as a human-biting species has implicated it as an auxiliary vector of LAC virus to man (Berry *et al.*, 1986). *Aedes albopictus* has been found naturally infected with LAC in North Carolina and in Tennessee (Gerhardt *et al.*, 2001), but as with other arboviruses, the actual importance of *Ae. albopictus* as a vector remains unknown.

La Crosse virus is of public health importance, particularly in the mid-western states of the USA. Though the number of reported cases is modest, the actual number of infections is probably quite large and infection with the virus may have serious consequences in children. As with most arboviruses, only supportive treatment is available. Control of the vector mosquito should be attempted in foci of the infection. Inasmuch as a risk factor identified from a survey of LAC virus case environments was the presence of discarded tyres, this suggests that removing these may be a more effective control measure than removing other kinds of containers, or filling tree holes.

Snowshoe hare virus

Snowshoe hare virus (SSH) is classified in the genus *Bunyavirus* of the family Bunyaviridae. It is an enveloped, single-stranded RNA virus, one of a large group of related arboviruses called 'California serogroup' viruses. This group includes Jamestown Canyon virus, the other California serogroup member associated with human disease in Canada. The California serogroup derives its name from the first virus of the group that was recognized – California encephalitis virus – which was isolated from mosquitoes in California in 1943. Snowshoe hare virus was first identified in a Snowshoe hare (*Lepus americanus*) in Montana in 1958. It was first recognized as a cause of human disease in 1978.

Snowshoe hare virus has been reported from all 10 provinces of Canada, the Yukon, the Northwest Territories and Nunavut. Its worldwide distribution includes Canada, some adjacent areas of the northern states, Alaska and some of eastern Asia. Surveys in Canada have found from 0.5–32% of people in the studied areas possess antibodies to one of two California serogroup viruses; in most of these people the virus would have been SSH. Disease in humans, when it occurs, takes the form of infection and inflammation of the brain (meningitis and encephalitis). Some cases of clinical illness due to SSH probably go unrecognized. However, it appears that infection is much more common than actual disease.

Snowshoe hare virus persists in a variety of different ecological associations of wild mammals and mosquitoes and is not limited to a small number of either vertebrate or invertebrate host species or to a narrow range of habitats. Mosquito vectors comprise several *Aedes* species and *Culiseta inornata* (Mclean, 1975). Snowshoe hare virus has been isolated from *Aedes stimulans* group mosquitoes in

Massachusetts (Walker *et al.*, 1993), and in the Yukon Territory it has been isolated from the larvae of *Ae. communis* and *Cs. inornata*.

Despite its wide distribution, SSH is of very limited public health importance in the area in which it is restricted in the northern USA and Canada.

Conclusions on the public health importance of arboviruses in North America

As the recent outbreak of WNV has shown, mosquito-borne arboviruses are of very considerable public health importance in the USA and Canada; the introduction and remarkable spread of WNV demonstrates that North America is at risk to the introduction of exotic arboviruses. St. Louis virus and Venezuelan equine encephalitis in periodic outbreaks have had serious effects on human populations. The reason for the periodic outbreaks is not always well understood and control is often difficult.

There have been serious epidemics of Rift Valley fever in southern Africa, Egypt and recently in Saudi Arabia. This virus, or others, could be introduced into North America with predictably grave effects on humans and domestic animal populations. Constant surveillance is necessary as is the preparation of plans for control. Clinicians should be aware of the diagnostic possibility of arboviruses especially among travellers returning from endemic areas.

22

Mosquito-borne diseases – malaria

As has been noted previously malaria was a severe scourge in much of the area of the southeastern states of the USA and as far north as southeast Canada. With the disappearance of endemic malaria, the importance of malaria for both countries is primarily as an imported disease. Since 1957, nearly all cases of malaria diagnosed in the USA have been imported. Approximately half the cases of imported malaria occur among USA-born civilians and half among foreign-born civilians, many of whom contract malaria while visiting their countries of origin. Malaria frequently occurs among military service men returning from endemic countries who have not complied with recommended chemoprophylaxis recommendations. In the USA, about 1200–1500 malaria cases are reported annually which are mostly attributed to imported malaria. These cases describe travellers or immigrants coming from malaria-endemic areas, where they were infected, and developing signs of malaria after arriving in the USA.

A total of 1337 cases of malaria, including 8 deaths, were reported for the year 2002 in the USA. Of these 1337 cases, all but 5 were imported, i.e., acquired in malaria-endemic countries. The USA Centers for Disease Control (CDC) report that in the USA, between 1957 and 2003, 63 outbreaks of locally transmitted mosquito-borne malaria have occurred; in such outbreaks, local mosquitoes become infected by biting persons carrying malaria parasites (acquired in endemic areas) and then transmit malaria to local residents. Of the 10 species of *Anopheles* mosquitoes found in the USA, the 2 species that were responsible for malaria transmission prior to eradication (*Anopheles quadrimaculatus* in the east and *An. freeborni* in the west) are still widely prevalent; thus there is a constant risk that malaria could be reintroduced into the USA. During 1963–1999, 93 cases of transfusion-transmitted malaria were reported in the USA; approximately two-thirds of these cases could have been prevented if the implicated donors had been deferred according to established guidelines (CDC, 2004d).

Imported malaria cases are not always diagnosed in a timely manner due to the unfamiliarity of many clinicians with symptoms of the disease; a delay in the diagnosis of *Plasmodium falciparum* may result in death. There were a total of 118 deaths due to malaria in the USA between 1979 and 1998 with an average of 5.9 deaths per year. Specific epidemiological data provided by the CDC regarding the 40 deaths that occurred between 1992 and 1998 yielded the following results. Deaths occurred in patients ranging from 9 months to 89 years of age (median, 53 years). Thirty-eight (95%) of these were due to *P. falciparum* and two (5%) due to *P. vivax*. Anti-malarial chemoprophylaxis was taken in 40% of cases, not taken in 45% of cases and unknown in 15% of cases. Twenty-four (60%) of the cases involved US travellers to malaria-endemic areas, of whom 59% travelled to Africa, 25% to South America, 8% to India, 4% to Haiti and 4% to unspecified areas. The remaining cases included 11 foreign travellers to the USA (27.5%), 3 induced cases (7.5%) and 2 undetermined cases (5%). Thirty-nine (98%) of the cases were diagnosed antemortem and only one case was known to have come to the attention of the medical examiner/coroner (Stoppacher & Adams, 2003).

The cost of treating malaria cases in persons that fall ill upon return to the USA can be very substantial and this will be discussed below. The failure of travellers to seek advice on malaria chemoprophylaxis and poor compliance with recommendations when such advice has been sought is perhaps the major cause of the occurrence of malaria cases after travel to endemic areas. The percentage of US travellers who make no use of recommended chemoprophylatic agents appears to be growing, in part because of a fear of side effects from the anti-malarial agents. Inasmuch as most of the imported cases, especially among travellers to Africa, are due to *P. falciparum*, the failure to comply may have serious consequences. Cave *et al.* (2003) carried out a study on a large number of foreign travellers to Nepal who originated from several different countries; the group included 1303 persons who responded to a questionnaire. Two hundred and eighty-eight respondents had taken chemoprophylaxis specifically for their trip to Nepal (22%), whereas 958 had not. Travellers from the UK and Denmark were significantly more likely, and those from the USA and Germany significantly less likely, to be taking chemoprophylaxis. Most of the travellers who responded had sought pre-travel advice (71%), and all sources were more likely to advise them not to take chemoprophylaxis than to take it. However, travellers advised by a family practitioner were significantly more likely to be taking chemoprophylaxis than those advised by a travel medicine specialist. The most common reason for choosing not to take chemoprophylaxis was either the occurrence of side effects or the fear of them (31%). Several studies have been carried out on immigrants to the USA and Canada who have travelled, usually with their families, to visit their country of origin. Only a very low percentage reported using chemoprophylaxis though most were aware of the risk of contracting malaria in the countries they visited. The increase in drug-resistant malaria has also been a factor in the limited use of chemoprophylaxis. Often health-care providers asked for guidance, when uncertain as to the most appropriate chemotherapeutic agents and dosages to be used for travel to malaria-endemic countries.

In Canada the number of imported cases of malaria is generally around 400 cases annually with a high of 1036 reported cases in 1997. However, it is estimated that only 30–50% of the actual cases are reported to public health agencies and the true number of imported cases into Canada is likely to be higher. Kain *et al.* (2001) noted that over the last decade there has been a marked increase in cases of drug-resistant and severe malaria in Canadian travellers. Risks for malaria infection include inappropriate recommendations for malaria prevention by health-care providers and lack of knowledge about or adherence to appropriate recommendations by the travelling public. Risks for death include delays in seeking medical attention, delays in diagnosis and inadequate care by physicians and hospitals.

The relative frequency of malaria species imported into Canada has changed over time, with a gradual increase in the proportion of *P. falciparum* cases from 20–30% in the early 1980s to 60–70% in the 1990s and 70–80% in the present decade (MacLean *et al.*, 2004); among Canadian travellers, sub-Saharan Africa was the region where most patients contracted malaria, with 353 case-patients (65%), followed by South Asia (23%), Southeast Asia (6%), Central America (5%) and South America (1%). However, India, with 110 cases (20%), was the single most frequent source. Tourists (29%), immigrants or refugees (29%) and foreign workers (24%) represented the categories most frequently reported.

Conclusions on the public health importance of malaria in North America

The fact that between 1957 and 2003, 63 outbreaks of locally transmitted mosquito-borne malaria occurred in the USA indicates that the country remains at risk of introduced malaria causing local transmission. Keeping in mind the extensive public health network in the USA, there is little likelihood that these small, local outbreaks might lead to the re-establishment of malaria transmission. The imported cases require costly treatment and greater attention should be given to ensuring that travellers to endemic countries are made aware of the risks and take appropriate chemoprophylaxis. There is also little concern about the possibility of re-establishment of malaria transmission in southern Canada where the potential vector would be *Anopheles quadrimaculatus* and the health network would quickly detect any local transmisssion.

The history of malaria in North America reinforces the viewpoint that although increased temperatures may lead to conditions suitable for the re-introduction of malaria to North America, socioeconomic factors such as public health facilities will play a large role in determining the existence or extent of such infections. The public health importance of malaria lies mainly in the substantial cost of the medical care and treatment of the large number of imported cases.

23

Mosquito-borne filarial infections

At one time bancroftian filariasis (*Wuchereria bancrofti*) was endemic in limited areas of the southern USA. Chernin (1987) suggested the hypothesis that bancroftian filariasis, which had been endemic since the early days of slavery in Charleston, South Carolina, disappeared around 1930 by virtue of the long-term effects of a municipal sewerage-water system begun in the 1890s or thereabouts. These public works, originally intended to abate typhoid and related diseases, helped to eliminate filariasis by reducing the availability of polluted domestic waters which are the preferred breeding sites for the urban vector, *Culex quinquefasciatus*. With the disappearance of this focus, the only filarial infection of man in the USA is the occasional case of dirofilariasis.

Canine heartworm disease is transmitted by mosquitoes and results from parasitization by the filariid worm *Dirofilaria immitis*. While presenting some danger to humans, the filariid has its greatest impact on canine populations and, to a much lesser extent, on feline populations. It is a potentially fatal infection in both dogs and cats. In recent years the disease has become established throughout much of the USA and part of Canada. Both *D. immitis* and *D. repens,* also found in dogs and cats, are found in Europe but in the USA the causative agents of human dirofilariasis may be *D. immitis* and *D. tenuis*; the natural hosts of the latter species are raccoons (*Procton lotor*) and opossums (*Didelphis* species). Heartworm is able to infect wild animals such as wild canines, ferrets (*Mustela nigripes*), muskrats (*Ondatra zibethica*), raccoons (*Procton lotor*), brown or grizzly bears (*Ursus arctos*), horses, coyotes (*Canis latrans*), wolves, foxes etc. Consequently, once heartworm is introduced in an area it is very likely to become endemic, as wild animals can not be treated with a preventive drug and transmission thus goes unchecked. Humans are dead-end hosts and microfilaraemia are not seen in the blood. The pathology of human dirofilariasis has already been described earlier. In the USA, the first case of subcutaneous dirofilariasis reported occurred in 1941 in New Orleans when *Dirofilaria* infection was discovered in a female cadaver (Shah, 1999). Before this it was not thought that *Dirofilaria* species were capable of infecting humans. The first case of human pulmonary dirofilariasis (HPD) was not reported until 1961. Fewer than 100 such cases have been reported in the USA, mostly from along the Atlantic Seaboard and Gulf Coast. The frequency and distribution of human infection is independent of dog ownership.

Diagnosis of human cases of dirofilariasis (HPD) is generally made by surgical resection, since it is often preoperatively presumed to be lung cancer. *Dirofilaria immitis* is known to cause solitary pulmonary nodules in humans. Human pulmonary dirofilariasis is more prevalent along the coastal regions of the USA, especially along the Mississippi River Valley. The incidence of this disease in Canada is relatively low. The four areas of Canada that are most dangerous are Southern Quebec, the region south of Winnipeg, the Okanagan Valley and Southwestern Ontario. Southwestern Ontario accounts for 90% of all cases in Canada.

Other species of *Dirofilaria* which only rarely have a relation to human disease include *D. tenuis* infection which is found only in raccoons (*P. lator*) in the southeastern USA, in the region between Florida and Texas, south of Arkansas; about 75% of all infections with this species have come from Florida. *Dirofilaria ursi* infection is found in the northern USA and Canada in bears (*U. arctos*). *Dirofilaria subdermata* infection is found in the northern USA and Canada in porcupines (*Erithizon dorsatum*). *Dirofilaria striata* infection is found in some species of wild cats including the puma (*Puma concolor*) on the American continent and *D. lutrae* infection is found in American otters (*Lontra canadensis*).

Dirofilaria ursi is transmitted by simuliid blackflies, and *D. tenuis* is transmitted by *Aedes taeniorhynchus* and *Anopheles quadrimaculatus*. *Aedes vexans* is probably the principal vector of *D. immitis* in the USA and Canada while *An. quadrimaculatus*, *An. punctipennis*, *Ae. trivittatus, Ae. sierrensis* and *Ae. albopictus* have been found naturally infected and are probably focal vectors.

Since the distribution of human dirofilariasis infection in the world roughly parallels the distribution of infection in the animal hosts, the most effective means of prevention of human disease is to reduce the incidence and prevalence of infection in the animal hosts. This strategy will be most effective in the species *D. immitis,* which is the infectious agent of heartworm in dogs, and to a lesser extent cats. There are drugs that can be given orally to domestic dogs and cats for the prevention of heartworm, including ivermectin, milbemycin, oxime and diethylcarbamazine (DEC). Use of these medications will be an effective means of reducing the rate of dirofilariasis in the parasite's natural hosts, and as a result, in humans as well. Treatment of animals should be carried out by a veterinarian.

The public health importance of dirofilariasis

The human public health importance of dirofilariasis in North America is relatively minor. Many cases go undiagnosed and others recover spontaneously without medical intervention. However, the veterinary costs associated with prevention and treatment of infected dogs and cats is considerable. Canine heartworm infection is widely distributed throughout the USA and heartworm infection has been found in dogs native to all 50 states. All dogs regardless of their age, sex or habitat are susceptible to heartworm infection. Most dogs infected with heartworm can be successfully treated.

24

Sandfly-borne diseases

The phleboviruses

The sandfly-transmitted viruses are all within the Bunyavirus group, the phleboviruses; globally, some 45 viruses are associated with sandflies. Sandfly fever, which is widespread in southern Europe and the Middle East, has not been found in the USA or Canada. In fact the only phlebovirus which is known to occur in the USA is Rio Grande virus which was isolated in Texas from pack rats (*Neotoma micropus*) (Calisher *et al.*, 1977). The infectious agent has not been reported from humans.

Most phleboviruses are associated with sandflies. A notable exception is Rift Valley fever which is found in Africa, Egypt and, most recently, in Saudi Arabia and Yemen. The infection, transmitted by culicine mosquitoes, causes severe illness and mortality in man and animals and, as has been noted earlier, the movement of the infectious agent should be kept under close surveillance to prevent its entry to the Americas.

Leishmaniasis

Leishmaniasis occurs focally in the Americas from Yucatan, Mexico, through Central and South America to the Peruvian Andes. The pathogenic New World species include the *Leishmania braziliensis* complex (*L. braziliensis, L. panamensis* and *L. guyanensis*), *L. mexicana* complex (*L. mexicana, L. amazonensis, L. venezuelensis*), *L. peruviana* and *L. chagasi*.

Several foci of endemic leishmaniasis have been reported in the USA. Autochthonous human cutaneous leishmaniasis (CL) caused by *Leishmania mexicana* appears to be limited to south central Texas and perhaps Oklahoma in the

USA; the infection is not found in Canada other than as an imported infection or, in a single instance, in a dog. Periodic cases of localized cutaneous and diffuse CL have been reported in southern Texas and Oklahoma. The usual reservoir host is the wood rat (*Neotoma micropus*) of the southern plains (Kerr *et al.*, 1995), but parasites have been identified in coyotes (*Canis latrans*) and domesticated dogs and cats. *Leishmania mexicana* is transmitted by the sandfly vector *Lutzomyia anthophora* (McHugh *et al.*, 1993). Cases are usually associated with exposure to the habitat of the wood rat. A case of visceral leishmaniasis (VL) with antibodies to *L. donovani* has also been diagnosed in a dog in Texas (Sellon *et al.*, 1993). *Leishmania mexicana* has been isolated from the white-throated woodrat (*Neotoma albigula*) near Tuscon, Arizona extending the distribution of enzootic leishmaniasis in the USA to the west (Kerr *et al.*, 1999).

In 1999 a disease started to kill fox hounds in the USA and subsequent laboratory investigations showed the cause to be leishmaniasis. None of the dogs had travelled to an area in which the disease was present, and the origin of the infection has not been found. Blood tests have since confirmed that about 12% of fox hounds in 29 states of North America, and in one state (Ontario) in Canada have been exposed to the parasite, but tests on other pet dogs and wildlife in the same areas have proved to be negative. The member of the *L. donovani* complex identified in dogs in the USA is *L. infantum* (Rosypal *et al.*, 2003). No humans have so far developed signs of the disease. The role of sandfly vector remains uncertain.

Travellers and military personnel from the USA are at risk overseas. Since 1969, over 450 US military personnel and dependents have been diagnosed with and treated for leishmaniasis. During the Persian Gulf War, 20 cases of CL and 12 cases of VL were reported. The Center for Disease Control and Prevention was notified of 129 cases between 1985 and 1990 involving travellers from the USA who acquired the disease abroad. Many hundreds of cases of CL are being treated among soldiers returning from Iraq and Afghanistan. A small number of cases of VL have also been imported.

The finding of previously unknown foci of autochthonous cutaneous and visceral leishmaniasis in the USA along with the finding of potential sandfly vectors in areas where they were not previously known (Ostfeld *et al.*, 2004) raises concern. Though not as yet a public health problem, the importance of the disease merits further study on its distribution and epidemiology in the USA.

25

Ceratopogonidae – biting midge-borne diseases

Members of this family of biting midges can be severe nuisances. They are vectors of bluetongue virus among cattle and sheep in the USA, transmitted by *Culicoides occidentalis* and *C. variipennis sonorensis* among other species, but are not known to be vectors of any human disease in North America. With the exception of a few species of *Leptoconops*, those species of biting midges attacking man and livestock in Canada all belong to the genus *Culicoides*. More than 50 species of this genus occur in Canada, most feeding either on mammals or on birds, although a few attack only reptiles or amphibians. Despite the large number of species found in Canada, relatively few are pest species. In the boreal forest, especially in eastern Quebec and the Atlantic Provinces, *Culicoides sanguisuga* and other species can make camping in forested areas intolerable in late June and July. In the Maritimes, adjacent to coastal salt marshes, *C. furens* can be abundant, although not the problem it is farther south in the USA. Main Drain serotype viruses in sheep in California have been associated with *Culicoides variipennis*. The group can be the cause of severe allergy in horses in the USA and Canada.

26

Dipteran-caused infections – myiasis

Autochthonous infestations by dipteran larvae or myiasis occur both in the USA and Canada. An aetiological classification of myiasis-causing flies has already been given in the section discussing myiasis in Europe. As has already been stated, most myiasis-causing flies belong to one of three major families: Oestridae, Sarcophagidae or Calliphoridae, although representatives of other families, such as Muscidae and Phoridae, also are known to cause human myiasis.

In the USA cases of human myiasis due to the following species have been reported: the secondary screwworm fly (*Cochliomyia macellaria*), the common cattle grub (*Hypoderma lineatum*), the green bottle (*Lucilia sericata*), the sheep botfly (*Oestrus ovis*), the black blow fly (*Phormia regina*), the stable fly (*Stomoxys calcitrans*) and the grey flesh fly (*Wohlfahrtia vigil* (= *W. opaca*)). Two cases of myiasis caused by *C. macellaria* and the little house fly (*Fannia canicularis*) have been reported from Canada.

The New World screwworm (*Cochliomyia hominivorax*) was at one time found throughout the southern USA, Central America and tropical regions of South America. As a result of massive state, federal and international eradication programmes, extant populations of *C. hominivorax* are no longer found in the USA or Mexico; isolated reports are often traced to the importation of infested animals from locations where the screwworm is still prevalent. Screwworm populations are found in Central and South America and in certain Caribbean Islands. Adult females deposit eggs in open wounds or discharging orifices, such as the nose. Larvae invade adjacent living tissue, including cartilage and bone. Because infestation of ears and nose provides the larvae with access to brain tissue, myiasis caused by *C. hominivorax* is more dangerous than infestation with other agents.

The occurrence of myiasis in hospitals, i.e. nosocomial myiasis, in Europe has already been discussed. Finlay *et al.* (1999) reported the case of a 55-year-old man from Massachusetts who contracted nasal myiasis while hospitalized with pneumonia and renal failure. About 100 larvae of *Lucilia sericata* were retrieved from his left nasal cavity by nasal-tracheal suction and by manual removal. The nasal trubinates were inflamed and copious amounts of bloody exudate were suctioned from the patient's nose and oropharynx. Sherman (2000) reviewed the epidemiological characteristics of human wound myiasis in the USA; of the 42 cases reported to him, 2 (5%) of the infestations were nosocomially acquired and not necessarily associated with patient neglect, both the patients being hospitalized at the time of their infestation. Host age averaged 60 years, with a male:female ratio of 5.5:1. Homelessness, alcoholism and peripheral vascular disease were frequent co-factors.

In a rather bizarre case of nasal myiasis, a large city hospital experienced an infestation of mice which was controlled, in part, by broadcasting poisoned baits. Months later there was an invasion of flies into the hospital, and two comatose patients in an intensive care unit were infested with nasal maggots. Adult flies were trapped and maggots removed from the nares of the second patient for identification; these were identified as the green bottle fly (*Lucilia sericata*). Recent downsizing of hospital personnel had led to the unintended and unrecognized loss of housekeeping services in the canteen food storage areas. A mouse infestation of the hospital occurred, with the epicentre in the canteen area. An attempt to control the mice was made by scattering poisoned bait and using rodent glue boards. This resulted in the presence of numerous mouse carcasses scattered throughout the building which attracted the blowflies. Adult gravid female flies trapped in the new intensive care unit (where mice were not present) laid eggs in the foetid nasal discharge of the two comatose patients. Live trapping of mice and removal of carcasses led to an abatement of the fly infestation. The cause-and-effect nature of the mouse carcasses and flies was underscored a year later when an outbreak of *L. sericata* occurred in the operating department and was linked to the presence of mouse carcasses on glue boards which had not been removed the previous autumn (Beckendorf *et al.*, 2002).

Many of the cases of cutaneous myiasis acquired in North America are attributable to *Cuterebra* species. Such cases are not common and only some 57 cases of cuterebrid myiasis have been reported in North America. Autochthonous cases are infrequently reported and are most often due to accidental infestation of humans by larvae of flies belonging to the genus *Cuterebra*, commonly known as rabbit bot flies or rodent bot flies. North American cuterebrid myiasis is rare, and the diagnosis is frequently delayed. Myiasis should be considered when treating patients with refractory furunculosis (Safdar *et al.*, 2003).

Flies of the family *Sarcophagidae* normally develop in decomposing tissue and are considered facultative parasites. Larvae of these flies parasitize wounds and other damaged tissues, and some species further invade living tissues adjacent to the wound. *Wohlfahrtia vigil*, the grey flesh fly, is incapable of penetrating adult skin; infestation occurs only in infants.

Myiasis is not a broad public health problem as infestations are almost always self-limited. The occurrence of nosocomial myiasis in hospitals is unfortunate and avoidable.

27

Flea-borne diseases

Plague

As has been noted previously, plague caused by *Yersinia pestis* first appeared in North America in San Francisco, in 1900. Plague is now endemic in North America from the Pacific Coast eastward to the western Great Plains and from British Columbia to Alberta, Canada and southward to Mexico. Most of the human cases occur in two regions; one in northern New Mexico, northern Arizona, and southern Colorado, another in California, southern Oregon and far western Nevada. Transmission of plague from person to person is uncommon and has not been reported in the USA since 1924. The last rat-borne epidemic in the USA occurred in Los Angeles in 1924–1925. Since then, all human plague in the country has been sporadic cases acquired from wild rodents or their fleas or from direct contact with plague-infected animals. Historically the plague bacillus has been demonstrated in western Canada during field studies involving rodents and fleas. More recent studies have resulted in the demonstration of antibodies in feral cats and dogs in southeastern Alberta and southwest Saskatchewan suggestive of ongoing plague activity in Canada. However, there have been no human cases identified in these areas.

In the USA during the 1980s plague cases averaged about 18 per year. During 1988–2002, a total of 112 human cases of plague were reported from 11 western states. The majority (87%) were from four states – New Mexico (48), Colorado (22), Arizona (16) and California (11). Approximately 80% of these exposures occurred in peridomestic environments, particularly those that provided abundant food and harbourage for flea-infested, plague-susceptible rodents (CDC, 2002) Most of the cases occurred in persons under 20 years of age. About one in seven persons with plague died. A map of the distribution of human plague cases

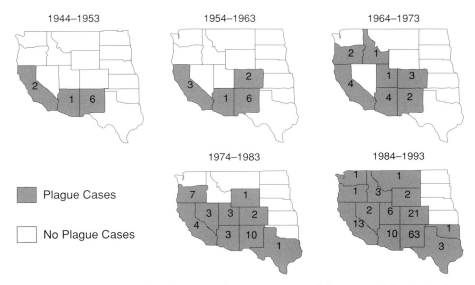

Figure 27.1 Number of human plague cases reported, by state and decade, in the USA, 1944–1993. Oklahoma Center for Family Medicine Research at the University of Oklahoma Health Science Center, 1(2), April 1999.

in the USA from 1944–1993 is shown in Figure 27.1. Plague was present in Hawaii for the period 1899–1957 and caused at least 370 fatalities. The simple biotic system in the islands – bacterium, three rodents and two fleas – appeared unable to maintain the disease over time. Improved sanitation, mechanized agriculture, gradients in rainfall and temperature, and the collapse of reservoir host and vector populations during drought are cited as probable factors in the extinction of plague from the state of Hawaii (Tomich *et al.*, 1984).

In the continental USA, movement of people to the suburbs has resulted in larger numbers of people living in or near active plague foci in the western part of the country and, consequently, increased peridomestic transmission. Surveillance of plague in rodents and rodent-consuming carnivore populations indicates that plague has spread eastward in the 1990s to areas that have been disease free since extensive animal surveillance started in the 1930s (Tikhomirov, 1999). American Indians have accounted for up to one-third of human plague cases due to their residence in areas of endemic foci and lifestyle (living in natural areas, shepherding and hunting).

Surveillance for plague in rodent and rodent-consuming carnivore populations during the 1990s indicates that plague has spread eastward to counties in areas (e.g. eastern Montana, western Nebraska, western North Dakota and eastern Texas) believed to be free of this disease since widespread animal surveillance began in the 1930s. For example, the potential for human plague cases in eastern Texas was demonstrated in 1993 when an infected roof rat (*Rattus rattus*) and two infected fox squirrels (*Sciurus niger*) were identified in Dallas. The continued expansion of human plague in the USA underscores the need to enhance plague surveillance and to increase efforts to prevent, detect and control human plague (CDC, 1994a).

Mortality among untreated cases of plague can range from 50–60% for bubonic plague to virtually 100% for septicaemic or pneumonic plague. Because the infection progresses rapidly and the organism grows slowly in culture, prompt empirical therapy with appropriate antibiotics for suspected cases is critical to reduce mortality and ensure successful recovery. Pneumonic plague has been rare in the USA but, as will be seen below, the number of cases of cat-associated pneumonic plague has been increasing.

Rodent and other reservoir hosts of plague in North America

Many species of prairie dogs, ground squirrels, chipmunks, rats and mice are highly susceptible to *Yersinia pestis* and are significant reservoir hosts of plague for humans in the western USA. The rodent species involved differ with the geographical area. During a period of active surveillance in 1970–1980, evidence of plague infection was found in 76 species of five mammalian orders in the USA. In Hawaii, the first infections of plague came from immigrant commensal rats, probably the roof rat, *Rattus rattus*, and the brown or Norway rat, *R. norvegicus*, on ships from the Orient. Both species were already established in Hawaii and became the widespread local carriers of plague, supplemented by the Polynesian rat, *Rattus exulans*, which had colonized the islands in ancient Polynesian times (Tomich *et al.*, 1984). *Rattus* species were probably involved in the first outbreaks of plague in San Francisco in the early twentieth century.

In the southwest the major reservoir hosts of *Y. pestis* are the prairie dog (*Cynomys gunnisoni*) and the rock squirrel (*Spermotophilus variegatus*). Devastating plague epizootics are common among prairie dog populations in the large colonies formed by this species. Epizootic outbreaks may kill 99% of the *C. gunnisoni* colonies and it may take 4–5 years for the colony to recover. Similar epizootics have been observed among *C. ludovicianus* (black-tailed prairie dog), *C. leucurus* (white-tailed prairie dog) and *C. parvidens* (Utah prairie dog).

On the Pacific coast of the USA, the most important reservoir host of *Y. pestis* is *Spermophilus beecheyi* (Californian ground squirrel); the chipmunks (*Eutamias* species), *Microtus californicus* (California voles) and *S. lateralis* (Wyoming ground squirrel) are also important reservoir hosts. There has been a single report of an epizootic in the domestic fox squirrel (*Sciurus niger*), in Colorado state. In the northern foci of plague ground squirrels including *S. beldingi* are important reservoir hosts (Gratz, 1999a). The black footed ferret (*Mustela nigripes*) has been found infected with plague in Wyoming which endangers the only known colony of this species.

There has been a significant increase in the number of cases of plague associated with domestic cats. Twenty-three cases of cat-associated human plague (5 of which were fatal) occurred in 8 western states from 1977 through 1998, which represent 7.7% of the total 297 cases reported in that period. No cat-related cases were reported prior to 1977. Bites, scratches or other contact with infectious materials while handling infected cats resulted in 17 cases of bubonic plague, 1 case of primary septicaemic plague and 5 cases of primary pneumonic plague. The 5 fatal cases were associated with misdiagnosis or delays in seeking treatment, which resulted in overwhelming infection and various manifestations of the systemic inflammatory response syndrome. Unlike infections acquired by flea bites, the occurrence of cat-associated human plague did not increase significantly during summer months. Plague epizootics in rodents also were observed less frequently at exposure sites for cases of cat-associated human plague than at exposure sites for other cases. The risk of cat-associated human plague is likely to increase as residential development continues in areas where plague foci exist in the western USA (Gage *et al.*, 2000). Although transmission by cats is often emphasized, ground and rock squirrels (*Spermophilus variegatus*) remain the most common animal source of human infections. Chipmunks, and other ground squirrels and their fleas, suffer plague outbreaks and some of these occasionally serve as sources of human infection. Deer mice (*Peromyscus maniculatus*) and voles (*Microtus* species) are thought to maintain the disease in animal populations but are less important as sources of human infection. Other less frequent sources of infection include wild rabbits, and wild carnivores that acquire infections from wild rodent outbreaks.

Flea vectors of plague in North America

Inasmuch as all recent cases of plague in the USA have occurred in rural areas, the most important flea vectors are ectoparasites of the principal reservoir hosts, rock squirrels particularly and, less frequently, prairie dogs. In California plague cases are generally transmitted by the most common flea ectoparasites of the California ground squirrel (*Spermophilus beecheyi*), which are mainly *Oropsylla montana* and to a lesser extent, *Hoplopsyllus anomalus*. *Oropsylla bacchi* occurs on the antelope ground squirrel (*Ammospermophilus leucurus*) in Arizona and New Mexico and has been responsible for several human plague infections (Craven *et al.*, 1993).

The public health importance of plague in North America

Although plague is no longer a particularly important public health problem in the USA, it may give rise to fatal disease if not promptly diagnosed and treated; physicians in the endemic states must be aware of the symptoms of plague to ensure timely treatment. Furthermore, on at least two occasions plague has been imported into the USA from an endemic area, once from Vietnam (Caten & Kartman, 1968) and once from Bolivia (CDC, 1990). On two other occasions plague acquired in New Mexico has been diagnosed after hospitalization in another state – once in Nebraska (Mann *et al.*, 1982) and once in New York City (CDC, 2003). Several other cases have been reported of plague among travellers in the USA between 1950 and 1980.

Much concern has been expressed about the possible utilization of *Yersinia pestis* as a bioterror agent, particularly in the form of pneumonic plague; its high virulence, and its ability to spread by aerosolization makes it a potential agent of biological warfare in the hands of terrorists (Krishna & Chitkara, 2003). Though an ancient disease, plague remains of global concern to this day. While the yearly number of infections in the USA is small, the potential virulence of an infection and potential fatal outcome of cases that are not diagnosed and treated in a timely manner, require a high degree of clinical awareness and public health preparation.

Flea-borne rickettsial diseases

Murine typhus (endemic or flea-borne typhus)

Murine typhus, whose causative agent is *Rickettsia typhi*, was first described in 1913 in the USA. By the mid-1940s, murine typhus was known to be endemic in 38 states though most prevalent in the southern states. Thousands of cases of the disease used to occur annually in the USA (Azad & Beard, 1998). Large scale DDT-dusting programmes were carried out in the southern USA in the late 1940s for rat-flea vector control resulting in a great diminution of transmission. As a result of dusting operations with DDT 10% dust, cases of murine typhus in the southern states decreased from 5338 cases in 1944 to an estimated 2200 in 1947 (Wiley & Fritz, 1948) and further decreased thereafter.

The classic cycle of *R. typhi* involves rats (*Rattus rattus* and *R. norvegicus*) and, mainly, the oriental rat flea (*Xenopsylla cheopis*) which, in most areas, is the main vector. Transmission is affected by contact with rickettsia-containing flea faeces or tissue during or after blood-feeding. Reported cases of murine typhus in the USA are, for the most part, from south and central Texas and the Los Angeles and Orange County area of California (Sorvillo *et al.*, 1993). Most of the cases are now associated with an opossum–cat flea cycle. Both opossums (*Didelphis* species) and domestic cats collected from the case areas were seropositive for

R. typhi antibodies. The cat flea (*Ctenocephalides felis*) which is also a competent vector of murine typhus, is the most prevalent flea species (97%) collected from opossums, cats and dogs in southern Texas; no fleas were recovered from rats in this area. In addition to *R. typhi, R. felis* (see next heading on ELB) was also found in opossums and their fleas; this finding, consistent with surveys in other areas of the country, further documents the reduced role of the rat and the flea *X. cheopis* in the maintenance of murine typhus within endemic-disease areas of the USA. The maintenance of *R. typhi* in the cat flea and opossum cycle is therefore of potential public health importance since *C. felis* is a very prevalent and widespread pest on cats and dogs and avidly bites humans.

The changing ecology of murine typhus in southern California and Texas over the past 30 years demonstrates the effects of suburban expansion. In suburbia the classic rat–flea–rat cycle of *R. typhi* has been replaced by a semidomestic or peridomestic animal cycle involving free-ranging cats and dogs, opossums, raccoons (*Procyon lotor*) and squirrels. Fleas found on these animals are picked up by household pets and brought into homes, and may carry rickettsiae.

Less than 100 cases of murine typhus a year are now reported in the USA. However, there are occasional outbreaks such as those that have occurred in Hawaii; while generally uncommon in the state, five cases of murine typhus were reported on southwestern Kauai, Hawaii in 1998 and in 2002 Hawaii reported 47 cases.

The ELB agent

A rickettsia later named the ELB agent, or *Rickettsia felis*, was identified in 1990, when cat fleas (*Ctenocephalides felis*) were examined as possible vectors of *R. typhi* in Texas; a novel *Rickettsia*-like organism was observed by electron microscopy in midgut epithelial cells of the fleas. The agent, named the ELB agent for the EL Laboratory (Soquel, CA), was detected in 1994 and 2000 by polymerase chain reaction (PCR) in four patients from Texas and Mexico. This flea-borne rickettsia has been implicated as a cause of human illness (Schriefer *et al.*, 1994a). In 1995, the name *Rickettsia felis* was proposed for the ELB agent on the basis of its phenotypic characteristics, as well as its clear genotypic differences from other known *Rickettsia* species. In 1997, the ELB agent was detected in two other flea species in the USA, *Ctenocephalides felis* and *Pulex irritans.* The organism has recently been reassigned to the spotted fever group of rickettsiae.

Studies in southern Texas revealed the presence of both *Rickettsia typhi* and the ELB agent in cat fleas and the first observation of ELB-infected vertebrates (opossums *Didelphis virginiana).* The persistence of the murine typhus disease focus appeared to be better accounted for by the presence of infected cat fleas, opossums and other non-rat hosts found in close association with human populations. Involvement of the ELB agent in the biology of murine typhus was suggested by its prevalence among suspected vectors and reservoir hosts (Schriefer *et al.*, 1994b). *Rickettsia felis* is now known to have a wide distribution both in the USA and in Central and south America, Europe and Australia. It is frequently present in cats and cat fleas which play an important role in the maintenance of *R. felis* and its transmission to humans with the involvement of an opossum/*Rickettsia*/cat flea triad in the flea-associated rickettsial transmission cycle of urban and suburban areas of south Texas and southern California (Boostrom *et al.*, 2002).

Infections due to *R. felis* represent an emerging disease associated with murine typhus and its flea vector as well as an emerging transmission complex involving opossums which had not been previously recognized. It seems likely that the infection will be found to be even more widespread; at present, much remains to be learned about its epidemiology.

Cat scratch disease

Cat scratch disease (CSD) commonly occurs as a result of infection by *Bartonella henselae*. The disease has emerged as a relatively common and occasionally serious zoonotic infection among children and adults in the USA. Some 80% of the cases occur among children. Although CSD occurs in persons of all ages, the highest age-specific incidence is among children < 10 years of age. Infection with *B. henselae* is one of the most common causes of chronic lymphadenopathy among children, and in some case series up to 25% of these infections result in severe systemic illness. Cat scratch disease is generally a mild, self-limited illness; the differential diagnosis often includes more serious conditions (e.g. lymphoma, carcinoma, mycobacterial or fungal infection, or neuroblastoma). Cat scratch disease was first described as a clinical syndrome in 1931, but it was not until 1983 that a bacterial aetiology was determined, and in 1992 the specific cause of CSD was identified as *B. henselae* (Koehler *et al.*, 1992).

Cat scratch disease is a feline-associated zoonotic disease, with an estimated annual incidence in the USA of at least 22 000 human cases, although Thompkins (1997) thought the number of cases annually in the USA was about 40 000. As has been noted, cat-to-cat transmission may occur by the cat flea *Ctenocephalides felis* but the exact role of fleas or other arthropods in the transmission of this pathogen to humans is not known. Scratches, licks and bites from domestic cats, particularly kittens, are important risk factors for infection. The infection is a paediatric problem in most parts of the USA. *Bartonella henselae* has been isolated from a high percentage of cats studied; Jameson *et al.* (1995) determined serum antibody titres to *B. henselae* for selected pet cats from 33 locations throughout the USA and several areas of western Canada. Seroprevalence paralleled increasing climatic warmth and annual precipitation; the warm, humid areas would have the largest numbers of potential vectors. The southeast USA, Hawaii, coastal California, the Pacific northwest and the south central plains had the highest prevalences (54.6%, 47.4%, 40%, 34.3% and 36.7% respectively). Alaska, Rocky Mountain–Great Plains region and the Midwest had low average prevalences of 5%, 3.7% and 6.7% respectively. Overall, 27.9% (175/628) of the cats tested were seropositive. Guptill *et al.* (2004) collected blood from 271 cats in four areas of the USA (southern California, Florida, metropolitan Chicago and metropolitan Washington, DC). Sixty-five (24%) cats had *B. henselae* bacteraemia, and 138 (51%) cats were seropositive for *B. henselae*. They also found that flea infestation, adoption from a shelter or as a stray cat, hunting, and being from Florida or California were significant risk factors for *B. henselae* seropositivity.

The association between *B. henselae* infections and fleas was also confirmed by Foley *et al.* (1998), who attempted to determine what factors were common to catteries with high *B. henselae* antibody prevalence compared with catteries with low prevalence; they noted the overall seroprevalence in 11 catteries from diverse geographical locations in North America in their study was 35.8%. They found that flea infestation was the most important risk factor for high *B. henselae* seroprevalence in the catteries they surveyed.

Although the existence of CSD has been known for some decades, it is only with the advent of more sensitive diagnostic tests that the true magnitude of infections can be seen. While, for the most part, the infection is benign in most patients, *B. henselae* infection has also been linked to bacillary angiomatosis, a vascular proliferative disease most commonly associated with long-standing human immunodeficiency virus infection or other significant immunosuppression; it has also been associated with bacillary peliosis, relapsing bacteraemia and endocarditis in humans. A study in a hospital in Huston, Texas emphasized that although CSD is generally a mild, self-limited illness, the differential diagnosis often includes more serious conditions (e.g. lymphoma, carcinoma, mycobacterial or fungal infection, or neuroblastoma) that might result in protracted hospital stays and lengthy treatments before diagnosis. Timely assessment of CSD is important, particularly when invasive diagnostic measures are being considered (CDC, 2002a). The public health burden of CSD is probably considerably higher than has been commonly thought and effective flea control on pet cats should be carried out as a measure of prevention.

28

The louse-borne diseases

Louse-borne or epidemic typhus

The epidemiological cycle of louse-borne typhus has been described in the section on this disease in Europe. The last louse-borne outbreak of the disease in the USA was in 1921 but sporadic cases of *Rickettsia prowazekii* are still reported. Between January 1976 and January 1979 serum specimens from 1575 individuals were received at the US Centers for Disease Control and tested for antibodies to rickettsiae. Of these, sera from 8 persons gave serological results indicative of recent infections with epidemic typhus rickettsiae. Five of the persons were from Georgia, and one each was from Tennessee, Pennsylvania and Massachusetts. The illnesses occurred during the winter, chiefly in persons living in a rural environment. The clinical picture was compatible with louse-borne epidemic typhus. There was no apparent contact with human body or head lice among the patients, and no cases occurred in patient contacts, indicating that infection was not associated with the classic man–louse–man cycle of epidemic typhus. Two of the eight patients had contact with flying squirrels (*Glaucomys volans*) suggesting that they became infected from this known non-human reservoir host of *R. prowazekii* (McDade *et al.*, 1980). Thus the infection is now known to exist as a zoonosis of flying squirrels in the USA. There have been 33 human infections diagnosed between 1976 and 1984; none of the cases has died. Most of the patients have either handled squirrels or squirrels have been present in the patient's home. *Rickettsia prowazekii* probably evolved from *R. typhi* but it is not known whether it first appeared in squirrels or humans (CDC, 1984).

McDade (1987) concluded that information gathered during the past decade indicates that the eastern flying squirrel, *G. volans*, is a zoonotic reservoir host of *R. prowazekii*. The sporadic cases of typhus that have occurred in the USA in association with flying squirrels provide evidence that the squirrels can transmit *R. prowazekii* infection to humans. Strains of *R. prowazekii* isolated from flying squirrels multiply readily in human body lice, but flying squirrel lice, although readily infected, are very host-specific and tend not to bite humans. The infection may be spread to humans in infective ectoparasite faeces aerosolized when the flying squirrels groom themselves. The author observed that current concepts of typhus epidemiology and control must be re-evaluated to take into account this zoonotic aspect.

The usual vectors of louse-borne typhus, body lice, are not common in the USA or Canada. It thus seems very unlikely that epidemic typhus could spread from the zoonotic reservoir hosts in flying squirrels and be the source of an epidemic outbreak of the disease and the threat to public health is thus minor.

Trench fever

The causative organism of trench fever is *Bartonella quintana*, transmitted by the body louse (*Pediculus humanus.*) Its great incidence during World War I in Europe has been described previously. *Bartonella quintana* is a gram-negative bacterium transmitted to humans by its only known reservoir host, the body louse. *Bartonella quintana* has been recently found in cat fleas though their vectorial role is unknown. Trench fever is characterized by a fever, headaches and leg pain, followed by relapses every 5 days which were the first clinical manifestation of *B. quintana* infection to be described during World War I. Since that date, additional manifestations of *B. quintana* infection, including endocarditis, bacillary angiomatosis and chronic bacteraemia, have been described. Reports of the re-emergence of *B. quintana* among urban homeless populations in Europe have been described above.

In the USA, between January and June 1993 in Seattle, Spach *et al.* (1995) isolated *B. quintana* from 34 blood cultures obtained from 10 patients not known to be infected with HIV; all of the patients were homeless, inner-city patients with chronic alcoholism. Body lice infestations were not documented among the patients. In a follow-up study in Seattle, 20% of the patients in a downtown clinic serving an indigent and homeless population, had a greater than 1:64 microimmunofluorescence titre to *B. quintana*, although most of these patients did not have symptoms of infection (Jackson *et al.*, 1996).

Shortly thereafter, the organism was identified in Oklahoma City and among inner-city intravenous drug users in Baltimore, Maryland, in the latter case along with *Bartonella elizabethae* and *B. henselae* (Comer *et al.*, 1996). In San Francisco, the cases of *B. quintana* infection were seen to be accompanied with body lice infestations (Thompkins, 1997). Lice infested, homeless people who are frequently also alcohol dependent are at greatest risk from *B. quintana* infections; the vector is the body louse (*Pediculus humanus*), but the reservoir host of the infectious agent remains unknown as well as where the organism persists before its reappearance. It appears that the cases of urban trench fever are often associated with HIV infections and that the major predisposing factors include poverty, low hygiene and chronic alcohol dependency.

Bartonella quintana is also present in Canada where, as in other countries, alcohol dependency and homelessness without previous valvular heart disease are risk or predisposing factors for *B. quintana* infection.

The reappearance of trench fever in cities far distant from one another in Australia, France, Germany, Greece, Russia, the UK and the USA after a long absence of any reports of the disease is a dramatic example of a resurging disease. Much work remains to be done on the epidemiology of this disease to elucidate, among other questions, what reservoir host has maintained trench fever over a period of several decades.

Body louse infestations

Body louse (*Pediculus humanus*) infestations in the USA and Canada mainly affect the homeless. There are indications that a recrudescence of body lice infestations is occurring as the numbers of individuals living under social conditions which predispose towards infestation has increased. Very little information is available on the prevalence of body louse infestations in either country as most infestations are among population groups that are found difficult to survey.

Head louse infestations

An estimated 6–12 million head lice (*Pediculus capitis*) infestations occur annually in the USA (West, 2004) with children of ages 3–12 most likely to be affected. The actual incidence may vary greatly but in both countries as much as 10% of the school children can be infested (Gratz, 1997). There are significant direct costs associated with treatment for the control of head lice and indirect costs due to lost time from school. Anecdotal reports suggest that direct costs of treatment are in the hundreds of millions of dollars annually (Hansen & O'Haver, 2004). Clore & Longyear (1990) reported that US $367 million were dispensed annually for head lice control in the USA including sums expended for the purchase of over the counter pediculicides and expense to the school systems. This will be further detailed below.

The presence of head lice among school children is generally viewed with much concern by parents and teachers. Many schools have an unfortunate 'no nits' policy which does not permit children to return to classes until their hair is free of nits or, at least, until they have a certificate of treatment (ignoring the fact that most infestations are actually contracted at the schools). This needlessly disrupts the child's education and ignores the fact that head lice are not a vector of disease.

Most cases of head lice in the USA, and elsewhere, are diagnosed and treated by non-physicians such as school nurses. Misdiagnosis may lead to treatment when no lice are present. Treatment failure may lead to repeated use of and improperly applied pediculicides, potentially resulting in overexposure to pesti-cides. These treatment failures are primarily due to the emergence of insecticide-resistant lice strains. In areas where resistant lice are common, patients or their families may self-treat numerous times with over-the-counter pediculicides before seeking advice and treatment from a physician (Burkhart, 2004).

Although head lice are not vectors of any disease, at most causing skin irrita-tion and infections due to scratching if left untreated, they are of public health importance because of the considerable sums spent for their control; as has been said earlier, these large expenditures detract from resources that would other-wise be available for other health-care problems or household expenditures.

Pubic louse infestations

Again it must be emphasized that pubic or crab lice (*Pthirus pubis*) are not vectors of disease. Nevertheless, they are of considerable public health impor-tance as their presence is frequently associated with other sexually transmitted diseases (STDs). It is worth repeating that physicians finding pubic lice should consider the possible presence of other venereal infections among infested patients as there is a high degree of correlation between the presence of *P. pubis* and other venereal infections. Furthermore, the possibility of sexual abuse having occurred should be considered in children who are found with pubic louse infestations.

The findings of Pierzchalski *et al.* (2002) described above bear repeating; the authors commented that health-care providers not skilled in the evaluation of sexually transmitted infections may treat pubic lice infestation without consider-ing the possible presence of other infections. They carried out a study comparing the rate of *Chlamydia* and gonorrhea infections in adolescents, with and without pubic lice, among a study group of 263 sexually active adolescents at a juvenile detention centre in the Midwest USA between July 1998 and June 2000. From the evidence of the study, they concluded that pubic lice infestation is predictive of a concurrent *Chlamydia trachomatis* infection in this population and emphasized that adolescents infested with pubic lice should be screened for other sexually transmitted diseases, including *Chlamydia* and gonorrhea.

As early as 1968, Ackerman observed that 'the concept that the crab louse is restricted to the poor is no longer applicable in the permissive America of the 1960s'. The incidence of *Pthirus pubis* like other venereal diseases has risen. Chapel *et al.* (1979) found that in a clinic for the treatment of sexually transmitted diseases 31.4% of the patients infested with *P. pubis* had other sexually transmitted diseases. As infestations with pubic lice do not have the requirement of notification as with other STDs, it is difficult to determine their incidence. However, The National Institute of Allergy and Infectious Diseases, in their Fact Sheet on STDs, states that there are some 3 million cases annually in the USA. Many infested persons hide their infestations and attempt to self-treat their crab lice infestations; it is felt, therefore, that the actual number of infestations is even greater than the 3 million cited.

29

Triatomine-borne diseases

The hemipteran triatomine bugs are vectors of *Trypanosoma cruzi*, the causative agent of Chagas disease in the western hemisphere in South and Central America.

American trypanosomiasis or Chagas disease

American trypanosomiasis, or Chagas disease, is caused by the protozoan parasite, *Trypanosoma cruzi*. The infection is usually transmitted via the faeces of blood-sucking reduviid bugs which belong to the Triatominae subfamily (kissing bugs). The infection is found in small mammals that serve as reservoir hosts in a sylvatic cycle and human disease results from the colonization of the human habitat by some vector species in a domestic cycle.

Vectorial transmission (via the faeces) is responsible for 80% of human infections. The entry of metacyclic trypomastigotes via the mucosal route (oral or ocular) is easy. Direct skin penetration is more difficult, and generally, the parasite enters through the site where the bug has taken a blood-meal through the lesions which result from scratching a bite. In South and Central America transmission by transfusion of infected blood (containing trypomastigotes) is responsible for 5–20% of the human cases of Chagas disease, mainly in urban centres.

The disease is found only in the American hemisphere. It is endemic in 21 countries; as a result of major control campaigns there is a decreasing trend in its prevalence and eradication of the disease has been achieved in Chile and Uruguay.

Nevertheless, 16–18 million people are currently infected in the hemisphere and 45 000 deaths are attributed to the disease each year by the WHO. Infection with *T. cruzi* is life-long, and 10–30% of persons who harbour the parasite chronically develop cardiac and gastrointestinal problems associated with the parasitosis. Although major progress has been made in recent years in reducing vector-borne and transfusion-associated transmission of *T. cruzi*, the burden of disability and death in persons chronically infected with the organism continues to be enormous.

Between 8 to 10 million persons born in countries in which Chagas disease is endemic currently reside in the USA, and epidemiologic and census data suggest that 50 000–100 000 are chronically infected with *T. cruzi* (Kirchhoff, 1993). The presence of these infected persons poses a risk of transmission of the parasite in the USA through blood transfusion and organ transplantation and several such cases have now been documented.

Trypanosoma cruzi is also endemic in parts of the USA; its distribution there includes approximately the southern half of the country. However, only five autochthonous insect-borne cases have been reported in humans in the USA (Herwaldt *et al.*, 2000).

The first naturally transmitted case in the USA was reported in Corpus Christi, Texas, in 1955 in a 10-month-old child residing in a triatomine insect-infested house (Woody & Woody, 1955). Further, serological evidence of *T. cruzi* infection was found in 6 (2.5%) of 237 residents and in 6 (60%) of 10 dogs from the vicinity of a patient with the disease in California (Navin *et al.*, 1985). A locally transmitted case in Tennessee was associated with serological evidence of *T. cruzi* infection in a domestic canine owned by the patient's family, as well as the finding of an infected reduviid insect within the household (Herwaldt *et al.*, 2000).

Chagas disease vectors

There are about 16 species and some 18 subspecies of *Triatoma* in the USA, distributed in the southern two-thirds of the country (Ryckman, 1984). Epidemiologically, the most important species are *Triatoma sanguisuga* in the eastern USA, *Triatoma gerstaeckeri* in the region of Texas and New Mexico, and *Triatoma rubida* and *Triatoma protracta* in Arizona and California. A factor limiting transmission is that North American *Triatoma* delay defecation until 20–30 minutes post-feeding. By this time, they are usually no longer in contact with the sleeping human (Kirchhoff, 1993).

Chagas disease reservoir hosts

Only mammals are susceptible to infection with *Trypanosoma cruzi*. Birds, amphibians and reptiles are naturally resistant to infection. In the domestic cycle, frequently infected mammals, besides humans, are dogs, cats, raccoons, (*Procyon lotor*), armadillos (*Dasypus* species), opossums (*Didelphis* species), mice, rats, guinea pigs (*Cavia porcellus*) and rabbits. Pigs, goats, cattle and horses only manifest transitory parasitaemia and do not play an important role in the transmission of infection.

Canines may be sentinels and/or reservoir hosts for human *T. cruzi* exposures. Seropositive dogs have been detected in both domestic and stray canine populations in Texas, Oklahoma, Louisiana, Virginia and other southern states. Estimates of seroprevalence in stray-canine populations in the lower Rio Grande Valley of Texas range up to 8.8%. Between 1994 and 1998, samples from 351 domestic dogs from Texas suspected of having Chagas disease were tested for antibodies to *T. cruzi*; 67 (19.1%) were seropositive (Shadomy *et al.*, 2004).

The public health importance of Chagas disease in the USA

Autochthonous cases of vector-borne Chagas disease in the USA are very rare. The infective agent is widespread in animal reservoir hosts in the southern part of the country and California but the competence of the vectors is limited. The presence of the infection among immigrants to the USA from Chagas disease-endemic countries represents a far greater medical and public health problem than local transmission.

30

Tick-borne diseases of the USA and Canada

Tick-borne viruses

Colorado tick fever

The causative agent of Colorado tick fever (CTF) is an orbivirus of the genus *Coltivirus* which is related to Eyach virus in Europe. It is the only member of this group found in the USA or Canada apart from Sunday Canyon virus which has been isolated from the tick *Argas cooleyi* in Texas.

The distribution of CTF is the same as that of its principal tick vector, *Dermacentor andersoni*. In the USA, the disease is found in California, Colorado, Idaho, Montana, Nevada, Oregon, Utah, Washington, Wyoming, and in Canada in British Columbia and Alberta. Many of the endemic areas are vacation destinations, frequented during the tick season (spring and early summer) by large numbers of tourists. Colorado tick fever virus has been isolated from humans vacationing or travelling through these states and physicians should be aware of the possibility of CTF in febrile patients returning from these regions (Calisher and Craven, 1998). Some 200–400 cases are reported each year. The disease is limited to elevations above 4000 feet. The number of reported cases is probably a small fraction of actual cases because many cases are either not diagnosed or are unreported. Campers, hikers and foresters are commonly infected. Infections occur mainly in April to June, when adult ticks are abundant. In 3–15% of infected children under 10 years of age, CTF virus invades the central nervous system (CNS) and causes encephalitis. Virus has been isolated from the cerebrospinal fluid of some of these patients. The disease is usually self-limiting and the prognosis is excellent, even in cases complicated by neurological symptoms. Only rare fatalities have been reported.

The virus overwinters in nymphal ticks, which feed on and infect small rodents in the spring. The rodents become viraemic and in turn infect larval ticks. The larvae metamorphose during the summer; they overwinter as infected nymphs and do not transmit virus to humans until reaching the adult stage, which may be 1 or 2 years after infection. Foci of infected ticks occur primarily in ecological zones favourable to large populations of ground squirrels and other small rodents.

Colorado tick fever replicates in *Dermacentor andersoni* ticks that are infected in the larval stage when they feed on blood of the golden-mantled ground squirrel (*Spermophilus lateralis*), or other rodents but with chipmunks (*Eutamias minimus*) and *S. lateralis* as the main hosts (Bowen *et al.*, 1981); it has been isolated from several other mammals including foxes, the western grey squirrel (*Sciurus griseus*) and porcupines (*Erethizon dorsatum*), among others. After a period of days or weeks, the virus appears in the tick's saliva. The tick, which remains infectious for life, feeds on humans during its adult stage. The virus does not pass transovarially in the tick.

The infection is certainly of public health importance within its extensive area of distribution in the USA and Canada; while mortality from CTF is rare, the illness produced by an infection can be serious among children. Physicians should be aware of the possibility of CTF infection in patients who have returned from the western USA and Canada with symptoms including abrupt onset and initial features of high fever, chills, joint and muscle pains, severe headache, ocular pain, conjunctival infection, nausea and occasional vomiting (Calisher & Craven, 1998). Treatment is primarily supportive.

Powassan virus

Powassan virus (POW) is a tick-borne and transmitted flavivirus which occurs in the USA and Canada as well as in Russia. Powassan virus was first isolated in 1958 from the brain of a 5-year-old child who died with encephalitis after having contracted the virus in a wilderness park near the town of Powassan, Ontario, Canada (McLean & Donohue, 1959); this was the first recognized case of disease caused by this virus though it is now known that a virus isolated from *Dermacentor andersoni* ticks collected in Colorado in 1952 was this same virus. The first recognized case of POW encephalitis in the USA occurred in New Jersey in 1970. Four cases of encephalitis attributed to POW were reported in Maine and Vermont between September 1999 and July 2001. These were the first cases in the USA since earlier reports in 1994 (CDC, 2001)

Powassan virus may be the cause of acute viral encephalitis but, generally, human disease from POW is not common. Probably many cases without severe symptoms are not recognized. Some patients who recover may have residual neurological problems. Powassan infection appears to be one of the least common causes of arbovirus encephalitis reported from the USA and Canada, ranking behind LaCrosse, St. Louis and eastern and western equine encephalitis. However, POW and eastern equine encephalitis have the dubious distinction of having the highest case-fatality rates and are associated with a very high incidence of neurologic sequelae (Ralph, 1999). Disease, when it does occur, takes the form of infection and inflammation of the brain (encephalitis and meningitis) but, as stated above, most infections do not result in disease. For example, among residents of northern Ontario from 1959–1961, studies of the pattern of occurrence of antibodies indicated that 3.3%, 0.5% and 1.9% of the population was infected by the virus in each of those 3 years. Tick bite is considered the major and possibly the only significant route of exposure for people, and risk of tick bite is the major risk factor for people within the geographic range of the virus.

In Canada, POW has caused clinical disease in people in Ontario, Quebec and New Brunswick, and has been isolated from wild mammals and ticks in Ontario. Antibodies against POW have been reported in wild animals from British Columbia, Alberta, Ontario and Nova Scotia. Powassan virus has been recognized at several different locations in the USA from coast to coast, and may occur in Mexico. It also has been recognized at multiple locations in southeastern Siberia, northeast of Vladivostok.

Powassan virus is maintained in cycles of transmission among several species of small wild mammals and several species of tick. The species of mammal and tick most important to the virus appears to vary from region to region. In eastern and central Canada, the woodchuck (*Marmota monax*) and the tick *Ixodes cookei* seem to be particularly important, but infection rates can be high in red squirrels (*Tamiasciurus hudsonicus*), grey squirrels (*Sciurus carolinensis*), eastern chipmunks (*Tamias striatus*), porcupines (*Erethizon dorsatum*), deer mice (*Peromyscus maniculatus*), voles (*Microtus* species), snowshoe hares (*Lepus americanus*), striped skunks (*Mephitis mephitis*) and raccoons (*Procyon lotor*). Powassan virus has been isolated from four species of tick in North America – *Ixodes cookei*, *I. marxi*, *I. spinipalpus* and *Dermacentor andersoni*. The virus can pass from larva to nymph to adult as the tick develops (that is transstadial transmission). It is not known whether or not it can be passed from an infected adult female to its eggs (that is, transovarial transmission). Clearly, however, the virus can pass through winter in dormant ticks at various stages in their life cycle, and be transmitted to mammalian hosts in the spring when the tick becomes active again. Thus, the virus appears to persist in a variety of ecological associations among a range of ticks and a range of mammalian hosts that include rodents, rabbits and hares and carnivores (Leighton, 2000).

Recently a POW-like virus was isolated from the deer tick (*Ixodes scapularis*). Its relationship to POW and its ability to cause human disease has not been fully elucidated but it appears to be a genotype of POW. It has been given the name deer-tick virus (Telford *et al.*, 1997).

Although the number of clinically recognized cases of POW in the USA and Canada is small, there is a grave risk associated with an infection both in terms of serious disease and a 10–15% case fatality rate. Persons who are likely to be exposed to tick-bites in endemic areas should take protective measures including the use of repellents and/or the use of insecticide-impregnated clothing.

As in Europe, a number of other viruses have been isolated from ticks in the USA and Canada whose role as pathogens to man is unknown and these will not be discussed.

Tick-borne bacterial infections

Tick-borne relapsing fever

Tick-borne relapsing fever (TBRF) was first recognized in North America in 1915 (Meador, 1915). The disease is caused by infection with *Borrelia hermsii* as well as *Borrelia turicatae* and, possibly, *B. parkeri*; these spirochaetes are maintained in zoonotic cycles involving rodents and soft ticks. The infection is acquired only within the geographic range of its specific tick vector, *Ornithodoros hermsi*. This tick has been found in southern British Columbia in Canada and in Washington, Idaho, Oregon, California, Nevada, Colorado and the northern regions of Arizona and New Mexico in the USA. Tick-borne relapsing fever occurs in scattered foci throughout much of the western USA and British Colombia in Canada. A single case of TBRF due to *B. turicatae* was diagnosed in Cincinnati in a child who had not travelled outside of Ohio, indicating the presence of *Borrelia* in this area (Linnemann *et al.*, 1978). Another case of TBRF, this time due to *B. hermsii* was reported in a man who had returned from the US Virgin Islands where the disease had not been previously reported (Flanigan *et al.*, 1991).

The true incidence of TBRF is unknown because it frequently goes undiagnosed and reporting is incomplete. The disease is currently reportable only in the states of California, Colorado, Idaho, Texas and Washington. Two hundred and eighty-five cases were reported in these states and in Arizona, New Mexico, Nevada, Oregon, Utah and Wyoming during the period 1985–1996 (CDC, unpublished data). Dworkin *et al.* (1998) found evidence that many cases of TBRF are not properly diagnosed nor reported to the public health authorities and concluded that TBRF is under-recognized and under-reported and may be falsely identified as Lyme disease.

The *Ornithodoros* vector ticks are nocturnal, transient feeders, and most patients are unaware of being bitten. Following an incubation period of about seven days (range 5–18 days), infected persons experience an abrupt onset of fever, shaking chills, headache, myalgias and arthralgias. The duration of the primary febrile attack averages 3 days followed by defervescence, usually accompanied by shaking chills. The afebrile interval averages 7 days and without appropriate antibiotic therapy, is followed by an average of three relapses.

Tick-borne relapsing fever is not an important public health problem in the USA or Canada. The number of cases is small and unlike TBRF in Africa, there is virtually no mortality and antibiotic treatment is efficacious. By taking preventive measures against tick bites such as the use of protective clothing when entering tick-infested areas or of repellents, the incidence could be reduced even further.

Lyme disease

Lyme disease (LD) is the most common vector-borne infection in the USA and Canada and is a major public health problem in both countries; its incidence is steadily increasing. Much of the material that follows is taken from a comprehensive review by Steere *et al.* (2004) and has already been partially covered in the description of Lyme disease in Europe.

In 1977, Steere *et al.* reported a new form of inflammatory arthritis, mainly in school children in the community of Lyme, Connecticut, USA, which they could also associate with preceding erythema and tickbites.

The disease was recognized as a separate entity in 1976 because of a geographic clustering of cases among children in Lyme. It then became apparent that Lyme arthritis was a late manifestation of an apparently tick-transmitted, multisystem disease, of which some manifestations had been recognized previously in Europe and America. Five years later, Burgdorfer (1982) isolated *Borrelia* spirochaetes from *Ixodes* ticks and in 1983, Steere *et al.* confirmed that the spirochaete isolated from the ticks was the causative agent of Lyme disease. The first reported case of erythema migrans acquired in the USA occurred in 1969 and in 1975 the full symptom complex now known as Lyme disease was recognized.

The earliest known American cases of what was later identified as Lyme disease occurred in Cape Cod in the 1960s. However, *B. burgdorferi* DNA has been identified by PCR in museum specimens of ticks and mice from Long Island which date from the late nineteenth and early twentieth centuries and the infection has probably been present in North America for millennia. During the European colonization of North America, woodland in New England was cleared for farming, and deer were hunted almost to extinction. However, during the twentieth century, ecological conditions changed in the northeastern USA, favouring the transmission of Lyme disease. As farmland reverted to woodland, deer (*Odocoileus virginianus*) proliferated, white-footed mice (*Peromyscus leucopus*) were plentiful, and the deer tick (*Ixodes scapularis*) thrived. Soil moisture and land cover, as found near rivers and along the coast, were favourable for tick survival (Guerra *et al.*, 2002). Finally, these areas became heavily populated with both humans and deer, as more rural wooded areas became wooded suburbs in which deer were without predators and hunting was prohibited.

Based on genotyping of isolates from ticks, animals and humans, the formerly designated *B. burgdorferi* has now been subdivided into multiple *Borrelia* species, including three that cause human infection as has been described previously. In the USA, the sole cause of Lyme disease is *B. burgdorferi*. In the northeastern US from Maine to Maryland and in the north central states of Wisconsin and Minnesota, a highly efficient, horizontal cycle of *B. burgdorferi* transmission occurs among larval and nymphal deer ticks (*Ixodes scapularis*) and certain rodents, particularly white-footed mice (*Peromyscus leucopus*) and chipmunks (*Tamias* species). This cycle results in high rates of infection among rodents and nymphal ticks and many new cases of human Lyme disease during the late spring and early summer months. The vector ecology of *B. burgdorferi* is quite different on the West Coast in northern California, where the frequency of Lyme disease is low. There, two intersecting cycles are necessary for disease transmission.

It has been estimated that Lyme disease is under-reported by a factor of 12 in the USA (Meek *et al.*, 1996). Current testing for the disease is non-standardized and often inadequate, and the diagnostic criteria for Lyme disease remain controversial. Nevertheless, Lyme disease is clearly a rapidly emerging vector-borne infectious disease in the USA and accounts for the majority of vector-borne illnesses reported in the country. In 2000, 17 730 cases of Lyme disease were reported to CDC (CDC, 2002) and in 2002, 23 763 cases were reported. In all, more than 157 000 cases have been reported to health authorities in the USA since 1982, when a systematic national surveillance was first initiated. The overall incidence rate of reported cases in the US is approximately 7 per 100 000 population, but there is probably considerable under-reporting. The CDC (2004a) carried out an analysis on the 40 792 cases of Lyme disease reported to CDC during 2001–2002. The results of that analysis indicated that the annual Lyme disease incidence had increased 40% during this period.

The disease occurs in distinct and geographically limited areas of the USA. The incidence in some of the most highly endemic communities may reach 1–3% per year. Persons of all ages and both genders are equally susceptible, although the highest attack rates are in children aged up to 14 years, and in persons

30 years of age and older. Although cases of Lyme disease have been reported from 49 states and the District of Columbia, significant risk of infection with the agent of Lyme disease, *B. burgdorferi*, is found in only about 100 counties in 12 states located along the northeastern and mid-Atlantic seaboard and in the upper north-central region, and in a few counties in northern California. However, since its identification nearly 30 years ago, Lyme disease has continued to spread, and there have been increasing numbers of cases in the northeastern and north-central USA.

In Canada, Lyme disease was first documented in 1984 (Doby *et al.*, 1987); however, the disease does not seem to be prevalent in the country. Between 1984 and the end of 1995, there was a total of 228 reported cases for all of the province of Ontario. In the same period, 14 cases were reported in the Thunder Bay district of the province: nine of these people had no history of travel outside the area. The disease is often concentrated in specific regions that have large populations of deer or other wildlife. In some areas where Lyme disease is common, blood tests in people show that up to 24% of the general population has been exposed to the infection. Long Point, Ontario appears to be the only region known to have a concentration of infected ticks and wild animals. Scattered cases of Lyme disease occur in other regions of the country and it has been reported from Alberta, British Colombia, Manitoba, New Brunswick, Newfoundland, Nova Scotia, Ontario and Quebec.

The vectors of Lyme disease in the USA and Canada

Steere *et al.* (1978b) first posed the hypothesis that erythema migrans and Lyme arthritis are tick-transmitted, specifically by *Ixodes dammini*. Anderson *et al.* (1983) in a survey in Connecticut found spirochaetes in small mammals parasitized by *I. dammini* which were serologically indistinguishable from isolates from patients and observed that this provided further support for the claim that a spirochaete is involved in the aetiology of Lyme disease. The species name *I. dammini* was later relegated to a junior subjective synonym of *I. scapularis* (Oliver *et al.*, 1993).

The deer tick (*Ixodes scapularis*) appears to have greatly extended its range and population densities since the 1940s; this increased abundance appears to be related to the increased abundance of deer, the preferred host of the adult stage of the tick. Its distribution includes the state of Florida in the southeastern USA, north to the provinces of Nova Scotia and Prince Edward Island, Canada, west to North and South Dakota and south to the state of Coahuila, in Mexico. The increased distribution of *I. scapularis* in both the USA and Canada has coincided with the increased distribution of Lyme disease. Though *I. scapularis* is present in the southern USA, it is a poor, although probably the only, vector of deer mice (*Peromyscus maniculatus*) parasites in that region and hence the vector of *B. burgdorferi*; about 90% of these immature ticks feed on lizards (Spielman *et al.*, 1984). While *I. scapularis* is the principal vector of *B. burgdorferi* throughout the eastern USA, *I. pacificus* is the vector in California; generally the rate of *B. burgdorferi* infection in *I. pacificus* is low as is the incidence of LD in California.

Several laboratory studies have compared the vectorial capacity of *I. scapularis* with other tick species; Sanders & Oliver (1995) compared the ability of three species of ticks in Georgia to transmit *B. burgdorferi* in the laboratory; *Amblyomma americanum* and *Dermacentor variabilis* did not transmit the infection from inoculated hamsters to mice and nymphal ticks of these species did not maintain this isolate transstadially. In contrast *I. scapularis* readily transmitted the infective agent thus adding support to the premise that *I. scapularis* is probably the main tick vector of *B. burgdorferi* in the southeastern USA as well. Similar results have been reported by several other authors confirming the role of *I. scapularis* as the principal, if not the only vector of *B. burgdorferi* in the eastern USA.

Reservoir hosts of Ixodes vectors of Lyme disease

Ixodes scapularis feeds on at least 125 species of North American vertebrates (54 mammalian, 57 avian and 14 lizard species), analysis of the US National Tick Collection holdings showed that white-tailed deer (*Odocoileus virginianus*), cattle (*Bos taurus*), dogs and other medium-to-large sized mammals are important hosts for adults as are native mice and other small mammals, certain ground-frequenting birds, skinks and glass lizards (*Ophisaurus attenuatus*) for nymphs and larvae (Keirans *et al.*, 1996). Horizontal transmission of pathogens is facilitated by a feeding pattern in which both the larval and nymphal stages feed on the white-footed mouse (*Peromyscus leucopus*) and mice serve as the principal reservoir host for the Lyme spirochaete (Spielman *et al.*, 1984). Bunikis *et al.* (2004) studied an enzootic site in Connecticut, and found that nearly the entire population of *P. leucopus* mice became infected with *B. burgdorferi* by late August, coinciding with the peak activity period of host-seeking larvae uninfected with the spirochaete *I. scapularis*, thereby perpetuating the agent through succeeding generations of ticks.

In the northeastern USA from Maine to Maryland and in the north-central states of Wisconsin and Minnesota, a highly efficient, horizontal cycle of *B. burgdorferi* transmission occurs among larval and nymphal *I. scapularis* ticks and certain rodents, particularly white-footed mice (*Peromyscus leucopus*) and chipmunks (*Tamias* species). This cycle results in high rates of infection among rodents and nymphal ticks and many new cases of human Lyme disease during the late spring and early summer months. White-tailed deer (*Odocoileus virginianus*), which are not involved in the life cycle of the spirochaete, are the preferred host of adult *I. scapularis*, and they seem to be critical for the survival of the ticks (Steere *et al.*, 2004).

Birds, both migratory and resident, are important in the epidemiology of Lyme disease in the USA and Canada, primarily for their role in supporting tick populations; they also disperse and introduce tick species into areas where they have not been previously present inasmuch as ticks readily parasitize birds. In a survey in Connecticut in 1983, Anderson & Magnarelli (1984), found spirochaete-infected *Ixodes scapularis* larvae and nymphs on eight and nine different species of birds, respectively. Spirochaetes grew in a cell-free medium inoculated with bloods from one northern mockingbird (*Mimus polyglottos*), one grey catbird (*Dumetella carolinensis*), two prairie warblers (*Dendroica discolor*), one orchard oriole (*Icterus spurius*), one common yellowthroat (*Geothlypis trichas*) and one American robin (*Turdus migratorius*). Anderson *et al.* (1986) indicated that *Borrelia burgdorferi* is unique among *Borrelia* species in being infectious to both mammals and birds and suggested that the cosmopolitan distribution of *B. burgdorferi* may be caused by long-distance dispersal of infected birds that serve as hosts for ticks. Many subsequent surveys have found *I. scapularis* infesting birds in large numbers and have considered the possibility that birds might also be reservoir hosts of LD spirochaetes. However, spirochaetaemia in most avian hosts may be short and therefore only certain species with concurrent infestations of nymphs and larvae of ticks may be effective reservoir hosts (Stafford *et al.*, 1995).

Lizards are important hosts for *Ixodes pacificus* in California but they are reservoir-incompetent for *Borrelia burgdorferi* though they serve as important hosts for the immature stages of *I. pacificus* in spring (Calisher *et al.*, 2002). Brush rabbits (*Sylvilagus bachmani*) and several species of rodents besides wood rats (*Neotoma fuscipes*) may contribute to the maintenance of *B. burgdorferi* in California because they harbour the spirochaete and are fed upon by competent enzootic vectors (Peavy *et al.*, 1997).

Conclusions on the public health importance of Lyme disease in the USA and Canada

Lyme disease is now the most common vector-borne disease in the USA. The chronic symptoms are of much importance; about 60% of untreated patients in the USA have intermittent attacks of joint swelling and pain, primarily in large joints and especially the knee, and the arthritis may persist for months or even years. Most of the symptoms are resolved after treatment. While mortality from infection is rare, it can occur. There is much concern and controversy related to the infection and mis-diagnosis is known to be common. Personal or

public expenditures for tick control and prevention against tick bites are largely
unknown.

Vaccination against Lyme disease appears only to be economically attrac-
tive for individuals who have a seasonal probability of *B. burgdorferi* infection
greater than 1% (Shadick *et al.*, 2001), but the only vaccine to prevent Lyme disease licensed in the USA has been discontinued due to what are reported to
be disappointing sales despite its good efficacy and tolerability. Lyme disease
will remain an important, and perhaps a continually growing, public health
problem in the USA and Canada unless an economic, safe and acceptable vaccine is developed that is readily available or greater measures are taken to control vector tick populations or ensure better protection is taken against tick
bites.

Southern tick-associated rash illness

Southern tick-associated rash illness, or STARI, is an infection with symptoms similar to the rash of Lyme disease; it has been described in humans residing in southeastern and south-central states; it is associated with the bite of
the lone star tick, *Amblyomma americanum*, and is apparently caused by infection with a spirochaete named *Borrelia lonestari*. It has also been referred to as
'Masters' disease' in recognition of the physician who first described its clinical
presentation.

In the southeastern and south-central United States, classic Lyme disease
occurs much less commonly than in the north-eastern states; the extent of the
existence of significant endemic Lyme disease transmission in the southern USA
was uncertain. In the early 1990s, despite mounting epidemiological evidence
that classic Lyme disease was not endemic in the South, physicians reported
patients with a clinical syndrome that appeared similar to acute Lyme disease. These patients presented with an expanding, circular skin rash (erythema
migrans) at the site of a tick bite accompanied by mild illness characterized by
generalized fatigue, headache, stiff neck and occasionally fever and other con-
stitutional signs; most cases responded to antibiotic treatment. Although apparently similar, the disease differed from classic Lyme disease in two key ways: (1)
when a tick was found biting it was identified as a lone star tick, *Amblyomma
americanum*, rather than an *Ixodes* vector of *Borrelia burgdorferi*, and (2) although
acute disease was recognized, chronic disease as is seen in classic Lyme disease
was not apparent.

Several studies have been made in the southern states of illnesses which
appeared to be Lyme disease but which serologically differed from *B. burgdorferi*. Serologic tests of 45 Missouri outpatients with annular rashes meeting a

surveillance case definition for erythema migrans and serologic tests failed to incriminate *B. burgdorferi* or selected other arthropod-borne pathogens. Skin specimens from the suspected erythema migrans lesions of 23 Missouri patients sampled prospectively in 1991–1993 were culture-negative for *B. burgdorferi*. The authors concluded that the tick bite-associated annular rashes in Missouri remained idiopathic. Possible causes were considered to include infection with a novel *Amblyomma americanum*-transmitted pathogen and an atypical toxic or immunological reaction to tick-associated proteins (Campbell *et al.*, 1995). Spirochaetes isolated from *A. americanum* from Missouri, Texas, New Jersey, New York and North Carolina were examined and phylogenetic analysis showed that the spirochaetes were a *Borrelia* species distinct from previously characterized members of this genus, including *B. burgdorferi* (Barbour *et al.*, 1996; Kirkland *et al.*, 1997). Eventually the illness STARI was shown to be caused by the newly described *Borrelia lonestari*.

Borrelia lonestari may be maintained in white-tailed deer (*Odocoileus virginianus*) as a reservoir host and transmitted between deer and to people by *A. americanum*. White-tailed deer are a preferred host for all three-life stages of *A. americanum*. Moore *et al.* (2003) evaluated blood samples from deer collected from eight southeastern states. Seven of 80 deer (8.7%) from 5 of 17 sites (29.4%) had sequence-confirmed evidence of a *B. lonestari* flagellin gene by PCR, indicating that deer are infected with *B. lonestari* or another closely related *Borrelia* species.

The tick vector *A. americanum* is abundant and is the most common cause of tick bites in southeast and south-central USA. Infection due to *B. lonestari* is being recognized in an increasing number of southern states and is becoming, in effect, the 'southern Lyme disease'. In this region of the USA, it is likely to spread to the areas where *A. americanum* is found in density.

Rocky Mountain spotted fever (RMSF)

Several species of spotted fever group (SFG) rickettsiae have been isolated from ticks in the USA; however, the only species which gives rise to human disease is *Rickettsia rickettsii*, which is the causative agent of Rocky Mountain spotted fever (RMSF). Rocky Mountain spotted fever was first recognized in 1896 in the Snake River Valley of Idaho. The disease was noted to be frequently fatal and affected hundreds of people in the area. By the early 1900s, the recognized geographic distribution of this disease included parts of the USA as far north as Washington and Montana and as far south as California, Arizona and New Mexico. It is now the most severe and most frequently reported rickettsial illness in the USA. The disease is distributed throughout the continental USA, as well as southern Canada, Central America, Mexico and parts of South America. Between

1981 and 1996, RMSF was reported from every USA state except Hawaii, Vermont, Maine and Alaska. North Carolina and Oklahoma account for one third of total cases reported. South Carolina, Tennessee and Georgia account for the third, fourth and fifth highest number of cases. Less than 2% of the total number of cases is found in the Rocky Mountain states.

In 2004, another agent, *Rickettsia parkeri* was also recognized as an infrequent cause of spotted fever in the southern USA and rarely in California. This rickettsia was first identified >60 years ago in Gulf Coast ticks (*Amblyomma maculatum*) (Paddock *et al.*, 2004). However, the vast majority of cases of spotted fever in the USA are due to *R. rickettsii* and *R. parkeri* is of little epidemiological importance.

The reported annual incidence of RMSF in the USA is 2.2 per million, but studies have suggested that human infection with *Rickettsia rickettsii* may be more common. From 1989–1996, more than 4700 cases of RMSF were reported in 46 states.

Between 1981–1992, the Centers for Disease Control collected and summarized 9223 cases of RMSF reported from 46 states and found that four states (North Carolina, Oklahoma, Tennessee and South Carolina) accounted for 48% of the reports (Dalton *et al.*, 1995). Paddock *et al.* (2002) carried out an assessment of the magnitude of fatal RMSF from 1983–1998 by an analysis of national surveillance reports; an estimated 612 RMSF-associated deaths occurred during the study period (median = 37 per year, range = 16–64), suggesting that approximately 400 fatal cases of RMSF went unreported to national surveillance during the period. Other studies comparing hospital records of cases of RMSF with official notifications have also shown that there is an under-reporting in endemic states.

Cases of RMSF are characterized by a rash and the infection may have a case-fatality rate as high as 30% in certain untreated patients. Even with treatment, hospitalization rates of 72% and case-fatality rates of 4% have been reported (CDC, 2004b). The mortality rate is now known to range from approximately 2% in children to 9% in elderly persons. The case-fatality rate is higher (6.2%) for persons whose treatment began more than 3 days after the onset of symptoms; for those treated within the first 3 days of illness mortality was 1.3%. The frequency of reported cases of RMSF is highest among males, Caucasians and children. Two-thirds of the RMSF cases occur in children under the age of 15 years, with the peak age being 5–9 years old. Individuals with frequent exposure to dogs and who reside near wooded areas or areas with high grass may also be at increased risk of infection.

Seasonal outbreaks of RMSF parallel the activity of the tick; 90% of cases are reported from 1 April to 30 September, with peaks in May and June.

Vectors of Rocky Mountain spotted fever

The vector of RMSF in eastern and southern USA is the American dog tick (*Dermacentor variabilis*), in the northwest USA, the wood tick (*D. andersoni*) and in the southwest, the lone star tick (*Amblyomma americanum*). In a survey in Mississippi, 18.9% of 884 brown dog ticks (*Rhipicephalus sanguineus*) taken from dogs contained spotted fever group rickettsiae (Sexton *et al.*, 1976). *Ixodes scapularis* has been found infected in Texas (Elliot *et al.*, 1990).

In Canada, the adult American dog tick (*D. variabilis*) is the second most common species encountered on humans and pets in Ontario. The preferred host of the adult tick is the dog, although it will also feed on horses and other large mammals including humans. In Canada, *D. variabilis* is found from Saskatchewan to the Maritime Provinces. These ticks sometimes enter buildings while attached to their hosts but they will not become established indoors; the species is the vector of RMSF in Canada.

Reservoir hosts of Rocky Mountain spotted fever

The disease is maintained in nature in Ixodidae ticks and their hosts. Small mammals are the natural reservoir hosts of RMSF in the wild. Dogs become infected when a tick harbouring *Rickettsia rickettsii* feeds on them but the dogs do not develop a sufficiently high rickettsaemia to act as a reservoir host in the transmission of the organism. On the other hand, dogs may act as sentinels to the presence of the disease, even though they cannot directly transmit infection (Comer, 1991).

Studies in Texas found antibodies to RMSF in 28% of the jack rabbits (*Lepus californicus*); four of 18 ticks (*Dermacentor parumapertus*) removed from 12 jack rabbits were positive for RMSF (Henke *et al.*, 1990). Burgdorfer *et al.* (1980) on the basis of laboratory transmission studies, concluded that cottontail rabbits do not serve as efficient reservoir hosts for infecting ticks. Antibodies to RMSF have been detected in chipmunks (*Tamias* species), raccoons (*Procyon lotor*), and white footed mice (*Peromyscus leucopus*), white-tailed deer (*Odocoileus virginianus*), skunks, squirrels, grey foxes (*Orocyon cinereoargenteus*), red foxes (*Vulpes vulpes*) and other mammals but their significance in the RMSF cycle often lies mainly as host animals for the tick vectors.

Conclusions regarding Rocky Mountain spotted fever

Although it is not widely realized, RMSF is the most prevalent rickettsial disease in the USA and is one of the most severe of all infectious vector-borne diseases in the USA. In Canada, the incidence cannot be obtained since RMSF is not a national notifiable disease. It remains an important infectious disease in the USA because of its prevalence, the difficulty of clinical diagnosis, potentially fatal outcome, and lack of a widely available sensitive and specific diagnostic test during the acute stage of the illness (Weber & Walker, 1991). It is important that physicians have a clinical suspicion of the possible presence of the infection among persons with the symptoms described above who live in or who have recently visited an endemic area or recall having been bitten by ticks.

Ehrlichiosis

As has been described earlier, ehrlichioses represent a group of emerging diseases caused by intracellular bacteria belonging to the family Rickettsiaceae, genus *Ehrlichia*. Several species previously known as *Ehrlichia* have been transferred to the genus *Anaplasma*, most notably *Ehrlichia phagocytophila* which is now named *Anaplasma phagocytophila*; however, it should be noted that many articles in the literature refer to *Anaplasma phagocytophilum*; the following section will follow the usage in the articles cited.

Two important ehrlichial human diseases recognized in the USA are human monocytic ehrlichiosis (HME) and human granulocytic ehrlichiosis (HGE). The causative agent of HGE in man is *Anaplasma phagocytophila*. Human monocytic ehrlichiosis is caused by *Ehrlichia chaffeensis*. The first case of HME in the USA was documented in 1986 (Maeda *et al.*, 1987). In 1991, *E. chaffeensis* was isolated from a military recruit stationed at Fort Chaffee, Arkansas (Anderson *et al.*, 1991).

Ehrlichia ewingii is the most recently recognized human pathogen of the group in the USA and was described in 1992 (Anderson *et al.*, 1992). Disease caused by *E. ewingii* has been limited to a few patients in Missouri, Oklahoma and Tennessee, most of whom have had underlying immunosuppression. *Ehrlichia ewingii*, was previously reported as a cause of granulocytic ehrlichiosis in dogs but this agent is not common. However, of 10 white-tailed deer (*Odocoileus virginianus*) tested from 20 locations in eight US states, six (5.5%) were positive for *E. ewingii*. In addition, natural *E. ewingii* infection was confirmed through infection of captive fawns. These findings expand the geographic distribution of *E. ewingii*, along with a risk for human infection, to include areas of Kentucky, Georgia and South Carolina. These data suggest that white-tailed deer may be an important reservoir host for *E. ewingii* (Yabsley *et al.*, 2002). Transmission of all members of this group of *Ehrlichia* is through the bite of infected ticks, which acquire the bacteria after feeding on infected animal reservoir hosts. Both

E. chaffeensis and *E. ewingii* are transmitted by *Amblyomma americanum*, the lone star tick.

More than 700 cases of HME have been recognized in the southeastern, south-central and mid-Atlantic regions of the USA since *E. chaffeensis* was first isolated. Tick bites, exposure to wildlife and golfing were associated with an increased risk of infection.

Human granulocytic ehrlichiosis was first described in 1994 and since then more than 600 cases have been described in the USA and Europe. In the USA, HGE has been documented or suspected in humans, animals or ticks in many states including California, Colorado, Connecticut, Florida, Georgia, Indiana, Massachusetts, Minnesota, Missouri, New York, Ohio, Virginia and Wisconsin. Gardner *et al.* (2003) reported on the data accumulated during the first 5 years of national surveillance for the human ehrlichioses in the USA and its territories, from its initiation in 1997 through 2001; reported cases of HME and HGE and cases of 'other ehrlichiosis' (OE) where the agent was unspecified, originated from 30 states. As anticipated, most HME cases were from the south-central and southeastern USA, while HGE was most commonly reported from the northeastern and upper-Midwestern region. The average annual incidences of HME, HGE and OE per million persons residing in states reporting disease were 0.7, 1.6 and 0.2, respectively. The median ages of HME (53 years) and HGE cases (51 years) were consistent with published patient series. Most (>57%) ehrlichiosis patients were male.

From 1986 through 1997, 1223 ehrlichiosis cases were reported by 30 state health departments in the USA. Data were reported from 19 states that considered ehrlichiosis notifiable as of August 1998, five that routinely collected information on cases, and six that occasionally received reports of ehrlichiosis cases. For states where ehrlichiosis was not notifiable, the designation routine reporting versus occasional reporting was based on the completeness of data provided. Because some states did not differentiate between probable and confirmed cases in their records, both categories were considered cases for the purposes of the report. Of the 1223 reported ehrlichiosis cases, 742 (60.7%) were categorized as HME, 449 (36.7%) as HGE and 32 (2.6%) as not ascribed to a specific ehrlichial agent. Using data from 20 states that reported information on deaths, the authors found case-fatality ratios of 2.7% (8 of 299) for HME and 0.7% (3 of 448) for HgE (McQuiston *et al.*, 1999).

In Canada, DNA extracts from *Ixodes scapularis*, collected by passive surveillance, were tested for the presence of HGE agent genome. A total of 576 *Ixodes scapularis* were screened for the HGE-agent using PCR. The HGE-agent genome was detected in ticks collected in Manitoba (5/203; 2.5%), Quebec (1/259; 0.4%) and Prince Edward Island (1/42; 2.4%). Although the prevalence of HGE-like pathogens in *I. scapularis* collected in Canada was much lower than those reported in tick-endemic localities in the US, Canadians clearly have a risk of exposure to this pathogen (Drebot *et al.*, 2001).

Clinically, HME and HGE are nearly indistinguishable and are characterized by one or more of the following symptoms: fever, headache, myalgia, thrombocytopenia, leukopenia and elevated liver enzyme levels. A rash occurs in approximately one-third of patients with HME but is less common in patients with HGE. Most cases of ehrlichiosis are characterized by mild illness. However, complications such as adult respiratory distress syndrome, renal failure, neurological disorders, and disseminated intravascular coagulation can occur. Case-fatality ratios may be as high as 5% for HME and 10% for HGE although more serious cases are probably over-represented in these estimates as they are more frequently reported. Other studies have reported case-fatality ratios of <5% for these diseases.

Canine ehrlichiosis caused by *Ehrlichia canis* was at first thought to be restricted to dogs but is now recognized as a human infection as well. Canine ehrlichiosis first was recognized in the USA in 1962. German shepherd dogs often develop fatal haemorrhage in the late chronic phase of the illness, whereas mixed breeds and beagles generally do not. One study found detectable antibodies to *E. canis* in 11% of military working dogs internationally and 57% of civilian dogs in the USA (Keefe *et al.*, 1982). *Ehrlichia canis* has been reported from humans in the southeastern USA; Conrad (1989) observed that although recovery from *E. canis* infection has been observed in humans without receiving treatment, prompt therapy with tetracycline is advised before obtaining results of serologic studies because the immunologically similar illness in untreated dogs has been lethal.

Vectors of ehrlichiosis

Human monoctyic ehrlichiosis due to *Ehrlichia chaffeensis* (HME) is primarily transmitted by the lone star tick, *Amblyomma americanum*.

Human granulocytic ehrlichiosis (HGE) is caused by species closely related or identical to the veterinary pathogens *E. equi* and *Anaplasma phagocytophila*, and is transmitted by the deer tick (*Ixodes scapularis*) and the western blacklegged deer tick (*I. pacificus*) in the western USA. *Ehrlichia ewingii* is also transmitted by *A. americanum* and has been isolated from *Dermacentor variabilis* as well in Missouri (Steiert & Gilfoy, 2002) where it appears to be the principal causative agent of canine ehrlichiosis. The probable role of *I. scapularis* as a vector in Canada has been mentioned above. The vector of *Ehrlichia canis* is the tick, *Rhipicephalus sanguineus*.

The public health importance of ehrlichioses has not, as yet, been well defined, largely because the group of diseases is relatively newly recognized. Various surveys have shown that both HME and HGE are under reported in the USA. The emergence of these newly recognized tick-borne infections as threats to human health may be due to increased clinical cognisance, but as in other emerging tick-borne infections, it is likely that the rapid increase in identified cases signals a true emergence of disease associated with changing vector-host ecology (Walker & Dumler, 1996).

Q fever

As described earlier, Q fever is a worldwide zoonosis caused by *Coxiella burnetii*, an obligate intracellular parasite. As with all rickettsiae, humans are dead-end hosts. The most common reservoir hosts are domesticated ruminants, primarily cattle, sheep and goats. Humans acquire Q fever typically by inhaling aerosols or contaminated dusts derived from infected animals or animal products. It is transmitted to humans mainly by the consumption of milk and meat from contaminated cattle or the inhalation of dried infected tick faeces by people working with cattle. The disease may be spread between infected rabbits, rodents and other wild animals by the bites of ticks. Vector transmission was identified when the organism was isolated from ticks in Montana in 1938. Humans are not usually infected by tick bites, although it is possible. Nevertheless, it is considered a tick-associated infection.

From 1948–1986, 1396 human cases were reported from almost every state (McQuiston & Childs, 2001) although it was only in 1999 that Q fever became a notifiable disease in the USA. During 2000–2001, a total of 48 patients who met the case definition of Q fever were reported to CDC in the USA. Only nine cases of Q fever were recorded in Canada in the 20 years prior to 1978. In the 18 months from August 1979 to January 1981 the disease was diagnosed serologically in six patients from the Maritime Provinces.

Up to 60% of initial infections are asymptomatic; acute disease can manifest as a relatively mild, self-limited febrile illness; more moderately severe disease is characterized by hepatitis or pneumonia. It manifests less commonly as myocarditis, pericarditis and meningoencephalitis. Chronic Q fever occurs in <1% of infected patients, months or years after initial infection. Mortality is not common.

As in Europe, improved working conditions on farms have reduced the number of infections. In view of the small number of cases of Q fever, most of which are not vector-borne, the disease can not be considered of public health importance in either the USA or Canada.

Babesiosis

Human babesiosis is a tick-transmitted protozoan zoonosis of the family Piroplasmoridae associated with *Babesia microti* in the USA. Even though bovine babesiosis has long been eradicated from the country, human babesiosis remains endemic in the northeastern and north-central regions of the country and is being recognized with increasing frequency and appears to be spreading. *Babesia microti* babesiosis in the USA is generally less severe than cases caused by *B. divergens* in Europe. *Babesia microti*, a babesial parasite of small mammals, has been the cause of over 300 cases of human babesiosis from 1969–2000, and was the cause of mild to severe disease, even in non-splenectomised patients. The disease can be fatal if it is untreated. Changing ecology favourable to the multiplication of tick populations has contributed greatly to the increase and expansion of human babesiosis in the USA which Kjemtrup & Conrad (2000) consider as an emerging disease in North America.

Between 1968–1993, more than 450 *Babesia* infections were confirmed in the USA by blood smears or serologic testing. Prevalence is difficult to estimate because of a lack of active surveillance, and because infections are often asymptomatic. Co-infection with *Borrelia burgdorferi* is being found more frequently as both infectious agents are carried by the same tick species and have the same rodent reservoir host. Further support to opinion that babesiosis is increasing in endemic areas of the USA is seen from the study of Krause *et al.*, (2003) carried out in Rhode Island (Block Island) and southeastern Connecticut. The incidence of babesial infection on Block Island increased during the early 1990s, reaching a level about three-quarters that of borrelial infection. The sera of approximately one-tenth of Block Island residents reacted against babesial antigen, a seroprevalence similar to those on Prudence Island and in southeastern Connecticut. Although the number and duration of babesial symptoms in people older than 50 years of age approximated those in people 20–49 years of age, more older adults were admitted to hospital than younger adults. Few *Babesia*-infected children were hospitalized. Babesial incidence at endemic sites in southern New England appears to have risen during the 1990s to a level approaching that due to borreliosis.

Hatcher *et al.* (2001) described 34 patients who were hospitalized with diagnosis of severe *Babesia* infection over the course of 13 years in Long Island, New York. Forty-one per cent of the patients with Babesia developed complications such as acute respiratory failure, disseminated intravascular coagulation, congestive heart failure and renal failure. Despite treatment, parasitaemia persisted for an average of 8.5 days and 3 of the patients died. There has been some southern extension of the infection and three patients with *Babesia microti* infections acquired in New Jersey have been reported (Eskow *et al.*, 1999). A new focus of transmission was detected in a woman in Wisconsin in 1983 and *B. microti* infected *Peromyscus leucopus* (white-footed mouse) were found on her property and elsewhere in the county in which she lived (Steketee *et al.*, 1985).

Human co-infection with Lyme disease and babesiosis has been described in Canada (dos Santos & Kain, 1999). The illness associated with concurrent babesiosis and LD is more severe than either infection alone (Krause *et al.*, 1996). The increasing number of co-infections being diagnosed in the USA is a serious clinical and diagnostic problem.

The agent of European human babesiosis, *Babesia divergens*, was detected in 16% of the cottentail rabbits (*Sylvilagus transitionalis*) sampled during 1998–2002 on Nantucket Island. The vector of *B. divergens* on Nantucket appears to be *Ixodes dentatus*, a rabbit- and bird-feeding tick that may feed on humans. Although the risk of human infection appears to be minimal, an autochthonously acquired Kentucky case due to this rabbit agent has been reported (Goethert & Telford, 2003). As this parasite may be the cause of severe haemolytic febrile syndromes, its discovery in the USA must be viewed with concern.

A previously unknown species of *Babesia* (WA-1) was isolated from a man in Washington State who had clinical babesiosis (Quick *et al.*, 1993).

Four patients in California were identified as being infected with a similar protozoal parasite. All four patients had undergone splenectomy; two of the patients had complicated courses, and one died. Serologic testing showed WA1-antibody seroprevalence rates of 16% (8 of 51 persons at risk) and 3.5% (4 of 115) in two geographically distinct areas of northern California. The clinical spectrum associated with infection with this protozoan ranges from asymptomatic infection or influenza-like illness to fulminant, fatal disease (Persing *et al.*, 1995).

Researchers have also described another, probably new, babesial species (MO1) associated with the first reported case of babesiosis acquired in the state of Missouri. Species MO1 is probably distinct from *B. divergens* but the two share morphologic, antigenic and genetic characteristics (Herwaldt *et al.*, 1996).

Reservoir hosts of babesiosis

Microtus pennsylvanicus host (the meadow vole) and *Peromyscus leucopus* (the white-footed mouse) are recognized as reservoir of *Babesia*. Deer are important hosts for all stages of the tick vectors but are not reservoir hosts of the infectious agent. As is noted below, *Microtus ochrogaster* (the prairie vole) may be the reservoir host in Colorado. Co-infections with *Borrelia burgdorferi* and *B. microti* in *P. leucopus* and *M. pennsylvanicus* are common.

Vectors of Babesia

The vector of *Babesia microti* spiroplasms in the eastern USA is *Ixodes scapularis*. *Ixodes spinipalpis* has been found infected with *B. microti* in Colorado in an area where 13 of 15 *Microtus ochrogaster* (the prairie vole) were positive as well (Burkot *et al.*, 2000). The vector of rabbit infections with *B. divergens* is *Ixodes dentatus*. Most cases of babesiosis are tick-borne, but cases of transfusion-associated and transplacental/perinatal transmission have been reported. The vector in Canada is also *I. scapularis*.

The public health importance of babesiosis

Babesiosis is an emerging disease in the USA; the number of cases being reported is increasing and the infection is spreading geographically. New species have been identified and *B. divergens*, a species with a more serious clinical picture, previously recorded only in Europe, has been identified in the USA. The wide clinical spectrum of human babesiosis extends to severe cases and occasionally death. The disease is of growing public health importance in the USA and probably in Canada as well. Growing leisure time has resulted in greater contact of humans with tick vectors and hence increasing infections. Inasmuch as the tick vector and rodent reservoir hosts are very widespread in the USA and Canada, it appears likely that the infection will continue to spread and the incidence increase.

Tularaemia

Tularaemia is a bacterial zoonosis that occurs throughout the temperate climates of the northern hemisphere. The causative organism, *Francisella tularensis*, is a small, gram-negative, non-motile coccobacillus. Tularaemia is spread by deer flies (*Chrysops discalis*) and a number of tick species in the USA; a large number of animal species are susceptible to the organism, especially the cottontail rabbit. Two known *F. tularensis* biotypes exist in the USA. Type A is more virulent than type B, but both can result in severe and sometimes fatal illness.

Francisella tularensis organisms are common in rabbits (cottontail, jack and snowshoe), and are frequently transmitted by tick bite. Biovar *palaearctica* strains are commonly found in mammals other than rabbits in North America. Approximately 150–300 human cases of tularaemia occur in the USA each year. The number of tularaemia cases reported in the country decreased during the latter half of the twentieth century; the incidence peaked in 1939, with 2291 cases. There were 1368 cases of tularaemia reported to the Centers for Disease Control from 1990 through 2000 with most cases occurring in Arkansas, Missouri and

Oklahoma. The disease has been reported from every state of the USA except Hawaii.

Francisella tularensis antibodies have been detected in humans in Alberta, Canada and transmission by *Dermacentor andersoni* has been documented in Saskatchewan (Gordon *et al.*, 1983).

The two biovars (*tularensis* and *palaearctica*) cause human disease of different severity. A third biovar, *F. philomiragia* is considered virulent only in immunocompromised individuals. *Francisella tularensis* biovar *tularensis* organisms are more virulent, with an untreated case-fatality rate of 5–15% primarily due to typhoidal or pulmonary disease. With appropriate antibiotic treatment, the case-fatality rate is negligible and early diagnosis and treatment can reduce that rate to 1–2%. *Francisella tularensis* biovar *palaearctica* organisms are less virulent and, even without treatment, few infections result in fatalities.

Tularaemia may present as several different clinical syndromes, including glandular, ulceroglandular, oculoglandular, oropharyngeal, pneumonic and typhoidal. The clinical course is characterized by an influenza-like attack with severe initial fever, temporary remission, and a subsequent febrile period of at least 2 weeks. Later, a local lesion may occur, with or without glandular involvement. Additional symptoms vary depending upon the method of transmission and form of the disease (Goddard, 1998).

In the summer of 2000, an outbreak of primary pneumonic tularaemia occurred on Martha's Vineyard, Massachusetts. The only previously reported outbreak of pneumonic tularaemia in the USA also occurred on the island in 1978. Of 15 patients with tularaemia, 11 were primary pneumonic tularaemia. *Francisella tularensis* type A was isolated from blood and lung tissue of the one man who died. Lawn mowing and brush cutting remained significant risk factors for infection. Of 40 trapped animals, one striped skunk (*Mephitis mephitis*) and one Norway rat (*Rattus norvegicus*) were seropositive for antibodies against *Francisella tularensis* (Feldman *et al.*, 2001).

There is concern about the possible use of tularaemia in biological warfare or bioterrorism. The organism could potentially be stabilized for weapons use by an adversary and theoretically produced in either a wet or dried form. In the USA, tularaemia was removed from the list of nationally notifiable diseases in 1994, but increased concerns about the potential use of *F. tularensis* as a biological weapon led to its reinstatement in 2000. A live, attenuated tularaemia vaccine is under investigation.

The vectors of tularaemia

The disease may be contracted in a variety of ways – food, water, mud, articles of clothing and particularly through arthropod bites. Arthropods involved in transmission of tularaemia include ticks, biting flies and possibly mosquitoes. In the original study of tularaemia in the USA, transmission by the deer fly (*Chrysops discalis*) was shown to occur. Fleas are poor vectors but may be important in maintaining tularaemia organisms in nature among small rodents. Ticks account for more than 50% of all cases, especially west of the Mississippi River and most transmission in the USA and Canada is by tick bite. The three major North American ticks involved in transmission of tularaemia organisms are the lone star tick (*Amblyomma americanum*); the wood tick (*Dermacentor andersoni*); and the American dog tick (*Dermacentor variabilis*). Both the lone star tick and the American dog tick are found over much of the eastern USA; the Rocky Mountain wood tick inhabits the West.

In Canada, Silverman *et al.* (1991) believed that mosquito vectors may be important in inducing the high seroprevalence of antibodies to *F. tularensis* above the tree line. *Dermacentor andersoni* was shown to be a vector in Saskatchewan.

The public health importance of tularaemia

The annual number of cases of tularaemia in the USA and Canada is not large and has, in fact, been decreasing. However, *F. tularensis* is one of the most infectious bacterial pathogens known. Fortunately human-to-human transmission is not known to occur. The pneumonic form of the disease that was seen in Martha's Vineyard occurs rarely but is the likely form of the disease should this bacterium be used as a bioterrorism agent. Thus the public health importance of tularaemia lies principally in the threat it may pose and continuing surveillance is required.

Tick paralysis

In the USA and Canada, tick paralysis may be caused by the attachment of one of several species of a female tick. A neurotoxin, secreted by the salivary glands during feeding, is responsible for the symptoms. The first symptom to occur is a numbness of the feet and legs. The patient also experiences difficulties in walking. This is followed by a numbness of the hands and arms. The throat muscles and tongue may become partially paralyzed. Death from respiratory failure may occur if the tick is not removed in time. Recovery is rapid if the tick is removed soon after attachment. Tick paralysis occurs worldwide and is caused by the introduction of a neurotoxin elaborated into humans during attachment of and feeding by the female of several tick species. If unrecognized, and the attached tick is not removed, tick paralysis can progress to respiratory failure and may be fatal in approximately 10% of cases.

In North America, tick paralysis occurs most commonly in the Rocky Mountain and northwestern regions of the USA and in western Canada. Most cases have been reported among girls aged less than 10 years during April–June, when nymphs and mature ticks are most prevalent. Only 10 cases were reported during 1987–1995.

In the USA, this disease is associated with *Dermacentor andersoni* (Rocky Mountain wood tick), *D. variabilis* (American dog tick), *Amblyomma americanum* (lone star tick), *A. maculatum, Ixodes scapularis* (black-legged tick) and *I. pacificus* (western black-legged tick). Onset of symptoms usually occurs after a tick has fed for several days. The pathogenesis of tick paralysis has not been fully elucidated and pathologic and clinical effects vary depending on the tick species. However, motor neurones probably are affected by the toxin, which diminishes release of acetylcholine (CDC, 1996).

The risk for tick paralysis is greatest for children in rural areas, especially in the northwest USA and western Canada, during the spring. The risk of infection may be reduced by the use of repellents on skin and permethrin-containing acaricides on clothing. Paralysis can be prevented by careful examination of potentially exposed persons for ticks and prompt removal of ticks.

Health-care providers should consider tick paralysis in persons, especially children, who reside in or have recently visited tick-endemic areas during the spring or early summer and who present with symmetrical paralysis. A rapid diagnosis and removal of the attached tick may avert death.

31

Mite-borne infections and infestations

Scabies

As has already been described, scabies is an intensely pruritic and highly contagious infestation of the skin caused by the mite, *Sarcoptes scabiei*. Although *Sarcoptes* mites do not transmit a disease, they are the cause of a disease condition (termed scabies) due to the extremely serious infestations that usually develop. As in many countries, there has been a noticeable increase of scabies in the USA and Canada. Scabies is a global problem and a significant source of morbidity in nursing home residents and workers because of its highly contagious nature. It is also a problem in hospitals that care for the elderly, the debilitated, and the immunocompromised. New outbreaks continue to occur, despite efforts in controlling the recurrent epidemics.

Severe infestations may lead to the development of a condition known as Norwegian scabies or crusted scabies. There are a number of differences between Norwegian scabies and typical scabies. One difference is that when Norwegian scabies affects immunocompromised patients, it results in a severe infestation of myriads of mites; in the healthy person the parasite numbers are far smaller (Clark *et al.*, 1992). The clinical manifestations are also usually different to those of typical scabies. Norwegian scabies may present as a generalized dermatitis with widely distributed burrows, extensive scaling and sometimes vesiculation and crusting. Typical scabies usually presents as lesions prominently around finger webs, anterior surfaces of the wrist, elbows, anterior folds and belt line and causes severe itching especially at night.

In developed countries, scabies epidemics seem to occur in 15-year cycles; however, the most recent epidemic began in the late 1960s and, for some unknown reason, continues today. In undeveloped countries, scabies infestation is pandemic, with millions affected worldwide.

There is an association between immunosuppressed persons and scabies infections. In 1975, Espy & Jolly reported the occurrence of a case of Norwegian scabies in a patient undergoing long-term immunosuppressive therapy after a kidney transplant. Altered host factors appeared to be the primary determinants in the pathogenesis of the disease. In 1989, Jucowics *et al.* reported the case of a 6-month-old infant with acquired immunodeficiency syndrome (AIDS) and typical scabies who subsequently developed Norwegian scabies, with deterioration of clinical status. The infestation spread to several health-care workers who were in close contact with the patient, despite standard isolation precautions. There is a rapidly growing hospitalized paediatric AIDS population. Kolar & Rapini (1991) concluded that patients with cognitive deficiency or an immunodeficiency disorder (including immunosuppressive therapy) are predisposed to developing crusted scabies. The infection often presents as generalized dermatitis with crusted hyperkeratosis on the palms and soles. Since the mid-1980s, worldwide reports confirm that scabies in individuals infected with the human immunodeficiency virus (HIV) result in a wide range of clinical manifestations which differ from those seen in immunocompetent patients. There is also general agreement that HIV-related scabies is more difficult to treat (Taplin & Meinking, 1997).

The treatment of patients with crusted or Norwegian scabies with lindane or permethrin may often fail due to the heavy mite burden in such persons and reinfestations are common. Fortunately, the systemic drug ivermectin is highly effective in most cases and appears to have little or no risk. Oral ivermectin is proving to be an alternative to topical scabicides, it appears effective as a local treatment for common scabies, so far there are few comparative studies. The best indications for ivermectin would be against outbreaks in institutions and the appearance of crusty or Norwegian scabies, but in association with topical treatment (Develoux, 2004). Nevertheless, treatment failures, recrudescence and reinfection can occur, even after multiple doses. Ivermectin resistance has also been induced in arthropods in laboratory experiments but until recently, resistance had not been documented among arthropods in nature. Currie *et al.* (2004) reported clinical and in vitro evidence of ivermectin resistance in two patients with multiple recurrences of crusted scabies who had previously received 30 and 58 doses of ivermectin over 4 and 4.5 years, respectively. Thus ivermectin resistance in scabies mites can develop after intensive ivermectin use. Other compounds have shown promise and may soon be available such as thiabendazole and flubendazole but further clinical, pharmacological and toxicological studies are required (Ena *et al.*, 1999).

In May 2002, the Centers for Disease Control included scabies in its updated guidelines for the treatment of sexually transmitted diseases (CDC, 2002b).

The public health importance of scabies

As the reporting of scabies is not obligatory in all areas, the overall incidence of infections in the USA and Canada is not known. From the growing number of reports in the literature, it is apparent that the trend of scabies infections is growing especially among immunosuppressed AIDS patients but with others as well. While scabies was at one time thought to be essentially a sexually transmitted disease, institutional infections are now a serious problem. Close surveillance and a high degree of clinical suspicion are essential as is the necessity of the immediate implementation of control measures along with strict measures of barrier nursing. Scabies must be considered to be a public health problem of importance even though its true magnitude is unknown.

Rickettsial pox

The causative agent of rickettsial pox is *Rickettsia akari*, transmitted by the house mouse mite (*Liponyssoides sanguineus*) from rodents, especially the house mouse (*Mus musculus*), to man. The disease was first identified in New York City in 1946 (Huebner *et al.*, 1946). During the next 5 years, approximately 540 additional cases were identified in New York City. During the subsequent 5 decades, rickettsial pox received relatively little attention from clinicians and public health professionals, few serum samples were submitted for identification and reporting of the disease diminished markedly. However, from February 2001 through August 2002, 34 cases of rickettsial pox in New York City were confirmed at the CDC from cutaneous biopsy specimens tested by using immunohistochemical (IHC) staining, PCR analysis and isolation of *R. akari* in cell culture, as well as an indirect immunofluorescence assay of serum specimens (Paddock *et al.*, 2003).

Small outbreaks of rickettsial pox have been recognized in several other cities in the USA, including Boston, Cleveland, Philadelphia, Pittsburgh and West Hartford. Most cases to date have occurred in large metropolitan areas of the northeastern USA and about half the described cases have occurred in New York City. Rickettsial pox is likely to be more common in the USA than is suggested by the relatively small number of reported cases during the past 50 years. Krusell *et al.* (2002) described a case of the disease from North Carolina which was confirmed by serologic testing. This case is the first to be reported from that region of the USA. Including rickettsial pox in the evaluation of patients with eschars or vesicular rashes is likely to extend the recognized geographic distribution of *Rickettsia akari*.

Single serum samples from 631 intravenous (i.v.) drug users from inner-city Baltimore, Maryland were tested for serologic evidence of exposure to spotted fever group rickettsiae. A total of 102 (16%) individuals had titres greater than or equal to 64 to *Rickettsia rickettsii*. Confirmation that infection was caused by *R. akari* was obtained by cross-adsorption studies on a subset of serum samples that consistently resulted in higher titres to *R. akari* than to *R. rickettsii*. Current intravenous drug use, increased frequency of injection, and shooting gallery use were significant risk factors for presence of group-specific antibodies reactive with *R. rickettsii*. There was a significant inverse association with the presence of antibodies reactive to *R. rickettsii* and antibodies reactive to the human immunodeficiency virus. This study suggests that intravenous drug users are at an increased risk for *R. akari* infections. Clinicians should be aware of the presence of rickettsial pox, as well as other zoonotic diseases of the urban environment, when treating intravenous drug users for any acute febrile illness of undetermined aetiology (Comer *et al.*, 1999).

The public health importance of rickettsial pox

The use of improved diagnostic methods has shown that both the incidence and distribution of rickettsial pox is greater than previously thought. Nevertheless, the total number of cases remains small; the disease is mild and self-limited and usually persists only for about a week and prognosis for full recovery from rickettsial pox is excellent. No deaths have ever been reported from this illness, and even the skin rash heals without scarring. Its public health importance is, therefore, small.

Mites and allergies

As has been observed, house dust mites are important sources of indoor allergens associated with asthma and other allergic conditions. In most temperate, humid areas of the world, house dust mites are a major source of multiple allergens in house dust. Mite allergens sensitize and induce perennial rhinitis, asthma, or atopic dermatitis in a large portion of patients with allergic disease. The most common dust mite species around the world include *Dermatophagoides pteronyssinus*, *Dermatophagoides farinae*, *Euroglyphus maynei* and *Blomia tropicalis*. The relationship of allergies to positive reactions to these mites was first observed in the 1960s. In many countries house dust mites are regarded as the most important, or one of the most important, causes of asthma, particularly among children.

Progressive changes in American housing and life styles have been associated with increased prevalence of allergen sensitization and asthma. Not only has there been a significant increase in the proportion of time spent indoors, but many of the changes made in houses are likely to increase exposure to indoor allergens. As an example, higher mean indoor temperatures, reduced ventilation, cool wash detergents and the widespread use of carpeting are all changes that could have increased the levels of allergens in homes in the USA and Canada. Over the past 15 years, dust mites, cockroaches and cats have been identified as major sources of indoor allergens. The combination of exposure and sensitization to one of these allergens is significantly associated with acute asthma. Furthermore, clinical studies have shown a direct quantitative correlation between dust mite allergen exposure and the prevalence of both sensitization and asthma (Platts Mills, 1994).

The prevalence of allergic reactions to the mite species *Dermatophagoides farinae* is extremely high in North America and Japan. In contrast to Europe and Australasia, *Dermatophagoides farinae* is generally the most abundant species within the USA (Wharton, 1976). Arlian *et al.* (1992) found that 12 of the 19 homes they examined within Ohio were infested with only *Dermatophagoides farinae*, whilst none contained *Dermatophagoides pteronyssinus* alone. In homes where both species were present *Dermatophagoides farinae* was generally dominant comprising between 74 and 93% of the total mite population with the remainder being *Dermatophagoides pteronyssinus*.

Arlian *et al.* (1992) determined the density and species prevalence of dust-mites at various times over a 5-year-period in 252 homes of dust-mite sensitive people with asthma who lived in eight geographic areas of the USA (Cincinnati, Ohio; New Orleans, Louisianna; Memphis, Tennessee; Galveston, Texas; Greenville, North Carolina; Delray Beach, Florida; San Diego and Los Angeles, California). The most common dust mites found in the homes were *Dermatophagoides farinae*, *Dermatophagoides pteronyssinus*, *Euroglyphus maynei*, and *Blomia tropicalis*. All homes in all locations contained *Dermatophagoides* species mites, but few homes were populated exclusively by either *Dermatophagoides farinae* or *Dermatophagoides pteronyssinus* alone. Most homes (81.7%) were co-inhabited by both *Dermatophagoides farinae* and *Dermatophagoides pteronyssinus*. In co-inhabited homes one species was predominant and usually made up at least 75% of the total mite population. Prevalence of the dominant or only species present varied between homes within a geographic area. *Euroglyphus maynei* occurred in significant numbers in 35.7% of homes in New Orleans, Memphis, Galveston, Delray Beach and San Diego. *Blomia tropicalis* occurred in these same cities but in low densities. For all dust samples, only 13 homes of the 252 sampled had 100 or fewer mites/g dust, which is considered to be the threshold for sensitivity. Most homes had average mite densities of 500 or more mites/g dust. The study suggests that there is a significant and widespread occurrence of both *Dermatophagoides farinae* and *Dermatophagoides pteronyssinus* in the USA.

In a later nationwide study by Arbes *et al.* (2003), carried out on the prevalence of dust mite allergen in the beds of homes in the USA, the authors attempted to identify predictors of dust mite allergen concentration; the percentages of homes with concentrations at or greater than detection, 2.0 μg/g bed dust, and 10.0 μg/g bed dust were estimated. Independent predictors of allergen concentration were assessed with multivariable linear regression. The percentages of the homes examined with dust mite allergen concentrations at or greater than detection, 2.0 μg/g and 10.0 μg/g were 84.2%, 46.2% and 24.2%, respectively. Independent predictors of higher levels were older homes, non-West census regions, single-family homes, no resident children, lower household income, heating sources other than forced air, musty or mildew odour and higher bedroom humidity. Most USA homes, therefore, have detectable levels of dust mite allergen in a bed and levels previously associated with allergic sensitization and asthma are common in bedrooms.

In Canada, over 2.2 million persons have been diagnosed with asthma by a physician (12.2% of children and 6.3% of adults). An estimated 10% of children and 5% of adults have active asthma (take medication for asthma or have experienced symptoms in the past 12 months). In Canada, approximately 20 children and 500 adults die each year from asthma and there has been an increase in the prevalence of asthma among children in the past 15 years. However, it is uncertain to what extent asthma cases in Canada can be ascribed to house dust mite allergens. Chan Yeung *et al.* (1995) found that the mean levels of both mite allergens in mattress and floor samples in homes in Vancouver and in Winnipeg were relatively low for all seasons; they considered that whether the low levels of mite allergens are partially responsible for the relatively low prevalence of childhood asthma in Canada, remains to be investigated.

Asthma prevalence among disadvantaged and minority children is disproportionately higher in inner-city populations in the USA. Overall, more than 80% of childhood asthmatics are allergic to one or more inhalant allergens. Environmental allergen exposure, particularly that of house dust mites and cockroaches, is known to contribute to asthma exacerbations in children. Kitch *et al.* (2000) evaluated whether socioeconomic status was associated with a differential in the levels and types of indoor home allergens. Dust samples for an ELISA allergen assay were collected from the homes of 499 families as part of a metropolitan Boston, Massachusetts, longitudinal birth cohort study of home allergens and asthma in children with a parental history of asthma or allergy. The proportion of homes with maximum home allergen levels in the highest category was 42% for dust mite allergen, 13% for cockroach allergen, 26% for cat allergen and 20%

for dog allergen. Homes in the high-poverty area (>20% of the population below the poverty level), were more likely to have high cockroach allergen levels than homes in the low-poverty area (51 vs. 3%); but less likely to have high levels of dust mite allergen (16 vs. 53%). Lower family income, less maternal education, and race/ethnicity (black or Hispanic vs. white) were also associated with a lower risk of high dust mite levels and a greater risk of high cockroach allergen levels. The costs surrounding environmental control of the arthropod causes of allergens may not be within the means of many families living in urban, inner-city environments (Kuster, 1996).

Humidity is a major reason for the presence or absence of dust mites in homes in the USA and elsewhere. Arlian *et al.* (2002) emphasized that the key factor that influences mite survival and prevalence is relative humidity. Mites are present in homes in humid geographical areas and are rare or absent in drier climates unless humidity in the homes is artificially raised. Generally speaking, dust mite allergen levels are low in public buildings and transportation compared to levels in homes. Due to low rainfall and high elevation, the ten Rocky Mountain states would not be expected to have the indoor humidity required to support the growth of significant numbers of house dust mites. Nelson & Fernandez Caldas (1995) obtained dust specimens from the homes of 58 bronchial asthma patients in Denver, Colorado. Forty-eight contained no mites, five had levels of mites considered clinically insignificant (less than 100 mites/g dust), three contained 100 mites/g of dust, a level at the threshold of significance, and two had high levels of mites (1000 and 3000 mites/g). Both of the homes with high mite levels had specific sources of excess moisture. Twenty-eight patients from adjacent states also submitted dust specimens. Significant levels of mites were encountered in specimens from these states reflecting the transition to zones of greater humidity and lower elevation. Thus, except in homes with unusual sources of humidity, significant levels of house dust mites are rarely encountered in homes in the ten Rocky Mountain states.

In fact, maintaining a relative humidity (RH) of less than 50% is one of the recommendations for reducing numbers of house dust mites and their allergens in homes. A study undertaken by Arlian *et al.* (2001) showed that it is practical to maintain an indoor RH of less than 51% during the humid summer season in a temperate climate, and, when done, this resulted in significant reductions in mite and allergen levels. Unfortunately, it may not be either possible or practicable for many families, especially those of limited economic means to adhere to this recommendation.

Because of the extent and gravity of the problem, many studies have been carried out in an effort to reduce or eliminate house dust mites in homes. Environmental measures have been proposed to reduce mite allergens in houses as well as including the reduction of humidity. However, there is debate as to the value of environmental control measures in the management of asthma, which is best illustrated in the case of house dust mite allergy. A meta-analysis (Gotzsche *et al.*, 1998) was unable to find a positive effect of house dust mite environmental control in patients with asthma. The authors noted that "current chemical and physical methods aimed at reducing exposure to allergens from house dust mites seemed to be ineffective". In that analysis, only five of 23 studies reviewed showed a significant fall in the concentration of house dust mite allergen, but four of the five studies were associated with some measured improvement of asthma. There is a need to very significantly decrease the concentration of house dust mite allergen to have an impact on the underlying asthma pathophysiology in patients allergic to house dust mites; environmental control measures may be effective where socioeconomic conditions permit the acquisition of protective mattress and pillow covers.

The public health importance of house dust mite allergies

Allergies caused by the allergens of house dust mites are a serious public health problem throughout most of the USA and to a lesser extent in southern Canada. Overall, the magnitude of the public health burden caused by house dust mite allergies has not been calculated. In view of the incidence of the condition and costs of treatment, costs to individuals and to the community must be considered as great.

32

Cockroaches and allergies

Along with house dust mites, allergies due to cockroach allergens are also a serious problem in the USA and Canada. Kang & Sulit (1978) carried out allergy skin tests in Illinois with cockroach antigen along with various common inhalant allergens on 222 atopic and on 63 non-atopic subjects. The most prevalent allergen producing a positive skin test was house dust mite antigen with a positive response of 72%, 78% and 57% in atopic adults, atopic children and non-atopic children, respectively. The next prevalent positive skin test was to cockroach antigen with 50%, 60% and 27%, respectively, of the three groups tested. Incidence of cockroach hypersensitivity was 58% among asthmatic adults and 69% among asthmatic children. The results indicate that cockroach hypersensitivity is very prevalent and that cockroach antigen is an independent agent from house dust mites as a cause of immediate hypersensitivity reaction.

A number of studies have established that cockroaches are an important and independent risk factor for allergic asthma, particularly among urban, inner city, Afro-American or Hispanic populations in the USA. The frequency of allergy to cockroaches in asthmatic patients of these populations is around 50% and is related to indoor exposure to cockroach allergens (Rosenstreich et al., 1997). Sarpong et al. (1996) observed that because cockroach populations are highest in crowded urban areas, it has been suggested that the increased asthma morbidity and mortality rates in inner cities could be related in part to cockroach allergen exposure; they examined cockroach allergen exposure in the homes of children with asthma in both urban and suburban locations and related the rates of exposure and sensitization to socioeconomic, racial and demographic factors. Based on the results of an inner-city study of 87 children with moderate to severe allergic asthma, aged 5–17 years, they concluded that African-American race and low socioeconomic status were both independent, significant risk factors for cockroach allergen sensitization in children with atopic asthma. Cockroach allergen is detectable throughout the house, including in the critical bedroom environment.

Allergy due to cockroach allergens is of somewhat lesser importance than that due to house dust mites and is more restricted in its occurrence. Cockroach allergen is more common in homes of relatively low socioeconomic status than in homes of relatively high status, whereas the opposite is true for cat allergen. This was seen in a study carried out among 499 families in Boston, Massachusetts by Kitch *et al.* (2000) which has been described above.

The public health importance of cockroach allergies

It is now realized that the public health importance of cockroach allergens is very considerable and second only to that of dust mite allergens. Cockroach allergens are much more linked to socioeconomic conditions than those of house dust mites and are an especially severe problem in provoking asthma and allergies in the inner cities of the USA and Canada. Homes in the high-poverty regions of a large metropolitan area are more likely to have high levels of cockroach allergen and, conversely, less likely to have high levels of dust mite or cat allergens. Well-conducted cockroach control will reduce allergens from this source and cockroach control in homes, not to mention food preparation establishments, should be assiduously supported.

33

Factors augmenting the incidence, prevalence and distribution of vector-borne diseases in the USA and Canada

From the evidence above, it can be seen that both in the USA and Canada, the incidence of many of the vector-borne diseases has been mounting and infections have often spread to geographical areas from which they have not previously been reported. It is important to understand the reasons why this is occurring as a basis for controlling these infections and attempting to prevent their further spread. In addition, new diseases are emerging, sometimes after having been introduced from elsewhere as has been the case with West Nile virus. In other cases, the emerging diseases may have been long-endemic and only recently recognized. The emergence may also be a consequence of changing ecological conditions. There are also instances of new clinical symptoms or conditions which have been reported or an increased virulence among infections long endemic.

The effect of ecological changes

Obviously, substantial urban and rural ecological changes have occurred in the USA and Canada, particularly since the mid-part of the twentieth century. These changes have included increases in human population densities in cities and, more recently, significant population movements to suburban areas resulting in greater exposure to vector populations. As in Europe, increased leisure time and the manner in which it is being used has increased exposure to vectors, especially to ticks. In most instances, the appearance of new diseases and syndromes and the resurgence of old can be associated with ecological changes that have favoured increased vector densities and exposure to them. In the USA, irrigation and other development projects, urbanization and movement to suburbs, and forestation have all resulted in changes in vector population densities that appear to have enabled the emergence of new diseases and the resurgence of old diseases. Greatly increased human travel has spread infectious agents, introducing them into areas in which they had been hitherto absent (Gratz, 1999b).

Perhaps the most dramatic example of an emerging vector-borne disease in the USA and Canada is that of Lyme disease; the mounting incidence and geographical spread of the disease is doubtlessly related to ecological changes. As Spielman (1994) stated

> This pattern of spread of Lyme disease and its vectors in the
> northeastern USA and Europe derives from the recent proliferation of
> deer, and the abundance of deer derives from the process of
> reforestation now taking place throughout the North Temperate Zone
> of the world. Residential development seems to favor small
> tree-enclosed meadows interspersed with strips of woodland, a
> 'patchiness' much prized by deer, mice, and humans. As a result,
> increasingly large numbers of people live where the risk of Lyme
> disease and babesiosis is intense. The agents of these infections that
> once were transmitted enzootically by an exclusively rodent-feeding
> vector have become zoonotic.

The possible effects of climate change

Over the past century, global mean temperature has increased by 0.7 °C and it is predicted that a further increase of 2.5 °C will occur by 2050. The possible effects of global warming on vector-borne diseases have already been considered in respect to Europe. In the USA and Canada, Patz & Reisen (2001) thought that global climate change might expand the distribution of vector-borne pathogens in both time and space, thereby exposing host populations to longer transmission seasons, and immunologically naïve populations to newly introduced pathogens. St. Louis encephalitis (SLE) viral replication and the length of the transmission season depend upon ambient temperature. Warming temperatures in the American southwest might place at risk migratory, non-immune elderly persons who arrive in early autumn to spend the winter. Warm temperatures might intensify or extend the transmission season for dengue fever. Reeves et al. (1994) estimated the possible effect if global warming occurs in California; temperatures may increase by 3–5 °C, precipitation patterns will change and the sea level may rise 1 m. The authors carried out studies on the effect of temperature changes on *Culex tarsalis*, the vector of western equine encephalitis (WEE) and SLE in two regions where the temperature differed by 5 °C. Daily mortalities of the vector increased by 1% for each 1 °C increase in temperature. At 25 °C only 5% of the *Cx. tarsalis* survived for 8 days or more, the time required for incubation of these viruses. Incubation time for the viruses shortened when temperatures were increased from 18°–25 °C. If the temperatures in the warmer regions increase by 5 °C, WEE virus may disappear but SLE would persist; WEE transmission would probably move more northwards. The geographic distribution of vector, human and animal populations would be altered with warming and North America would be more receptive to tropical vectors and diseases.

Aedes albopictus represents a potential risk for transmission of dengue and other arboviruses in the USA. The mosquito strain that has established itself in the country is the diapausing form, and it appears to have adapted well to the climate in more northern states. Now in both Asia and North America, there is a potential that with global climate change the vectors of dengue will move perhaps much further north than their current distribution.

For the time being, there does not appear to have been any obvious change in the pattern of transmission of vector-borne infections in the USA or Canada that is associated with global warming trends. Such changes have, however, occurred in some parts of Europe and may well occur in North America as well. Warmer temperatures would extend the distribution and length of seasonal activities of mosquitoes and other vectors increasing the risk of tranmission.

Vector-borne disease problems associated with introduced vectors

The introduction and successful establishment of exotic arthropod species into countries or areas where they have not been previously found, is always a cause for concern; the newly introduced species, whether mosquitoes, ticks or other groups, may turn out to be vectors or potential vectors of existing diseases and may even be more efficient vectors than indigenous species of the country into which they have been introduced. The most dramatic instance of such an occurrence followed the introduction of the African malaria vector, *Anopheles gambiae* into Brazil in 1930. The species rapidly spread and resulted in a very considerable increase in the incidence and mortality from malaria; a large-scale and very costly campaign directed by F. L. Soper and covering 54 000 km^2 finally eradicated the species from Brazil.

Species may be introduced into a new country or territory by arriving aboard a ship or an aircraft. The increase in international trade has resulted in great increases in the number of ships and aeroplanes arriving in the USA and Canada; during the fiscal year 1997 more than 83 500 vessels and 414 000 aircraft arrived at USA ports of entry from foreign locations. Some carry exotic pests such as *Aedes albopictus* and *Musca autumnalis*, both of which have become established in North America. Screwworms (*Cochliomyia hominivorax*) have been detected in infested animals and were found in the wound of a human patient airlifted to the USA from Central America (Bram & George, 2000). From at least the late 1940s, aircraft arriving from foreign countries located between 45° north and 45° south were subjected to disinfection. Planes arriving from points south of the 35th parallel were subjected to disinfection on a yearly basis. During a 13-year period ending in June 1960, about 250 000 insects, dead and alive were recovered aboard inspected aircraft. Among these insects, there were more than 20 000 mosquitoes found on aircraft arriving in the USA; 51 of the 92 species of mosquitoes found were not known to occur in the continental USA, Hawaii or Puerto Rico (Hughes, 1961).

Introduced species may spread rapidly in the territory into which they have been introduced, as has occurred following the introductions of *Ae. albopictus* and *Ae. japonicus* into the USA. Wherever *Ae. albopictus* has been established, it has become a serious pest and is a potential or actual vector of disease (Gratz, 2004).

Over the last two decades *Ae. albopictus* has spread from the western Pacific and South-East Asia to Europe, Africa and the Middle East, north and South America and the Caribbean. In 1985, breeding of *Ae. albopictus* was found for the first time in the USA in Harris County, Texas (Sprenger & Wuithiranyagool, 1986). Since that time it has spread very widely throughout eastern and central USA and in 2001 was found breeding in California where over-wintering populations have been established (Linthicum *et al.*, 2003). The species may have become established in California after arriving from China as larvae in standing water in plastic boxes containing the Lucky Bamboo plant, *Dracaena* species.

While disease transmission has not as yet been associated with *Ae. albopictus* in the USA, evidence suggests that the species may be implicated in the ecology of West Nile virus based on the isolation of the virus from the species in nature (Holick *et al.*, 2002). It is possible that *Ae. albopictus* could serve as bridge vectors from virus-infected birds. Vertical transmission of West Nile virus in the laboratory has been demonstrated on several occasions with the species but for the time being, there is no indication of *Ae. albopictus* being involved in the field transmission of this rapidly spreading arbovirus in the USA. However, as other infective agents such as Potosi virus (Francy *et al.*, 1990) and *Dirofilaria immitis* (Comiskey & Wesson, 1995) have been isolated from *Ae. albopictus* populations in the USA, the risk of the increased transmission of these, and other agents, can not be excluded. The urban-suburban distribution, aggressive behaviour and broad viral susceptibility of *Ae. albopictus* may lead to the transmission of viruses and other infective agents of known public health importance and perhaps of viruses hitherto not transmitted to humans because suitable vectors were not present.

Whatever the actual or potential role of *Ae. albopictus* as a disease vector in the USA, its introduction and rapid spread and its presence, often in high densities, has resulted in the species becoming a serious nuisance and added considerably to mosquito control costs in the country.

Aedes japonicus japonicus is an Asian mosquito that was recognized for the first time in the USA in 1998 (Peyton *et al.*, 1999). It is now established in several states east of the Mississippi River and has been reported from the state of Washington on the west coast (Roppo *et al.*, 2004). The species is suspected of being a vector of Japanese encephalitis virus to swine in northern Japan. In the laboratory *Aedes japonicus* is also a competent experimental vector of West Nile virus. West Nile virus has been detected in this species in New York State. An avid biter on humans, its introduction and presence in areas where several arboviruses are endemic risks an aggravation of their transmission.

Small, but stable, populations of *Aedes togoi*, also an Asian species, exist in the northern San Juan Island region of Puget Sound, Washington. Geological formations in this region are conducive to rock holes and support populations of this species which probably arrived as eggs or larvae in tyres imported from Asia. Between 18 May and 4 December 1986, 79 seagoing containers and their contents of 22 051 used tyres were inspected for the presence of adult mosquitoes, eggs or larvae. Of the total inspected, 5507 tyres (25%) contained significant amounts of water. No adults or eggs were found. Fifteen tyres contained mosquito larvae that were identified as *Aedes albopictus, Aedes togoi, Culex pipiens* complex, *Tripteroides bambusa* and *Uranotaenia bimaculata*. The infestation rate for all species was 6.8 infested tyres per 10 000 tyres (wet and dry) inspected. *Aedes albopictus* larvae were most frequently collected, occurring at a rate of 20 infested wet tyres per 10 000 inspected (Craven *et al.*, 1988).

Aedes bahamensis which was originally described from the Bahamas, was introduced into Florida, apparently in 1986, and has been spreading from the area in which it was first found in the southern part of the state. In the USA, larvae and pupae of *Aedes bahamensis* have been frequently found in association with immature *Aedes aegypti*. Fortunately, this species does not appear to have yet been implicated as a vector of any disease.

As has been observed above, infected mosquito vectors may be carried from one country to another by aircraft; such introductions may result in local transmission of the infective agent and there have been many instances of 'airport malaria' in Europe (Gratz *et al.*, 2000). There is less risk of airport malaria occurring in the USA than in Europe as far more direct flights arrive in Europe from Africa than arrive in the USA or Canada. There is a degree of risk for the USA in view of the numbers of mosquitoes, including anophelines, which have been found aboard aircraft arriving in the USA from malaria-endemic countries in Latin America.

Among the other exotic dipteran species which have been introduced into the USA are the oriental latrine fly (*Chrysomya megacephala*), which has been introduced to several countries including the USA, and has been reported to be the mechanical carrier of food-borne pathogens including *Shigella* and *Salmonella* species; it is an old world calliphorid species discovered in Brazil in 1977 and now is widespread in Latin America. In the USA it has become established in California and Texas (Wells, 1991).

Cases of myiasis are occasionally seen among travellers returning from tropical areas of Latin America and Africa and may be caused by the screwworm fly (*Cochliomyia hominivorax*), the tumbu fly (*Cordylobia anthropophaga*), the tropical botfly (*Dermatobia hominis*) and the black soldier fly (*Hermetia illucens.*) Most physicians have never encountered myiasis and may attribute a patient's complaints to an insect bite or skin infection that will heal without treatment. However, the diagnosis of furuncular myiasis should be considered by remembering the basic elements of this condition: recent travel history to the tropics and a sterile, persistent furuncle with sensations of movement and pain (Robert & Yelton, 2002).

Cockroaches are frequently transported from country to country aboard cargo ships or aircraft. *Blatta lateralis, Ischnoptera bilunata* and *Ischnoptera nox* are species known to have been introduced into the USA. The Surinam cockroach (*Pycnoscelus surinamensis*) was first recorded in Canada in 1938 having been introduced sometime before that date.

The most common flea species imported into the USA and Canada is *Tunga penetrans*, a flea that burrows into human skin, causing the condition called tungiasis. The parasite is not endemic in the USA or Canada but patients may present with this disease upon returning from Africa and Central and South America. Infestations with this flea may lead to secondary bacterial infections. Little information is available on any other species which may have been imported although the plague vectors *Xenopsylla cheopis* and *X. vexabilis* were probably both introduced into Hawaii along with *Rattus rattus* and *R. norvegicus* (Tomich *et al.*, 1984).

Exotic tick species and tick-borne diseases are serious threats to livestock, companion animals, wildlife and humans in the USA and there have been recurring introductions of exotic tick species into both the USA and Canada. A review of the literature and unpublished records from the US National Tick Collection on the importation of ticks from foreign lands revealed that at least 99 exotic tick species assignable to 11 genera had been either detected and destroyed at ports of entry or inadvertently imported into the USA in the last half of the twentieth century. This number included four argasid and 95 ixodid species, some of which are important vectors of agents that cause disease to both man and animals. If one includes *Aponomma, Hyalomma* and *Rhipicephalus* species, the total exceeds 100 taxa (Keirans & Durden, 2001). Humans may also return from travel with attached ticks; *Rhipicephalus simus* was removed from a patient suffering from a rickettsial infection who had just returned to the USA from Kenya (Anderson *et al.*, 1981). Ticks are commonly found on migratory birds and may be introduced from one area to another in this manner, sometimes over very considerable distances.

The risk of the establishment of introduced vector-borne diseases

The rapid speed of the dispersion of West Nile virus in the USA and Canada following the first appearance of the virus in New York in 1999, graphically illustrates the extent of risk that may be associated with an introduced vector-borne infection. Whether the virus was introduced by a viraemic human, an infected mosquito aboard an aircraft or a bird remains uncertain but the dramatic consequences of the introduction are clear and the virus is now endemic in both the USA and Canada. It is prudent to consider what other vector-borne diseases might be introduced into North America and what factors could lead to their establishment. After the terrorist attacks in New York and Washington, DC in 2001, the possibility that such introductions may be the result of a deliberate attack must also be considered; a number of vector- and rodent-borne diseases are high among the infectious agents which could be used in such bioterrorism.

Whatever the manner of their introduction, whether intentional or unintentional, the consequences can be dramatic if conditions in the recipient country are suitable for indigenous transmission of the infective agent as was obviously the case for WNV. In 2002, WNV virus caused the largest recognized epidemic of neuroinvasive arboviral illness in the Western hemisphere and the largest epidemic of neuroinvasive WNV ever recorded.

The presence of a competent indigenous local vector or vectors and maintenance reservoir host(s) are the most obvious factors which will determine whether a given, introduced, infective agent may become established. In the case of WNV in the USA and Canada, many different species of mosquitoes have been found to be actual or potential vectors of the virus. Turell *et al.* (2005) in a summary review of the north American vectors of WNV reported that all the *Culex* species from the USA that were tested by his group were competent vectors in the laboratory and varied from highly efficient vectors (e.g. *Culex tarsalis*) to moderately efficient ones (e.g. *Culex nigripalpus*). Virtually all of the *Culex* species tested could serve as efficient enzootic or amplifying vectors for WNV. Several container-breeding *Aedes* and other *Aedes* species were highly efficient vectors under laboratory conditions, but because of their feeding preferences, would probably not be involved in the maintenance of WNV in nature but could be potential bridge vectors between the avian-*Culex* cycle and mammalian hosts. Most of the pool-breeding *Aedes* species tested were relatively inefficient vectors under laboratory conditions and would probably not play a significant role in transmitting WNV in nature. The authors emphasized that in determining the potential for a mosquito species to become involved in transmitting WNV, it is necessary to consider not only its vector competence in the laboratory but also its abundance, host-feeding preference and its involvement with other viruses with similar transmission cycles. This observation is equally applicable to any investigation on vector potential.

The extent to which there is a risk that an introduced infectious agent may become established will obviously vary with the disease and the potential vectors present in the area of introduction. As an example, a number of potential *Anopheles* vectors of malaria are present in the USA and Canada; there have been many instances of autochthonous transmission of malaria by local vectors that had become infected after feeding upon malaria-infected persons who had returned from travel to malaria-endemic areas. However, the outbreaks have been limited by the early diagnosis and treatment of infected persons following epidemiological investigations to determine the source of the parasites.

When no suitable vector is present to transmit an introduced vector-borne disease the health risk is limited to the infected individual. Risks of epidemic outbreaks in the USA and Canada are constrained by the highly developed public health systems in both countries that will usually succeed in quickly detecting and reporting the presence of an exotic infection and then take appropriate measures to deal with it.

The greatest threat remains the possible introduction of an arbovirus not endemic in North America; the introduction and rapid spread of West Nile virus shows the degree of risk which may follow the introduction of an exotic infectious agent when competent local vectors are widely present. Small outbreaks of dengue have occurred in Texas in areas bordering dengue-endemic regions in Mexico and the disease is a serious problem in Puerto Rico. The potential vectors of dengue in the USA, *Aedes aegypti* and *Ae. albopictus*, are widespread, particularly in the southern states and introduction of the virus by viraemic returning travellers could lead to outbreaks in other states. For the time being, the sporadic, autochthonous cases of dengue in the continental USA have been limited to Texas and the outbreak in Hawaii. The infection is frequently found in travellers returning from dengue-endemic areas of the world including Puerto Rico.

There are many exotic pathogens whose introduction into the USA or Canada might have a serious detrimental impact on livestock, agricultural economy and public health. Many of these pathogens are arthropod-borne and potential vectors are present in both countries. An infection that poses a serious risk of being introduced into the USA and Canada is Rift Valley fever virus (RVF), a phlebovirus of the Bunyaviridae family. The virus can be transmitted from animal to animal or from animals to humans by mosquitoes. Transmission can also occur by contact with body fluids and during the butchering of infected animals. Infection may cause serous illness and death in humans and approximately 1% of humans that become infected with RVF die of the disease. Case-fatality proportions are significantly higher for infected animals. The most severe impact is observed in pregnant livestock infected with RVF, which results in abortion of virtually 100% of fetuses. Although vaccines against this virus are available, their use is limited because of deleterious effects or incomplete protection.

Rift Valley fever has been the cause of several very large epizootics and epidemics with a high morbidity and mortality among animals and humans. It has spread from southern Africa and widely throughout the continent to West Africa, Kenya, Zimbabwe, Zambia, Somalia, the Sudan, Tanzania, Egypt, Madagascar and recently has spread from the African continent to Saudi Arabia and the Yemen in 2000 with subsequent outbreaks in 2001 and 2004/2005. The infection spread to Egypt in 1977, probably imported with animals from Sudan, and was the cause of at least 18 000 human cases and some 600 deaths; a second, smaller outbreak occurred in 1978 with subsequent outbreaks with human infections in 1986 and 1993. An epizootic occurred in Egypt in 1997.

The epidemic which broke out in Saudi Arabia and Yemen in 2000–2001 was the first reported outside the African continent or Madagascar. In Saudi Arabia there were a reported 886 human cases with a mortality rate of 13.9%. During the same period in Yemen, well over 1000 human cases were estimated to have occurred with 121 deaths. Some 89 000 human cases were estimated to have occurred in an outbreak in East Africa in 1997; the biological features of RVF appear to be different with each outbreak and the virus has been successful in spreading and adapting to different ecological conditions and has a significant potential of further emergence (Sall *et al.*, 1998).

Rift Valley fever could be introduced into North America by an infected mosquito or by an infected animal; it is thought that infected animals were the means by which the virus was introduced into Egypt and across the Red Sea into Saudi Arabia. In the Arabian Peninsula, the virus was transmitted by *Aedes vexans arabiensis* and *Culex tritaeniorhynchus*, both of which were found infected in nature and transmitted the virus in the laboratory (Jupp *et al.*, 2002).

Several studies have been carried out on the vector-potential of North American species of mosquitoes and their ability to transmit this virus in the laboratory. Gargan *et al.* (1988) evaluated field populations of *Aedes canadensis, Aedes cantator, Aedes excrucians, Aedes sollicitans, Aedes taeniorhynchus, Aedes triseriatus,*

Anopheles bradleyi-crucians, Culex salinarius, Culex tarsalis, and *Culex territans*. While all species, with the exception of *Anopheles bradleyi-crucians* transmitted virus, *Aedes canadensis, Aedes taeniorhynchus* and *Culex tarsalis* had the highest vector potential of the species tested. Following inoculation of approximately 10 plaque-forming units of virus, 100% of the mosquitoes of each species became infected. Turell *et al.* (1988) carried out transmission studies on a strain of *Aedes albopictus* from Houston, Texas and concluded that it is a potential vector of RVF if the virus were introduced into the USA. The results of these studies is a cause of concern; following the introduction of WNV into the USA and Canada a large number of native species have been found to be efficient vectors of this introduced virus and had a role in the transmission of the virus in the field. Should RVF be introduced into North America, it might be as readily transmitted by native species as was the case with WNV with even more serious consequences for human and animal health.

Among the other arboviral diseases that would pose a threat to public health were they to be introduced and established in North America, are Ross River virus and Japanese encephalitis. Laboratory transmission studies have shown that a number of North American mosquito species could transmit both of these infections were they to be introduced.

Whether local transmission of an introduced exotic infectious agent occurs depends on many factors in addition to the presence of potential vectors. Nevertheless, the risk of introduction of exotic vector-borne infections is a very real one and this possibility has been dramatically illustrated by the introduction and widespread transmission and spread of WNV as well as by the earlier incursions of Venezuelan equine encephalitis into the USA. It is essential that effective surveillance systems are in place which will enable imported infections to be rapidly detected and effective control measures to be rapidly undertaken to prevent spread of such introduced infectious agents. Infectious agents know no borders and once pathogens cross their known geographic limits, they tend to adapt to the local ecology in order to survive and maintain transmission (Fagbo, 2002).

34

The rodent-borne diseases of the USA and Canada

As was with Europe, only those infectious diseases transmitted directly from rodents to humans will be considered below.

The hantaviruses

The distribution of the hantaviruses has been described in the section dealing with this group of infections in Europe. Hantaviruses are a diverse group of RNA arboviruses in the Bunyaviridae family and are parasites of small mammals, predominantly peridomestic and commensal rodents. In most cases human infection is manifested by one of a variety of acute illnesses involving haemorrhagic fever and renal disease. Adult rodents show persistent infection without any clinical manifestations and secrete virus for prolonged periods. Following inoculation, a viraemia develops in the rodent during which the virus is disseminated throughout the body. The virus is found in the lungs, spleen and kidneys for long periods in the rodent, perhaps for life. Saliva appears to play an important role in the horizontal transmission of the virus between rodents.

Man is probably infected through the respiratory route via aerosols of virus particles excreted by rodents in their lungs, saliva, urine and faeces. Transmission has also been documented following bites by rodents. Horizontal transmission among humans has not been documented, although blood and urine are infectious during the first 5 days of illness. It is generally accepted that arthropods are not involved in transmission of the viruses between rodents and from rodents to man. Many hantavirus isolates have been isolated from humans and rodent hosts and are typed according to their serological cross-reactivity. There are at least 14 subtypes.

The viruses in the genus *Hantavirus* are shown in Table 34.1.

Table 34.1 *Viruses of the genus* Hantavirus, *family Bunyaviridae*

Virus	Abbr.	Original source	Location	Geographic distribution of rodent host[a]	Human disease	Isolated in cell culture
Murinae subfamily-associated virus						
Hantaan	HTN	*Apodemus agrarius*	Korea	Asia, Europe	HFRS	yes
Seoul	SEO	*Rattus norvegicus, R. rattus*	Korea	Asia, Europe, the Americas	HFRS[b]	yes
Dobrava-Belgrade	DOB	*Apodemus flavicollis*	Slovenia	Europe, Middle East	HFRS	yes
Thai-749	THAI	*Bandicota indica*	Thailand	Asia	unknown	yes
Arvicolinae subfamily-associated viruses						
Puumala	PUU	*Clethrionomys glareolus*	Finland	Europe, Asia	HFRS	yes
Prospect Hill	PH	*Microtus pennsylvanicus*	Maryland	N. America	unknown	yes
Tula	TUL	*Microtus arvalis*	Russia	Asia	unknown	yes
Khabarovsk	KBR	*Microtus fortis*	Russia	Asia	unknown	yes
Topography	TOP	*Lemmus sibiricus*	Siberia	Russia, Asia, N. America	unknown	yes
Isla Vista	ISLA	*Microtus californicus*	California	N. America	unknown	no
Sigmodontinae subfamily-associated viruses						
Sin nombre	SN	*Peromyscus maniculatus*	New Mexico	N. America	HPS[c]	yes
New York	NY	*Peromyscus leucippus*	New York	N. America	HPS	yes
Black Creek Canal	BCC	*Sigmodon hispidus*	Florida	The Americas	HPS	yes
Bayou	BAY	*Oryzomys palustris*	Louisiana	South eastern United States	HPS	yes
Cano Delgadito	CANO	*Sigmodon alstoni*	Venezuela	S. America	unknown	yes
Rio Mamore	RM	*Oligoryzomys microtis*	Bolivia	S. America	unknown	yes
Laguna Negra	CHP	*Calomys laucha*	Paraguay	S. America	HPS	yes
Muleshoe	MULE	*Sigmodon hispidus*	Texas	The Americas	unknown	no

Table 34.1 *(Cont.)*

Virus	Abbr.	Original source	Location	Geographic distribution of rodent host[a]	Human disease	Isolated in cell culture
El Moro Canyon	ELMC	*Reithrodontomys megalotis*	California	N. America	unknown	no
Rio Segundo	RIOS	*Reithrodontomys mexicanus*	Costa Rica	Mexico, Central America	unknown	no
Andes	AND	*Oligoryzomys longicaudatus*	Argentina	S. America	HPS	yes
Insectivore-associated virus Thottapalayam	TPM	*Suncus murinus*	India	Asia	unknown	yes

[a]Given as approximate distribution: many rodent species occur focally, while many others have widespread distributions.

[b]HFRS, haemorrhagic fever with renal syndrome.

[c]HPS, hantavirus pulmonary syndrome.

The prototypic New World hantavirus is Prospect Hill virus (PHV) of the meadow vole *Microtus pennsylvanicus*. It has not been associated with human disease, but is widespread among voles in the USA and Canada (Yanagihara *et al.*, 1987). After the observations that *Rattus* species in Asia harbour hantaan-like viruses associated with haemorrhagic fever with renal syndrome, Lee *et al.* (1982) examined sera from rodents captured in Alaska, California and Virginia; samples from wild meadow voles (*M. pennsylvanicus*) trapped in Virginia and Alaska, from California voles (*M. californicus*) trapped in California and from the red-backed vole (*Clethrionomys rutilis*) from Alaska were found to contain antibody which appeared to be identical to the agent causing Korean haemorrhagic fever.

LeDuc *et al.* (1984) carried out investigations to clarify the possibility that similar viruses might be present in the USA. Rats were captured at major port cities, their sera were examined for hantaviruses and tissues from selected antibody-positive rats were examined for hantavirus antigen. Rats positive for both antibody and antigen to HTN virus were found in Philadelphia, Pennsylvania, and Houston, Texas. Infected rats were found clustered in discrete foci where a significant proportion was antibody and antigen positive. Viruses isolated from lung tissues of Norway rats (*Rattus norvegicus*) captured at Philadelphia and Houston were shown to be closely related to each other.

By 1990, three antigenically distinct hantaviruses had been isolated from Norway rats (*Rattus norvegicus*), house mice (*Mus musculus*), and meadow voles (*Microtus pennsylvanicus*) in the USA and serologic evidence of a hantavirus enzootic has been found in several other indigenous rodent species. Although hantaviruses had been isolated from rodents in the USA and serological studies have documented human infections with hantaviruses, acute disease associated with infection by pathogenic hantaviruses was not reported in the Western hemisphere. Seoul virus was identified in Korea as a hantaan-like virus whose natural urban host is the Norway rat. Serologic surveys have detected its presence worldwide, including in the USA; seroprevalence rates of 12% were detected in urban rats in Philadelphia and of about 44% in Norway rats (*R. norvegicus*) in Baltimore with viruses that are serologically indistinguishable from disease-causing hantavirus strains isolated from rats in the Far East (Diglisic *et al.*, 1999). Infection with human disease caused by Seoul virus is largely restricted to Asia. Only three suspected clinical cases have been reported in the USA. Overall mortality associated with Seoul virus infection is probably <1%. Despite the widespread distribution of hantavirus-infected rodents, confirmed cases of haemorrhagic fever with renal syndrome had not been recognized in the USA at the time. Moreover, the overall risk of hantavirus infection in humans in the USA was considered low, even among individuals who have frequent exposure to commensal and wild rodents (Yanagihara, 1990).

In 1993, a previously unknown human infectious disease was recognized in the western USA. A cluster of cases of an illness characterized by fever, chills and severe myalgia was noted in rural Navajo Indian residents of the Four Corners region of the USA, where the states of Colorado, New Mexico, Utah and Arizona share a common border. The infection mainly affected the lungs and had a fatality rate of 60% or more. The causative agent was identified and partially characterized. It was shown to be a newly recognized and genetically distinct hantavirus later designated as sin nombre virus (also previously called Muerto Canyon virus) and was named Hantavirus pulmonary syndrome (HPS).

In 1993 there were a total of 52 confirmed cases of which 32 died; the groups most at risk were American Indians who represented half the cases and 42% were non-hispanic whites (CDC, 1993). The oral history of local American Indian healers describes clusters of similar deaths occurring over three cycles during the twentieth century in association with identifiable ecological markers. The abrupt introduction to western medical practitioners of a disease long recognized by indigenous healers through illness occurring among a cohort of patients seeking care from medical officers parallels western medical recognition of previous human illnesses associated with hantaviral infections through disease outbreaks among military troops (Chapman & Khabbaz, 1994). Nevertheless, the outbreak was surprising as serious disease due to a hantavirus had not been previously reported in the USA.

Sin nombre virus (SNV) is unique but genetically similar to the non-pathogenic Prospect Hill serogroup found in North American rodents for decades. Several other serotypes that cause Hantavirus pulmonary syndrome (HPS) have subsequently been recognized. Hantavirus pulmonary syndrome has turned out to be a newly identified, but not a new disease. In fact, the earliest case of a serologically confirmed SNV infection was in a person who developed an HPS-compatible illness in July 1959 and was found to have IgG antibodies in September 1994 (Frampton *et al.*, 1995). The earliest case of HPS was a 38-year-old Utah man who had died from an illness compatible with hantavirus in 1959. Researchers located his lung tissue and were able to isolate SNV in 1994. The earliest case diagnosed by immunohistochemistry in postmortem tissues was from a patient who died in 1978.

Sin nombre virus was first isolated from rodents collected on the premises of one of the initial HPS patients in the Four Corners region. Additional viral strains which were likely the cause of HPS have also been isolated from the deer mouse (*Peromyscus maniculatus*) associated with a fatal case in California and from the white-footed mouse (*P. leucopus*) from the vicinity of probable infection of a case in New York. Black Creek Canal virus was isolated from cotton rats (*Sigmodon hispidus*) collected near the residence of a human case in Dade County, Florida. Closely related strains (Bayou and Black Creek Canal viruses), found in the southeastern USA, may produce a variant of the syndrome that is characterized by a greater degree of renal failure. The New York virus is the cause of cases of HPS in New York and Rhode Island and is genetically distinct from sin nombre virus. Bayou virus has been reported in Louisiana and Texas and is carried by the marsh rice rat (*Oryzomys palustris*). Blue River virus has been found in Indiana and Oklahoma and seems to be associated with the white-footed mouse. Monongahela virus, discovered in 2000, has been found in Pennsylvania and is transmitted by the white-footed mouse.

Shortly after the appearance of cases in the Four Corners region in 1993, cases of HPS were reported from Louisiana and Florida; by mid 1994, 83 cases and 42 deaths were reported (54% case fatality rate) with 96% of the cases having been identified east of the Mississippi River (CDC, 1994b). In early 1994, two cases of HPS probably acquired in California were also the first two confirmed cases in that state. Both persons died; both cases were in rural areas and both had been in contact with rodents.

The CDC reported that a total of 387 cases of HPS had been reported in the USA by 3 May 2005. The case count started when the disease was first recognized in May 1993. Thirty-six per cent of all reported cases have resulted in death. Cases have been confirmed in 31 states. Of persons ill with HPS, 62% have been male, 38% female. The mean age of confirmed case patients is 38 years (range: 10–83 years).

In Canada cases have been reported from British Columbia, Alberta, and Saskatchewan. Hantavirus pulmonary syndrome was made a nationally notifiable disease as of 1 January 2000; provincial and territorial public health authorities had previously reported confirmed cases. The first case of HPS recognized in Canada during active surveillance was in 1994 in British Columbia. Subsequently cases have been identified retrospectively with the earliest case dating back to 1989 in Alberta (CCDR, 2000). Since 1989, there have been 31 cases of hantavirus in Alberta, and 9 deaths. In May 2005 there were three cases in central Alberta including a woman, who died, the first fatality in the province since 2002. About 57 cases have been reported in Canada, with at least 19 deaths. All of the cases have occurred in the four western provinces. Infected mice have been found in every province except Nova Scotia and Prince Edward Island, suggesting that the potential exists for human hantavirus cases to emerge in other parts of the country. In Quebec a case was contracted in a forest just north of Trois-Rivieres in 2004 and confirmed by Health Canada.

Rodent reservoir hosts

The striped field mouse (*Apodemus agrarius*) and the Norway rat (*Rattus norvegicus*) are members of the family Muridae. The majority of hantavirus isolates have come from these rodents. Older classifications of rodents recognize the families Muridae and Cricetidae. All of the New World rodents that have been identified as vectors of HPS are confined within the subfamily Cricetinae but more recent classification recognizes Cricetinae as a subfamily of the Muridae. This family includes voles, hamsters, gerbils, deer mice and harvest mice, and has members that are hosts to Puumala virus, Four Corners virus, harvest mouse hantavirus, Black Creek Canal virus and Prospect Hill virus.

Phylogenetic analysis of Old World and American hantaviruses indicates that the relationship among hantaviruses corresponds with the phylogeny of their rodent hosts. Viruses of rodents belonging to the subfamily Murinae are monophyletic as are hantaviruses of arvicoline and sigmodontine rodents. Viruses found in Old World murine rodents, including hantaan virus (HTNV), Seoul virus (SEOV) and Dobrava virus (DOBV), are associated with haemorrhagic fever with renal syndrome in Eurasia. Viruses carried by New World sigmodontine rodents, including sin nombre virus, Black Creek Canal virus and Bayou virus, are associated with hantavirus pulmonary syndrome in the Americas.

Hantaviruses do not cause overt illness in their reservoir hosts. Although infected rodents shed virus in saliva, urine and faeces for many weeks, months or for life, the quantity of virus shed can be much greater approximately 3–8 weeks after infection. The demonstrated presence of infectious virus in saliva of infected rodents and the marked sensitivity of these animals to hantaviruses following intramuscular inoculation suggest that biting might be an important mode of transmission from rodent to rodent. Field data suggest that transmission in host populations occurs horizontally, more frequently among male rodents, and might be associated with fighting, particularly, but not exclusively, among males (Hinson *et al.*, 2004).

The deer mouse (*Peromyscus maniculatus*) was implicated as the host for the virus causing HPS in New Mexico, Colorado and Arizona (Nichol *et al.*, 1993, Childs *et al.*, 1994). Black Creek Canal virus, which was recovered from the cotton rat (*Sigmodon hispidus*) in Florida, represents a new hantavirus distinct from the previously known serotypes (Rollin *et al.*, 1995). The same species was also found infected with Muleshoe virus in Texas, a virus closely related to Bayou hantavirus. In the same area, El Moro Canyon virus was detected in western harvest mice (*Reithrodontomys megalotis*) and in deer mice (*P. maniculatus*) (Rawlings *et al.*, 1996). Antibody to sin nombre virus has also been found in the marsh rice rat (*Oryzomys palustris*) and in the chipmunk (*Tamias minimus*). Monongahela virus has been isolated from the white-footed mouse (*Peromyscus leucopus*) in New York. Brush mice (*P. boylii*), and cactus mice (*P. eremicus*) have been found seropositive for sin nombre virus in Arizona (Kuenzi *et al.*, 1999). The Pinyon mouse (*P. truei*) is among the reservoir hosts in Arizona, California, Colorado and Nevada.

Peromyscus maniculatus is the primary rodent vector for the sin nombre virus, and was the source of the first recognized epidemic of HPS in the southwest USA in 1993; in view of the large number of rodent species in the endemic areas and their high population densities, it seems likely that other rodent hosts will be found and possibly other causative viruses of HPS.

Conclusions on the public health importance of the hantaviruses

Although the number of reported cases and deaths due to HPS is not great, the infection must be considered as one of considerable public health importance due to the severity of the disease and the high mortality among patients. Hantavirus pulmonary syndrome is now known to have a wide distribution in the USA and Canada and a number of countries in Latin America. Cases of HPS have been confirmed in South and Central America in the following countries: Argentina, Bolivia, Brazil, Chile, Paraguay, Panama and Uruguay.

The CDC has reported that rodents carrying hantavirus have been found in at least 20 national parks and that it is possible that the virus is in all of the parks in the USA. However, most people who have been exposed have come into contact with rodent droppings in their own homes. An effective treatment for hantavirus is not yet available and even with intensive therapy, over 50% of the diagnosed cases have been fatal. The list of distinct hantaviruses associated with HPS is growing. The burgeoning human population is causing disruption of natural habitats as more and more land is cleared for commercial and residential purposes. Many rodents readily adapt to life in human settlements, where they generally benefit from reduced predation and where they sometimes proliferate to high numbers. Although often referred to as emerging pathogens, HPS-associated hantaviruses emerge through increased exposure of humans to rodents and their excreta, not through genetic drift or reassortment of the viral genome. Based on current human population growth and development trends, hantavirus diseases will become more common in the near future unless public health measures are taken to curtail or eliminate rodents from human communities (Lednicky, 2003).

Arenaviruses

Arenaviruses cause chronic infections in rodents and zoonotically acquired disease in humans. In 1934, lymphocytic choriomeningitis virus (LCMV) was the first *Arenavirus* isolated in the USA during serial monkey passage of human material that was obtained from a fatal infection during the first epidemic of St. Louis encephalitis. Lymphocytic choriomeningitis virus was the first recognized cause of aseptic meningitis in humans. Other arenaviruses from South America and Africa are classic causes of viral haemorrhagic fever syndrome. Each virus is associated with either one species or a few closely related species of rodents, which constitute the virus' natural reservoir host.

Arenaviruses have been divided into two groups based on whether the virus is found in the Eastern hemisphere or the Western hemisphere; LCMV is the only arenavirus to exist in both areas. Of the 15 known New World arenaviruses, four (Junin, Machupo, Sabia and Guanarito) have been associated with haemorrhagic fever in humans. The arenaviruses known to occur in North America are lymphocytic choriomeningitis virus (LCMV), Tamiami virus (TAM) and Whitewater Arroyo virus (WWA). The LCM virus was doubtlessly introduced into the Americas along with its principal rodent host, the house mouse (*Mus musculus*); TAM virus is known only from the cotton rat (*Sigmodon hispidus*) in southern Florida. Whitewater Arroyo virus was originally recovered from the white-throated wood rat (*Neotoma albigula*), collected from northwestern New Mexico.

Lymphocytic choriomeningitis

The following information is taken, in part, from the website of the Special Pathogens Branch of the CDC in Atlanta, Georgia, USA (http://www.cdc.gov/ncidod/dvrd/spb/mnpages/dispages/lcmv.htm accessed 29 April, 2005). As has already been described, lymphocytic choriomeningitis, or LCM, is a rodent-borne viral infectious disease that presents as aseptic meningitis (inflammation of the membrane, or meninges, that surrounds the brain and spinal cord), encephalitis (inflammation of the brain), or meningoencephalitis (inflammation of both the brain and meninges). Its causative agent is the lymphocytic choriomeningitis virus (LCMV). Although LCMV is most commonly recognized as causing neurological disease, as its name implies, asymptomatic infection or mild febrile illnesses are common clinical manifestations. Additionally, pregnancy-related infection has been associated with abortion, congenital hydrocephalus and chorioretinitis and mental retardation.

Lymphocytic choriomeningitis infections have been reported in Europe, the Americas, Australia and Japan, and may occur wherever infected rodent hosts of the virus are found. However, the disease has historically been under-reported, often making it difficult to determine incidence rates or estimates of prevalence by geographic region. Several serologic studies conducted in urban areas have shown that the prevalence of LCMV infection among humans ranges from 2–10%.

Humans become infected by inhaling infectious aerosolized particles of rodent urine, faeces or saliva, by ingesting food contaminated with virus, by contamination of mucous membranes with infected body fluids, or by directly exposing cuts or other open wounds to virus-infected blood. Infection has also been documented among staff handling infected hamsters. Person-to-person transmission has not been reported, with the exception of vertical transmission from an infected mother to foetus.

Lymphocytic choriomeningitis virus is naturally spread by the common house mouse (*Mus musculus*). Once infected, these mice can become chronically infected by maintaining virus in their blood and/or persistently shedding virus in their urine, a common characteristic of other arenavirus infections in rodents. Chronically infected female mice usually transmit infection to their offspring, which in turn become chronically infected. However, hamsters are also a host of the virus. An outbreak of LCM infections occurred among medical centre personnel at the University of Rochester, New York, in 1972–1973. A total of 48 infections was discovered – the source of the infection was Syrian hamsters (*Mesocricetus auratus*) which were used in tumour research (Hinman *et al.*, 1975). Several outbreaks have occurred in the USA among the personnel and clients of pet stores selling hamsters.

The geographical distributions of the rodent hosts are widespread both domestically and abroad. However, infrequent recognition and diagnosis, and therefore under-reporting, of LCM, have limited ability to estimate incidence rates and prevalence of disease among humans. Understanding the epidemiology of LCM and LCMV infections will help to further delineate risk factors for infection and develop effective preventive strategies. Increasing physician awareness will improve disease recognition and reporting, which may lead to better characterization of the natural history and the underlying immunopathological mechanisms of disease, and stimulate future therapeutic research and development.

In a survey in Baltimore between 1986–1988, 1180 sera were examined from persons visiting a sexually transmitted disease clinic in an inner-city area; antibodies to LCMV were detected in 54 individuals (4.7%). Self-reported human–rodent contact indicated more exposure to house mice (*Mus musculus*) than rats, most likely, *Rattus norvegicus*, within residences, although rats were more commonly sighted on streets (Childs *et al.*, 1991). Among 480 mice (*M. musculus*) trapped in urban sites in Baltimore, Maryland from 1984–1989, 9.0% of them were LCMV antibody positive; prevalence of antibody among the mice varied from 3.9–13.4% and the location with the highest prevalence was an inner city residential site (Childs *et al.*, 1992). In cities where environmental conditions have improved, the prevalence of LCM has fallen.

The public health importance of lymphocytic choriomeningitis

Although the yearly number of cases of LCM in the USA and Canada is relatively small, the infection is an under-diagnosed foetal teratogen; the public and health-care professionals should be aware of the hazard that wild, pet and laboratory rodents pose to pregnant women. Cases are not rare and many are almost certainly not recognized and the disease is probably under-reported.

Whitewater Arroyo virus

Three isolates of a novel arenavirus were collected from two arenavirus antibody-positive white-throated wood rats (*Neotoma albigula*) collected from Whitewater Arroyo in McKinley County, New Mexico. Serologic tests indicated that Whitewater Arroyo virus (WWA) is antigenically distinct from other arenaviruses but most closely related to Tamiami virus (Fulhorst *et al.*, 1996).

Infection with this virus was found to be the cause of death of a small number of patients in California and Texas. Studies indicate that the southern plains wood rat (*Neotoma micropus*) is a principal host of WWA virus in southern Texas. Analyses of viral gene sequence data revealed substantial genetic diversity among WWA virus strains isolated from the wood rats, suggesting that multiple variants of the virus can coexist in a single wood rat species in a small geographic area. The virus has been recovered from *N. albigula* in New Mexico and Oklahoma, *N. cinerea* in Utah, *N. mexicana* in New Mexico and Utah, and *N. micropus* in Texas. The present-day association of WWA virus with the genus *Neotoma* represents a long-term shared evolutionary relationship between virus and rodent host; the geographic range of the virus may extend far beyond the southwestern USA (Fulhorst *et al.*, 2001). So far it is known to be present in California, Colorado, Oklahoma, Texas and Utah.

Only a small number of cases of WWA have been recognized; 3 cases were reported from California between June 1999 and May 2000; 2 patients lived in southern California and 1 lived in the San Francisco Bay area. All 3 cases were fatal. It is found in North America among wood rats (*Neotoma* species) and has not previously been known to cause disease in humans. Of 20 *Neotoma* with species status, 9 occur in the USA. The geographic range of these species incorporates most of the country and at least 5 of the 9 species may harbour the virus; however, complete description of its distribution requires further study. The abundance and habits of wood rats suggest that potential contact between *Neotoma* spp. and humans is limited (CDC, 2000c).

Tamiami virus

Tamiami virus (TAM) is a member of the Tacaribe complex of arenaviruses. It was isolated from hispid cotton rats (*Sigmodon hispidus*) in the Everglades National Park in south Florida and is the only Tacaribe complex arenavirus known to occur in North America (Calisher *et al.*, 1970). The virus is antigenically and genetically most closely related to Whitewater Arroyo virus. *Sigmodon hispidus* seropositive for TAM virus have been found only in Florida. In 1996, Kosoy *et al.* examined the prevalence of arenavirus antibodies in rodents from the USA. They found wood rats (*Neotoma* species) seropositive for Tamiami virus antigen from several localities in the southwest, including San Diego and Ventura counties in southern California. They also demonstrated that no other species of rodent in the southwestern USA was seroreactive to Tamiami virus antigen.

The virus induces severe disease and death in suckling mice but does not appear to cause disease in humans in the USA.

Leptospirosis

The overall aetiology and epidemiology of leptospirosis have already been described under the section dealing with this infection in Europe.

Leptospirosis is a spirochaetal infection caused by contact with an infected animal's urine or tissue or indirectly by contact with contaminated water or soil. It is also the cause of Weil's disease, a severe form of leptospirosis. Rats are the most common reservoir hosts for this spirochaetal zoonosis, although farm animals and livestock can also harbour the infection and in the USA, dogs and occasionally cats may also carry it. In humans, leptospirosis is caused by *Leptospira interrogans*, a species with some 250 serovars. *Leptospira ballum* and *L. icterohaemorrhagiae* are associated with rats and mice and *L. canicola* is associated with dogs. The clinical course of leptospirosis is highly variable, and severity depends on the serovar. In 85–90% of patients the disease is self-limited and mild. The first case in the USA was reported in 1907, in Louisiana. Urban residential exposure is now on the rise, most notably in crowded inner city locations with rat infestations.

The reported incidence of leptospirosis is 100–200 cases per year in the USA with most (50–100 cases) occurring outside the continental USA in Hawaii where it was recognized as an occupational disease of sugar plantation workers at the beginning of the twentieth century. Leptospirosis is likely under-diagnosed in the USA, with reported incidence depending largely upon clinical index of suspicion. The true incidence is unknown because leptospirosis is not nationally notifiable in Canada or the USA.

In Hawaii the Norway rat (*Rattus norvegicus*), had the highest infection rate, 33.3%, and the predominant (72.2%) organism in these infections was *Leptospira icterohaemorrhagiae* in a survey carried out in the early 1970s (Higa & Fujinaka, 1976).

Sporadic outbreaks of leptospirosis in the continental USA have occurred in the East, Midwest and in Texas in the last decade and recently in California where the overall incidence of leptospirosis appears to be on the rise (Meites *et al.*, 2004). Human cases and cases in dogs appear to be increasing in northeastern USA. Human leptospirosis infection is probably considerably more important than is generally recognized in inner-city areas; *L. interrogans* was present in 19 of 20 Norway rats (*R. norvegicus*) trapped in alleys in Baltimore, Maryland, where the patients may have acquired *L. interrogans* (Vinetz *et al.*, 1996). Leptospirosis is a re-emerging infectious disease in California where it has been a reportable disease since 1922. From 1982 to 2001, most reported California cases occurred in previously healthy young adult white men after recreational exposures to freshwater contaminated by cattle urine infected by leptospires.

The actual incidence of leptospirosis in the USA and Canada is unknown since, as noted above, it is not a reportable disease on a national level. In 1995, the Council of State and Territorial Epidemiologists and the CDC removed leptospirosis from the US list of notifiable diseases. Since reliable diagnostic testing was not readily available and organized reporting had not resulted in implementation of methods to control the disease, many states have stopped reporting incidences of leptospirosis.

Occupational exposure probably accounts for 30–50% of human cases. The main occupational groups at risk include farm workers, veterinarians, pet shop and dog owners, field agricultural workers, abattoir workers, plumbers, meat handlers and slaughterhouse workers, coal miners, workers in the fishing industry, military troops, milkers and sewer workers. To an increasing extent, recreational exposure has become important during canoeing, hiking, kayaking, fishing, swimming, wading, riding trail-bikes through puddles, white-water rafting and other outdoor sports played in contaminated water. As has been remarked above, incidence is increasing among inner-city inhabitants exposed to rodent urine.

The public health importance of rodent-borne leptospirosis

Inasmuch as there appears to be a resurgence of leptospirosis, it is a disease of public health importance. The possibility of infection should be considered among patients who may have had occupational or recreational exposure and reporting will provide more accurate information on the degree of risk and special attention should be given to the possibility of infection among inner-city inhabitants.

Rat-bite fever

The causative agent of this disease in North America is *Streptobacillus moniliformis*. Rat-bite fever is also called Haverhill fever. It is generally transmitted by rat-bite. Groups of people can occasionally be infected by contaminated food, milk or water; an outbreak in Haverhill, Massachusetts in 1926 caused by contaminated ice cream gave rise to the name Haverhill fever. Rat-bite fever is rare in the USA, occurring mostly in children and laboratory personnel exposed to infected rodents. Infection usually produces a mild, protracted illness that has either a favourable response to antibiotic therapy or spontaneously resolves. The case-fatality rate for RBF may approach 10% for untreated cases. Infection with *S. moniliformis* can cause fulminant sepsis and death in previously healthy adults and several deaths have been reported recently in the USA.

As the disease is not reportable either in the USA or Canada, no true measure of its incidence exists. Of 14 cases of RBF on record since 1958, 7 originated from the bite of laboratory rats (CDC, 1994c). Bites by wild rodents (rats, squirrels) can also transmit the infectious agent.

The rate of nasopharyngeal carriage of *S. moniliformis* by healthy laboratory rats has been reported to vary between 10–100%. In view of the likely high rates of exposure of laboratory personnel to *S. moniliformis*, three possible explanations for the rarity of diagnosis of RBF are: a true low incidence of disease in spite of common exposure, a low index of suspicion of attending physicians, and the strict growth requirements of the organism (CDC, 1994c).

Rat-bite fever is generally rare and its public health importance limited; two quite disparate groups seem to be at the greatest risk, i.e. people who handle rodents in pet stores and in laboratories and inner-city inhabitants exposed to wild rats, usually *Rattus norvegicus*.

Salmonellosis

An estimated 1.4 million persons in the USA contract salmonellosis annually leading to approximately 14 800 hospitalizations and 415 deaths; most of these cases are related to contaminated food (CDC, 2005). In addition, a common form of transmission is through food contaminated by rat or mouse faeces that contain *Salmonella* (especially *S. typhimurium*) organisms. The importance of rodents in the spreading of *Salmonella* has already been reviewed above.

Salmonella-contaminated animals and chickens are not uncommon in the USA and Canada. Most outbreaks of salmonellosis in the USA and Canada are food-borne and there is little information on the role of wild rodents in the epidemiology of the infections. Rodents have been implicated in outbreaks in a dairy herd in Canada (Tablante & Lane, 1989), and in beef herds in the USA; rodents are serious pests in poultry farms and can transmit *Salmonella* which may eventually lead to human outbreaks. *Salmonella* are not always readily found in surveys of rodent populations as infections are generally fatal to rats.

Recently, hamsters and mice purchased from pet shops in South Carolina and Minnesota have been associated with human illness; a search of the National *Salmonella* Database in 2004 revealed 28 matching human case-isolates of serotype *typhimurium* from 19 states; patient illness onset dates ranged from December 2003 to October 2004. Of 22 patients interviewed, 13 (59%) had been exposed to rodents purchased from retail pet stores. The 15 patients with primary or secondary rodent exposure were from Illinois, Kentucky, Missouri, Pennsylvania and South Carolina (2 cases each), Georgia, Michigan, Minnesota, New

Jersey and North Carolina (1 case each). The source of infected rodents for this multistate outbreak is unknown. Public health practitioners should consider pet rodents a potential source of salmonellosis and, when indicated, should obtain cultures from pet rodents during an investigation (CDC, 2005).

The use of *Salmonella*-based rodenticides risks the spread of the infectious agent to humans as has occurred in Europe. *Salmonella*-based rodenticides are still produced and used in Central America, South America, and Asia, especially a preparation called Biorat manufactured by Labiofam, Cuba, which is a *Salmonella enterica*-based rodenticide; it is made by coating rice grains with a combination of *Salmonella enteritidis* serotype and warfarin. In July 2001, US custom authorities seized a shipment of Biorat destined for distribution in the USA (Painter *et al.*, 2004). Rodenticides containing salmonellae were evaluated during a plague outbreak in San Francisco in 1895; they were found to have no definable impact on the rodent population, but they caused illness and death in humans who prepared and handled them (Friedman *et al.*, 1996). *Salmonella*-based rodenticides can not be registered or legally used in either the USA or Canada.

While salmonellosis is of very considerable public health importance in the USA and Canada, the role of rodents as reservoir hosts is now limited. The frequency of infections being acquired from rodents purchased from pet shops has grown but the incidence of such infections remains low.

Toxoplasmosis

The parasite *Toxoplasma gondii* is very widespread in the USA and Canada as it is in Europe. Although many species of animal may be infected, members of the cat family are the definitive hosts for the organism; domestic cats play a major role in transmission of the disease and only members of the cat family shed oocysts. Cats become infected by ingesting either oocysts from faecal contamination or tissue cysts present in the flesh of eaten animals, including rodents or birds or by eating flesh scraps of other infected animals such as pigs, cattle or sheep. Digestive enzymes release the organisms, which invade the feline small intestine.

The parasite affects 10–20 out of every 100 people in North America by the time they are adults. The disease is less frequent in areas where the environment is unfavourable for the oocysts, such as at the extremes of temperatures and at higher altitudes. Although the rate of seropositivity for *T. gondii* is high, clinical manifestation of the disease is relatively rare. Clinical toxoplasmosis occurs in as many as 40% of patients with AIDS. In the USA, as many as 40% of adult humans have serologic evidence of subclinical infection with *T. gondii*. However, the vast majority of infections are without clinical symptoms.

Approximately 85% of women of childbearing age in the USA are susceptible to acute infection with the parasite. Transmission of *T. gondii* to the foetus can result in serious health problems, including mental retardation, seizures, blindness and death. Some health problems may not become apparent until the second or third decade of life. An estimated 400–4000 cases of congenital toxoplasmosis occur in the USA each year. About one half of the cases of toxoplasmosis in the USA are caused by eating undercooked infected meat (Jones *et al.*, 2003).

The public health importance of toxoplasmosis

The actual reported incidence of toxoplasmosis in the USA and Canada is low, although up to 6000 congenital cases occur annually in the USA. The risk for pregnant women is significant. The disease is also of growing importance among AIDS patients. It is a disease of public health importance. With improved rodent prevention and control on pig farms, the extent of the importance of rodents as reservoir hosts has probably declined.

Cestode infections

Hymenolepiasis

As has been mentioned earlier, *Hymenolepsis nana* (also known as the dwarf tapeworm) is a cestode of rodents occasionally seen in humans and frequently found in rodents; humans or rodents (mice primarily) are the definitive hosts and flour beetles (*Tribolium* species) and grain beetles (*Oryzaephilus* species) are the intermediate host. It is the only cestode that parasitizes humans without requiring an intermediate host and is the most common of all cestode infections in man. While the life-span of the adult worm is short, no more than 4–6 weeks, internal autoinfection allows the infection to persist for years. The parasite is most frequently found in areas of poor sanitation. However, *H. nana*-infected Norway rats (*Rattus norvegicus*), house mice (*Mus musculus*) and prairie dogs (*Cynomys ludovicianus*) have also been found in pet shops. Both *Hymenolepis nana* and *H. diminuta* are present in the southern USA. Infections with *H. nana* are usually asymptomatic and are more of a nuisance than a problem.

Hymenolepsis diminuta, the rat tapeworm, occurs throughout the world. Its principal definitive hosts are rodents. Humans are an accidental host and rarely may become infected through the ingestion of an infected arthropod. *Hymenolepsis diminuta* was reported in a 17-month-old child in North Carolina in 1990 (Hamrick *et al.*, 1990); the source of the infestation was unknown. The authors observed that the incidence of *H. diminuta* infections in humans is rare and that only 200 cases have been reported worldwide and none in the USA since 1965. The prevalence of *H. nana* is much higher. No records have been found of this species in Canada.

Fleas may also serve as the intermediate hosts of *H. nana* and harbour the infective larvae (cysticercoids) and when dogs lick their coats or children kiss and fondle dogs, fleas may be swallowed by dogs and children and they thus become infected.

Unless worm burdens are unusually heavy, neither of these cestodes is a public health problem and diagnosed infections can be readily treated.

Echinococcosis (hydatid disease)

Echinococcosis or hydatid disease results from infections with larvae of the tapeworms *Echinococcus granulosus, E. multilocularis* or *E. vogeli*. *Echinococcus granulosus* is found most commonly in dogs that have consumed viscera of infected sheep, but infections can also be found in coyotes (*Canis latrans*), wolves (e.g. *Canis lupus*), dingos (*Canis lupus dingo*) and jackals (e.g. *Canis aureus*). This species occurs worldwide, typically in rural areas of Africa, the Middle East, southern Europe, Russia, China, Australia and South America (especially Argentina and Uruguay). *Echinococcus multilocularis* is found in foxes (e.g. *Vulpes vulpes*), coyotes, dogs and cats. The intermediate hosts are predominantly rodents or other small mammals, or, accidentally, humans. *Echinococcus multilocularis* causes alveolar hydatid echinococcosis and occurs only in the northern hemisphere. *Echinococcus vogeli* has been identified only in Central and South America. Both *Echinococcus granulosus* and *E. multilocularis* have been reported in Canada.

Infection with *E. granulosus* results in the formation of cysts in the liver, lungs, kidney and spleen. This condition is also known as cystic hydatid disease and can usually be successfully treated with surgery. Infection with *E. multilocularis* results in the formation of parasitic tumours in the liver, lungs, brain and other organs. If untreated, this infection may be fatal, and is more likely to result in death than the disease caused by *E. granulosus*. It is also called alveolar hydatid disease. Infection is generally asymptomatic for 10 to 20 years, until the cyst grows large enough to cause problems.

The public health importance of echinococcosis

Human echinococcosis is prevalent in Alaska and northern Canada and is present in parts of the USA. The widespread occurrence and expanding range of *E. multilocularis* in north-central USA and south-central Canada, point to an increasing public health importance of alveolar hydatid disease. Its importance lies in the possible fatal outcome if untreated and the difficulty of treatment. It is thought that *E. granulosus* and *E. multilocularis* together infect at least 2.7 million people worldwide. Alveolar echinococcosis especially is very important from a public health standpoint because it can be lethal in up to 100% of untreated patients, and treatment is difficult and costly, with therapy costing many tens of thousand of dollars per patient.

Nematode infections -- trichinosis

Trichinosis or trichinellosis is an infection due to *Trichinella* nematodes, most commonly *Trichinella spiralis*. Rats (usually the Norway rat, *Rattus norvegicus*) and other rodents maintain the infection in nature. Pigs and bears may eat the rodents, which, in turn, may infect humans when their meat is consumed.

Occurrence in the USA is now largely limited to sporadic cases or small clusters of infections related to consumption of home-processed meats usually from non-commercial farm-raised pigs and wild game. Trichinosis has been a reportable disease since 1966. The Centers for Disease Control and Prevention (CDC) surveillance system has data as far back as 1947 demonstrating a significant decrease in cases from a peak of nearly 500 in 1948 to averages of fewer than 50 over the past several years. From 1991–1996, an annual average of 38 cases per year were reported in the USA. The number of cases has decreased because of legislation prohibiting the feeding of raw meat garbage to hogs, increased commercial and home freezing of pork, and the public awareness of the danger of eating raw or undercooked pork products. Through an aggressive programme of meat inspection, the incidence of trichinosis in pigs in the USA has been lowered to less than 1%. The US Department of Agriculture conducts periodic surveillance of farm-raised pigs. In a 1999 study, the major risk factor for seropositivity in tested pigs was access to live wildlife or wildlife carcasses.

Rattus norvegicus populations may maintain a high intensity of infection through cannibalism (Leiby *et al.*, 1990); effective rodent control is essential in sites where infections are known to occur in rat populations.

Trichinosis in the USA and Canada has been declining in importance with only sporadic outbreaks and is now of little public health importance.

Conclusions

The most important change in the frequency of rodent-borne diseases in the USA and Canada has been the emergence of hantavirus pulmonary syndrome, an infection with a case fatality rate of more than 50%. First recognized in southwestern USA, in 1993, in the Four Corners Region, it has now been widely reported in the USA and Canada as well as in Central and South

America, and due to changing ecological conditions favourable to rodent populations is likely to continue spreading the disease in the Americas (Lednicky,
2003).

The persistence of domestic rat populations, mainly of Norway rats (*Rattus
norvegicus*), particularly in inner cities, ensures that a number of rodent-borne
diseases persist in urban areas; the rat populations carry the risk of spreading
introduced or newly emerged infections.

35

The economic impact of vector- and rodent-borne diseases in the USA and Canada

Human and animal diseases transmitted by arthropods cause almost immeasurable social and economic losses. Many studies have been done on the socioeconomic costs of one or another of this group of diseases in the tropics but similar information for the USA and Canada is scattered, incomplete and not easily found. Most of the studies which have been carried out in the USA or Canada, have concentrated on the costs of health-care and cost of work loss for the patients. However, the cost of vector and reservoir host control activities to deal with outbreaks of vector-borne diseases is also very substantial and must also be taken into account. Though poorly documented, the economic costs associated with vector- and rodent-borne diseases are growing in the USA and Canada. Some examples are given here.

Eastern equine encephalitis

A study was undertaken to ascertain the economic burden imposed on some residents of Massachusetts who had survived eastern equine encephalitis (EEE) infections in the late 1980s. Transiently affected persons mainly required assistance of direct medical services; the average total cost per case was US $21 000. Those who suffered persistent sequelae remained at home and seemed likely to live a normal life span, but without gainful employment, and they represented costs that ranged as high as US $0.4 million but levelled off at US $0.1 million after 3 years. Hospital costs approached about US $0.3 million a patient. The cost associated with persistent sequelae, which included medical expenses, education, institutionalization, and loss of income, was approximately US $3 million per case. Insecticidal intervention to avert EEE outbreaks cost between US $0.7 and 1.4 million depending upon the extent of the area to be treated. This is less than the US $3 million imposed on one person suffering residual sequelae (Villari *et al.*, 1995). Eastern equine encephalitis also imposes serious costs on the equine industry despite the availability of vaccines.

Among other states, EEE virus continues to cause losses in Florida horses. A conservative estimate of the cost of EEE to the state equine industry in that state was greater than US $1 million per year (Wilson *et al.*, 1986); the costs have probably increased by now.

La Crosse encephalitis

A study was carried out in North Carolina to assess the economic and social impacts of the illness; 25 serologically confirmed La Crosse encephalitis (LAC) case patients and/or families were interviewed to obtain information on the economic costs and social burden of the disease. The total direct and indirect medical costs associated with LAC over 89.6 life-years accumulated from the onset of illness to the date of interview for 24 patients with frank encephalitis totalled US $791 374 (range = US $ 7521–175 586), with a mean per patient cost of US $ 32 974 +/− 34 793. The projected cost of a case with lifelong neurological sequelae ranged from US $48 775−3 090 798; thus the socioeconomic burden resulting from LAC is substantial in the endemic area of the state (Utz *et al.*, 2003).

St. Louis encephalitis

St. Louis encephalitis (SLE) was the most important cause of epidemic viral encephalitis in the USA until the introduction of West Nile virus. The disease is widespread in the USA and has been responsible for recurring epidemics with substantial mortality. The true impact of SLE during epidemics is difficult to assess, since there are typically several hundred mild or asymptomatic cases generated for every diagnosed case. A study of the 1966 SLE epidemic in Dallas, Texas (172 cases, 20 deaths) estimated the cost of a mosquito-borne disease epidemic in the USA (Schwab, 1968). The total costs of that epidemic were an estimated US $796 500 in 1966 dollars. Adjusting each cost component by the appropriate Consumer Price Indices for all items or for medical care, the total epidemic cost was US $5.4 million in 2002 dollars, of which the largest share was for epidemic control expenditures (US $348 500 in 1966 dollars (US $1.9 million in 2002 dollars). The 1977 and 1990 SLE epidemics in Florida resulted in considerable disruption of normal activities of permanent residents and retarded tourism to the state. Economic loss to the state has not been well documented, but the 1990 epidemic alone is likely to have been responsible for millions of dollars of direct and indirect losses.

West Nile virus

The introduction and rapid spread of West Nile virus (WNV) into the USA and Canada has led to extraordinary costs both for individuals and for health services. During 1999, the State of New York incurred approximately US $14 million in costs to prevent the spread of the WNV for activities including virus surveillance, vector mosquito control, mosquito larvicides and adulticide, laboratory procedures and insect repellent purchases (news release, Department of Health New York State, 4 August 2000).

A number of studies have been carried out attempting to determine the costs of outbreaks of the virus; generally they show that there is a serious economic impact as a consequence of outbreaks. In 2002, in the course of an outbreak of WNV in a single county, Cuyahoga, in Ohio, USA, nearly US $4 million was expended for health care for WNV cases. At least 150 county residents with life-threatening complications of the disease spent a combined 1800 days in the hospital. Each case typically cost more than US $25000 to treat when hospitalized. The cost of treating the large number of non-hospitalized patients was not included nor was the lost productivity resulting from the sick days. In 2002, the epidemic of WNV illness, which focused in the Midwestern USA resulted in 4156 reported cases; 2942 cases had central nervous system illness (meningitis, encephalitis, or acute flaccid paralysis) and 284 died.

A total of 329 persons with WNV disease were reported in Louisiana State, with illness onsets from June to November. Among these, 204 had illnesses involving the CNS; 24 died. To estimate the economic impact of the 2002 WNV epidemic in Louisiana, data were collected from hospitals, by a patient questionnaire, and various public offices. Hospital charges were converted to economic costs by using Medicare cost-to-charge ratios. The estimated cost of the Louisiana epidemic was US $20.1 million from June 2002 to February 2003, including a US $10.9 million cost of illness (US $4.4 million medical and US $6.5 million non-medical costs) and a US $9.2 million cost of public health response. Severe cases were said to cost US $51 826 per patient to treat. These data indicate a very substantial short-term cost of the WNV disease epidemic in Louisiana (Zohrabian *et al.*, 2004).

In 2002, 378 and 1100 equine cases of WNV infections were confirmed in Colorado and Nebraska, respectively. These outbreaks of WNV cost equine owners in Colorado and Nebraska more than US $1.25 million, while preventive measures are estimated to have cost an additional US $2.75 million. The study estimated that the total cost attributable to death or euthanasia of equids from WNV infection was US $600 660, the estimated revenue lost by owners because of lost equine use due to WNV was US $163 659, and the estimated cost attributable to treatment in the two states was US $490 844.

There have been many other estimates of the overall costs of the epidemics of WNV in the USA including the costs of medical care, increased vector-control costs, costs to the equine industry, etc. Some of the estimates go as high as a billion dollars from 1999 to 2004. The CDC has estimated that WNV fever cost the USA, during 2002 alone, at least US $20.1 million in costs of illness; hospitalization costs were estimated at approximately US $26 000 per patient. In 2003, 7386 people were diagnosed with WNV, of whom 155 died. The CDC estimated that the 2003 cases cost an estimated US $139 million in hospitalization, doctor's bills and such things as loss of wages, need for extra child-care and transportation and loss of productivity. The cost to families and to the communities of the fatal cases must also be added to this figure.

In February 2005 the region of Waterloo, Ontario, Canada where WNV has been present since 2001, released a 'West Nile Virus 2005 Contingency Plan and Report'. Based on costing for the 2004 WNV programme (CAN $585 000) which represented three rounds of larviciding, and comparative costing with health units of similar size (and circumstances), the report anticipated that an annual cost of CAN $600 000 will be required for the minimum activities of monitoring, larviciding and education and noted that additional contingency funds may be requested in the event of certain risk factors, such as high WNV activity, an emergency, or a human outbreak. Costs are doubtlessly similar for other affected regions of Canada.

The total health-care costs and added costs of vector mosquito control caused by this introduced disease remain to be calculated but it is certain that they have been very great both in the USA and Canada. As has been stated before, the very considerable expenditures as a result of the introduction and spread of WNV, reduces the availability of funding available to deal with other health problems.

Dengue

In most of the highly dengue-endemic countries in the world, the economic impact of dengue and dengue haemorrhagic fever (DHF), is enormous, and causes significant economic burdens to affected communities. The impact may include loss of life, medical expenditures for hospitalization of patients, loss in productivity of the affected workforce, strain on health-care services due to sudden high demand during an epidemic, the expenditures for large-scale emergency control actions, and loss of tourism as a result of negative publicity (Meltzer *et al.*, 1998). Von Allmen *et al.* (1979) estimated the cost of the

1977 dengue epidemic in Puerto Rico to be between US $6 and US $16 million. The authors estimated the direct costs (medical care, epidemic control measures) of the epidemic were between US $2.4 and US $4.7 million, while indirect costs (days of work lost by ill workers and parents of ill children) were between US $3.7 and $11 million. A study by Torres (1997) in Puerto Rico found that the impact of a dengue outbreak was greater for poor communities in the urban and semi-rural areas, and particularly for women who described themselves as housewives and mothers, and their children. Social expectations and the family's demands for these women to fulfill the role of caretaker superseded their own sick role. In addition they experienced the greatest loss of time as a consequence of the outbreak. The main effect of the outbreak on work activities not traditionally remunerated with money such as housework, was the inability of adult females in the household to perform their routine activities to maintain family life. The monetary costs of health care absorbed a significant percentage of the household weekly income. When hospitalization was required for serious cases of dengue or DHF, the median duration was 5 days.

Imported malaria

Bloland *et al.* (1995) collected data by phone interview on the treatment of US citizens reported to the CDC as having acquired *Plasmodium falciparum* malaria in 1988–1989 while travelling in sub-Saharan Africa. These data were used to derive a relative index of illness severity, to estimate the costs of malaria-specific therapy, and to assess adherence to existing therapy recommendations. All monetary values are expressed in US dollars. Of 142 patients, 110 (77%) were classified as having mild, 21 (15%) as having moderate, and 11 (8%) as having severe infections. Two (1.4%) deaths were reported. Overall, the mean cost of treatment per case was $2743.51 (range $191.75–$79 801.73). Estimated with relation to severity, the median cost for treatment per case was $467.54 for mild, $2701.16 for moderate and $12 515.52 for severe infections. Forty-two (30%) of these patients had at least one element of therapy that was inconsistent with recommendations current at the time of the study; 27(19%) received chloroquine; 12 (9%) received primaquine unnecessarily; 8 (6%) received inappropriate dosages of pyrimethamine/sulfadoxine (Fansidar); and 3 (2%) received potentially inappropriate dosing regimens of quinine.

Most patients from Europe, the USA and Canada who have contracted malaria during visits to endemic countries have no protective immunity; when infected by malaria they may develop a rapidly severe, even fatal disease. Health-care providers are often unfamiliar with the symptoms of malaria, and this can cause delayed or incorrect diagnosis and treatment of the disease and considerable medical costs.

Lice infestations

Body lice (*Pediculus humanus*) are rare in the USA and Canada and almost all costs related to lice infestations are those for the control of head and pubic lice (*P. capitis* and *Pthirus pubis* respectively). While neither species is a vector of disease, expenditures on their control probably detracts from resources available for other health care. Head lice are a major community public health problem in the USA with an estimated 6–12 million cases annually. It has already been remarked that this results in about US $367 million a year in consumer costs, loss of parental wages and school system expenses (Clore & Longyear, 1990). The costs involved in the control of pubic lice are not known.

Scabies infestations

Nosocomial outbreaks of scabies caused by *Sarcoptes scabiei*, in hospitals, prisons and especially nursing homes for elderly or incapacitated persons are not uncommon. Because of the highly contagious nature of scabies, early diagnosis and early treatment is essential to prevent the rapid spread of the infection in hospital and nursing home settings. In a recent outbreak of scabies in an extended care unit attached to an acute care hospital in Canada the infection had spread so widely that 78 residents and over 100 staff and family members had to be treated at a cost of more than US $20 000 (Jack, 1993). As many institutional outbreaks involve even greater numbers of infected persons among both the patients and the staff, the cost of isolation and treatment would be even greater as in a large hospital, hundreds of staff and patients may require prophylactic treatment at considerable cost.

Lyme disease

During the period 1980–2005, a total of 246 247 cases of Lyme disease (LD) (caused by *Borrelia burgdorferi*) were reported to the CDC; LD is now the most frequently acquired vector-borne disease in the USA. It was designated a nationally notifiable disease in 1991. However, it has been estimated that only 1 in 10 cases of Lyme disease is actually reported to the CDC – so there may have been as many as 2.46 million cases of LD since 1980. Studies from the early 1990s suggested that LD cases were under-reported by 6–12-fold in some areas where LD is endemic

As has been observed above, there is some controversy in the USA as to the accuracy of the diagnosis of LD and hence as to its real incidence; in any event, it is likely that the total incidence since reporting started is around 2 million cases. Factoring in the costs of diagnosis, treatment and lost wages, a recent study estimated the cost of each acute case of LD to be US $161, while just the economic costs to treat the sequelae of early and late disseminated LD increases the total costs >10 to 100-fold. Social costs, including losses in tourism and recreation, pain and suffering, disability and the like further add to the toll that vector-borne diseases impose. In the supporting information provided for a bill introduced into the House of Representatives seeking to establish a Tick-Borne Disorders Advisory Committee, it is stated that LD costs the USA between US $1000 million to US $2000 million each year in increased medical costs, lost productivity, prolonged pain and suffering, unnecessary testing and costly delays in diagnosis and inappropriate treatment. Due to both medical and public concern with the disease there may be undue tests and treatment for patients suspected of having LD which add to the national cost. Maes *et al.* (1998) conducted a study to establish the medical and economic burden of LD in the overall US population, which included determining its endemicity in high-risk states and counties, describing current treatment patterns, measuring direct and indirect costs, and defining the cost burden by age group. Using an annual mean incidence of 4.73 cases of LD per 100 000 population in the decision–analysis model yielded an expected national expenditure of US $2.5 billion (1996 dollars) over 5 years for therapeutic interventions to prevent 55 626 cases of LD sequelae. This estimate included both direct medical and indirect costs.

The probability of infections with *B. burgdorferi* in Canada are low outside of British Colombia and Ontario and the costs to the community are proportionally lower.

Less than 20 years elapsed between the 1982 report of the identification and isolation of *B. burgdorferi* and the licensure and marketing in the USA of a prophylactic vaccine against this pathogen. However, the manufacturer removed the vaccine from the market under 4 years after its release. The low demand undoubtedly was the result of limited efficacy, need for frequent boosters, the high price of the vaccine, exclusion of children, fear of vaccine-induced musculoskeletal symptoms and litigation surrounding the vaccine (Hanson & Edelman, 2003). As the LD vaccine is no longer available greater attention should be given in the future to the prevention of tick bites and tick control.

Mosquito control

There are a substantial number of mosquito abatement and vector control organizations in the USA along with commercial control companies that carry out mosquito control under contract. The latest information made available to the American Mosquito Control Association in an unpublished survey conducted in 1999, listed 345 mosquito abatement agencies, including a small number of private companies. However, the actual number may be as high as 700 (personal communication, J. Conlon, American Mosquito Control Association).

The main objective of these agencies in carrying out mosquito control is to reduce the mosquito populations as a cause of annoyance and discomfort. In doing so they also serve, to some extent, to reduce the likelihood of mosquito populations transmitting infectious diseases. When disease outbreaks occur, such as West Nile virus, these organizations carry out targeted vector mosquito control, often by large-scale space spraying.

Information on the estimated overall annual expenditures for mosquito control in the USA and to a lesser extent Canada, gleaned from proprietary information obtained from market analyses, estimates that approximately US $550 million is expended annually on mosquito control products. This is for the products themselves (insecticide sprays, traps, repellents) and does not include equipment and personnel expenditures which are very considerable. It also does not include expenditures by individuals on household sprays, repellents etc.

The funds expended on organized mosquito control are very substantial and emergency funding during outbreaks of disease, often federal funding, can also be large. It is difficult to separate mosquito pest control and disease control and the two are at times identical.

Conclusions

The overall expenditures for health-care costs, epidemiological investigations and disease surveillance as well as the losses of income and productivity due to illness and death from both autochthonous and introduced cases of vector-and rodent-borne diseases are far greater than are generally recognized by public health authorities or the public. If the costs of vector-and rodent control measures are also taken into consideration, the total economic burden of this group of infections and the diseases they cause is very great indeed. Effective preventive measures through the control of arthropod vectors and rodent reservoir hosts would reduce the overall costs of some of the diseases in the group. As an example greater awareness among travellers of the necessity of taking chemoprophylactic drugs would significantly reduce the health-care costs due to imported malaria. The use of protective clothing and repellents in tick-infested areas might very likely have an impact on the incidence of Lyme disease.

36

Conclusions on the burden of the vector-and rodent-borne diseases in Europe, USA and Canada

The actual incidence of this group of diseases and their public health importance is not always recognized by the public, public health authorities or the medical profession. Travellers to vector-borne disease endemic areas all too frequently fail to comply with measures of personal protection to prevent contracting an infection such as malaria and often fall ill on their return from travel; frequently diagnosis is delayed as the patient may not have been asked or volunteered information about recent travel or because of a lack of familiarity with the symptoms of tropical diseases by physicians being consulted. As a result, treatment may also be delayed with serious consequences for the patient. As regards the vector-and rodent-borne diseases endemic to Europe, the USA and Canada, the necessary clinical suspicion to ensure an accurate diagnosis must be based on awareness of distribution of the infections and the risk that travellers may have occurred or the risk that inhabitants of endemic areas face.

Several of this group of vector-and rodent-borne infections have emerged in recent years as diseases of considerable and widespread importance, perhaps foremost among them Lyme disease and West Nile virus. As an example, the incidence of Lyme disease in Germany has risen to an estimated 60 000 cases a year. Overall, Lyme disease has become the most commonly reported arthropod-borne illness in American and European countries.

West Nile virus, newly introduced into the USA in 1999, has spread from its point of introduction in New York City and by 2004 had reached California and Canada and caused large arboviral meningoencephalitis outbreaks. The fact that many different species of mosquito are efficient vectors of the infective agent has facilitated its rapid spread. Romania suffered an epidemic involving some 800 cases hospitalized with viral meningoencephalitis caused by WNV in

1996 and outbreaks of WNV appear in Russia, some with an increased degree of virulence.

The introduction and spread of WNV in the USA and Canada highlights the threat from other arboviral infections that may be introduced such as Rift Valley fever and Ross River virus. Active, informed surveillance programmes are necessary to cope with such possible introductions.

Usutu virus, a flavivirus closely related to WNV, was apparently confined to the African continent until 2001, when it was isolated unexpectedly from encephalitic birds in Vienna. It has rapidly spread throughout the country causing a severe mortality among birds. Antibodies to this virus have been detected in birds in the UK. The appearance of this virus and its dramatic spread illustrate the risk of the introduction of exotic vector-borne infections to Europe and North America.

Despite the availability of an effective vaccine, tick-borne encephalitis (TBE) persists in Europe. The TBE reporting system is inadequate and in some countries there are only incomplete data, low awareness and underestimation of the risk of infection. In TBE-endemic countries with a low vaccine coverage rate, the number of reported TBE cases is increasing and new TBE foci are emerging in a number of European countries. TBE-infected ticks are even found in urban parks. It is predicted that with climate change the transmission of TBE will likely move further to the north.

The emergence and spread of Lyme disease has been seen to be due to changing ecological conditions and it appears that other tick-borne diseases are similarly affected and are emerging such as Southern Tick-Associated Rash Illness or STARI in the USA. There are indications that still other tick-borne diseases are emerging. Diseases such as Rocky Mountain spotted fever, the most severe tick-borne infection in the USA, persist and in fact are spreading and may cause mortality if not timely diagnosed and treated. Lyme disease has become the most common vector-borne disease in the USA, in the course of only three decades. Ticks infected with *Borrelia burgdorferi* are being found with increasing frequency in urban parks in Europe feeding on Norway rats (*Rattus norvegicus*) and yellow-necked mice (*Apodemus flavicollis*) thus increasing the risk of transmission to humans. A similar phenomenon has been seen in the USA.

The last few decades have been marked by the emergence of a number of tick-borne diseases not previously known in Europe or North America; Mediterranean spotted fever (MSF) due to *Rickettsia conorii* was thought for a long time to be the only tick-borne rickettsial disease prevalent in Europe but five more spotted fever rickettsiae have been described as emerging pathogens in the last decade. Further, cases of infection due to *Anaplasma phagocytophila*, the agent of human anaplasmosis (previously known as human granulocytic ehrlichiosis) have been reported throughout Europe. *Ehrlichia chaffeensis*, the agent of human monocytotropic ehrlichiosis (HME) in the USA, was only first recognized in 1986; it may present as meningoencephalitis and may lead to death.

Cases of visceral leishmaniasis (VL) (*Leishmania donovani* s.l.) co-infection in HIV-positive individuals have been reported from most areas of the world where the geographical distributions of the two infections overlap. The majority of the co-infected cases that have been recorded, however, live around the Mediterranean basin. In AIDS patients, VL is a recurrent disease that is highly prevalent. Highly active antiretroviral therapy (HAART) has not reduced the frequency of VL relapses in patients receiving HAART which remains high. Relapses of VL are observed only in individuals with uncontrolled HIV replication and/or poor immunological responses. The appearance of sporadic cases of autochthonous leishmaniasis in animals and humans in Germany is ominous and may indicate that transmission via a sandfly vector species is already occurring.

Louse-borne trench fever caused by *Bartonella quintana* was long thought to have disappeared after World War II as it had after World War I; yet it has reappeared after a long absence, among homeless people in Europe and the USA with little or nothing known about its reservoir hosts.

Among the rodent-borne diseases, hantavirus pulmonary syndrome, which causes a high degree of mortality among infected individuals, was first recognized in 1993 and thought to have a limited distribution; it is now seen to be spreading geographically within the USA. While the number of cases of this infection may be small, any disease that may cause 60% mortality must be taken account of. The hantaviruses appear to have spread in Europe and while they usually show a benign course, some more virulent strains have been observed. The number of cases is increasing and more cases are appearing in urban areas.

The costs of vector- and rodent-borne diseases to individuals and to the community have been discussed above. On both continents, the costs are very great indeed. It is important to quantify these costs in terms of treatment, hospitalization and absence from work, the cost of mortality and costs associated with essential control operations to control transmission of these infective agents. The magnitude of these costs will be seen to justify improved surveillance and control.

Vector- and rodent-borne diseases know no frontiers. Not only are they constantly introduced in large numbers with travellers from disease-endemic tropical countries but they are also present in both Europe and North America, from the Arctic to warmer southern regions. Ecological changes have favoured the recrudescence and spread of diseases long known to be present in the northern hemisphere and appear to have enabled the emergence of infectious agents and diseases previously unrecognized. A rise in mean temperature of 2–3 °C is expected in Europe within the next 50 years, suggesting a significant increase of autochthonous cases of vector-borne, non-endemic diseases as well as the spread of already endemic vector-borne infectious diseases. Such warming will also be likely to occur in North America and may result in the increased distribution of this group of diseases to the north. The introduction of potential vector species of mosquitoes and ticks may have serious consequences on both continents.

Knowledge of this group of diseases and their burden on public health is essential both to public health planners and administrators and to individual physicians who must diagnosis and treat the extremely large number of diseases in this group.

References

Aberer, E., Kehldorfer, M., Binder, B., Schauperi, H. (1999) The outcome of *Lyme borreliosis* in children. *Wien Klin. Wochenschr* **111**(22–23):941–4.

Abranches, P., Lopes, F. J., Fernandes, P. S., Gomes L. T. (1982) Kala-azar in Portugal. I. Attempts to find a wild reservoir. *J. Trop. Med. Hyg.* **85**(3):123–6.

Abranches, P., Santos-Gomes, G. M., Campino, L. (1993) Epidemiology of leishmaniasis in Portugal. *Arch. Inst. Pasteur Tunis* **70**(3–4):349–55.

Abranches, P., Silva-Pereira, M. C. D., Conceicao-Silva, F. M., Santos-Gomes, G. M., Janz, J. G. (1991) Canine leishmaniasis: pathological and ecological factors influencing transmission of infection. *J. Parasitol.* **77**(4):557–61.

Acedo Sanchez, C., Martin Sanchez, J., Velez Bernal, I. D. *et al.* (1996) Leishmaniasis eco-epidemiology in the Alpujarra region (Granada Province, southern Spain). *Int. J. Parasitol.* **26**(3):303–10.

Ackerman, A. B. (1968) Crabs – The resurgence of *Phthirus pubis*. *N. Engl. J. Med.* **278**(17):950–1.

Ackermann, R. (1976) Tick-borne meningopolyneuritis (Garin-Bujadoux, Bannwarth). *MMW Munch. Med. Wochenschr.* **118**(49):1621–2.

Ackermann, R., Kabatzki, J., Boisten, H. P. *et al.* (1984) *Ixodes ricinus* spirochete and European erythema chronicum migrans disease. *Yale J. Biol. Med.* **57**(4):573–80.

Adhami, J., Murati, N. (1987) Presence of the mosquito *Aedes albopictus* in Albania. *Rev. Mjekesore* **1**:13–16.

Adhami, J., Reiter, P. (1998) Introduction and establishment of *Aedes* (*Stegomyia*) *albopictus* Skuse (Diptera: Culicidae) in Albania. *J. Am. Mosq. Control Assoc.* **14**(3):340–3.

Aeschlimann, A., Buttiker, W. W., Eichenberger, G. (1969) Les tiques (Ixodoidea) sont-elles des vecteurs de malidae en Suisse? *Bull. Soc. Entomol. Suisse* **48**(1–2):69–75.

Afzelius, A. (1921) Erythema chronicum migrans. *Acta Dermato-Venereologica* **2**:120–5.

Aitio, A. (2002) Disinsection of aircraft. *Bull. Wld Hlth Org.* **80**(3):257.

Albanese, M., Bruno-Smiraglia, C., Di di Cuonzo, G., Lavagnino, A., Srihongse, S. (1972) Isolation of Thogoto virus from *Rhipicephalus bursa* ticks in western Sicily. *Acta Virol.* **16**(3):267.

Alberdi, M. P., Walker, A. R., Urquhart, K. A. (2000) Field evidence that roe deer (*Capreolus capreolus*) are a natural host for *Ehrlichia phagocytophila*. *Epidemiol. Infect.* **124**(2):315–23.

Aleksandrov, E., Teokharova, M., Runevskaya, P., Dimitrov, D., Kamarinchev, B. (1994) Rickettsial diseases, health and social problems. *Infectology* **31**(2):3–7.

Alekseev, A. N., Dubrinia, H. V., Van De Pol, I., Schouls, L. M. (2001) Identification of *Ehrlichia* spp. and *Borrelia burgdorferi* in *Ixodes* ticks in the Baltic regions of Russia. *J. Clin. Microbiol.* **39**(6):2237–42.

Alston, J. M., Broom, J. C. (1958). *Leptospirosis in Man and Animals*. Edinburgh: E.° & °S. Livingstone.

Amaducci, L., Inzitari, D., Paci, P. *et al.* (1978) Encefalite da zecche (TBE) in Italia. *Giorn. Malattie Infett. Parassit.* **30**(9):693–9.

Ambroise-Thomas, P., Quilici, M., Ranque, P. (1972) Reappearance of malaria in Corsica. Value of seroepidemiologic studies. *Bull. Soc. Path. Exot.* **65**(4):533–42.

Amela, C., Mendez, I., Torcal, J. M. *et al.* (1995) Epidemiology of canine leishmaniasis in the Madrid region, Spain. *Eur. J. Epidemiol.* **11**(2):157–61.

Amundson, T. E., Yuill, T. M. (1981) Natural La Crosse virus infection in the red fox (*Vulpes fulva*), gray fox (*Urocyon cinereoargenteus*), raccoon (*Procyon lotor*), and opossum (*Didelphis virginiana*). *Am. J. Trop. Med. Hyg.* **30**(3):706–14.

Anda, P., Sanchez-Yebra., W., del Mar Vitutia, M. *et al.* (1996) A new *Borrelia* species isolated from patients with relapsing fever in Spain. *Lancet* **348**:(9021)162–5.

Anderson, B. E., Dawson, J. E., Jones, D. C., Wilson, K. H. (1991) *Ehrlichia chaffeensis*, a new species associated with human ehrlichiosis. *J. Clin. Microbiol.* **29**(12):2838–42.

Anderson, J. F., Johnson, R. C., Magnarelli, L. A., Hyde, F. W. (1986) Involvement of birds in the epidemiology of the Lyme disease agent *Borrelia burgdorferi*. *Infect. Immun.* **51**(2):394–6.

Anderson, J. F., Magnarelli, L. A. (1984) Avian and mammalian hosts for spirochete-infected ticks and insects in a Lyme disease focus in Connecticut. *Yale J. Biol. Med.* **57**(4):627–41.

Anderson, J. F., Magnarelli, L. A., Burgdorfer, W., Barbour, A. G. (1983) Spirochetes in *Ixodes dammini* and mammals from Connecticut. *Am. J. Trop. Med. Hyg.* **32**, 818–24.

Anderson, J. F., Magnarelli, L. A., Burgdorfer, W., Casper, E. A., Philip, R. N. Importation into the United States from Africa of *Rhipicephalus simus* on a boutonneuse fever patient. *Am. J. Trop. Med. Hyg.* **30**(4):897–9.

Angelov, L., Aeschlimann, A., Korenberg, E. I. *et al.* (1990) Data on the epidemiology of Lyme disease in Bulgaria. *Med. Parazit* (*Mosk.*) **4** (Jul.–Aug):13–14.

Angelov, L., Arnaudov, D., Rakadzhova, T., Kostova, E., Lipchev, G. (1993) Study on the epizootiology of Lyme borreliosis in Bulgaria. *Infectology* **30**(5):12–14.

Angelov, L., Dimova, P., Berbencova, W. (1996) Clinical and laboratory evidence of the importance of the tick *D. marginatus* as a vector of *B. burgdorferi* in some areas of sporadic Lyme disease in Bulgaria. *Eur. J. Epidemiol.* **12**(5):499–502.

Angolotti, E. (1980) Yellow fever. History and current status. Yellow fever in Barcelona in 1821. *Rev. Sanid. Hig. Publ.* (*Madr*) **54**(1–2):89–102.

Anic, K., Soldo, I., Peric, L., Karner, I., Barac, B. (1998) Tick-borne encephalitis in eastern Croatia. *Scand. J. Infect. Dis.* **30**(5):509–12.

Anthonissen, F. M., De Kesel, M., Hoet, P. P., Bigaignon, G. H. (1994) Evidence for the involvement of different genospecies of *Borrelia* in the clinical outcome of Lyme disease in Belgium. *Res. Microbiol.* **145**(4):327–31.

Antipa, C., Girjabu, E., Iftimovici, R., Draganescu, N. (1984) Serological investigations concerning the presence of antibodies to arboviruses in wild birds. *Virologie* **35**(1):5–9.

Antoniadis, A., Alexiou-Daniel, S., Malissiovas, N. *et al.* (1990) Seroepidemiological survey for antibodies to arboviruses in Greece. *Arch. Virol.* Suppl. 1:277–85.

Antykova, L. P., Kurchanov, V. I. (2001) Tick-borne encephalitis and Lyme disease epidemiology in Saint Petersburg. *EpiNorth 2* **1**.

Aranda, C., Eritja, R., Roiz, D. (2006) First record and establishment of the mosquito *Aedes albopictus* in Spain. *Med. Vet. Entomol.* **20**(1): in press.

Arbes, S. J. Jr, Cohn, R. D., Yin, M. *et al.* (2003) House dust mite allergen in US beds: results from the First National Survey of Lead and Allergens in Housing. *J. Allergy Clin. Immunol.* **111**(2):408–14.

Arcan, P., Topciu, V., Rosiu, N., Csaky, N. (1974) Isolation of Tahyna virus from *Culex pipiens* mosquitoes in Romania. *Acta Virol.* **18**(2):175.

Arlian, L. G., Bernstein, D., Friedman, S. *et al.* (1992) Prevalence of dust mites in the homes of people with asthma living in eight different geographic areas of the United States. *J. Allergy Clin. Immunol.* **80**(3-part 1):292–300.

Arlian, L. G., Morgan, M. S., Neal, B. S. (2002) Dust mite allergens: ecology and distribution. *Curr. Allergy Asthma Rep.* **2**(5):401–11.

Arlian, L. G., Neal, J. S., Morgan, M. S. *et al.* (2001) Reducing relative humidity is a practical way to control dust mites and their allergens in homes in temperate climates. *J. Allergy Clin. Immunol.* **107**:99–104.

Arlian, L. G., Rapp, C. M., Fernandez-Caldas, E. (1993) Allergenicity of *Euroglyphus maynei* and its cross-reactivity with *Dermatophagoides* species. *J. Allergy Clin. Immunol.* **91**(5):1051–8.

Arnedo-Pena, A., Bellido-Blasco, J. B., Gonzalez-Moran, F. *et al.* (1994) Leishmaniasis in Castellon: an epidemiological study of human cases, the vector and the canine reservoir. *Rev. Sanid. Hig. Publ. (Madr.)* **68**(4):481–91.

Arteaga, F., Garcia-Monco, J. C. (1998) Association of Lyme disease with work and leisure activities. *Enferm. Infett. Microbiol. Clin.* **16**(6):265–8.

Artsob, H. A. (1990) Arbovirus activity in Canada. *Arch. Virol.* (suppl. 1):249–58.

Artursson, K., Gunnarsson, A., Wikstrom, U. B., Engvall, E. O. (1999) A serological and clinical follow-up in horses with confirmed equine granulocytic ehrlichiosis. *Equine Vet. J.* **31**(6):473–7.

Asokliene, L. (2004) Tick-borne encephalitis in Lithuania. *Eurosurveillance Weekly* **8**(26).

Autorino, G. L., Battisti, A., Deubel, V. *et al.* (2002) West Nile virus epidemic in horses, Tuscany region, Italy. *Emerg. Infect. Dis.* **8**(12):1372–8.

Azad, A. F., Beard, C. B. (1998) Rickettsial pathogens and their arthropod vectors. *Emerg. Infect. Dis.* **4**(2):179–86.

Azad, A. F., Radulovic, S., Higgins, J. A., Noden, B. H., Troyer, M. J. (1997) Flea-borne rickettsioses: some ecologic considerations. *Emerg. Infect. Dis.* **3**(3):319–27.

Babenko, L. V., Prisyagina, L. A., Smirnov, O. V. *et al.* (1975) Epidemiology of tick-borne encephalitis in the Latvian SSR. *Med. Parazit. (Mosk.)* **44**(5):533–41.

Bacellar, E., Beati, L. França, A. *et al.* (1999) Israeli Spotted Fever Rickettsia (*Rickettsia connorii* complex) associated with human disease in Portugal. *Emerg. Infect. Dis.* **5**(6):835–6.

Bacellar, F., Dawson, J. E., Silveira, C. A., Filipe, A. R. (1995) Antibodies against rickettsiaceae in dogs of Setubal, Portugal. *Cent. Eur. J. Publ. Hlth.* **3**(2):100–102.

Bacellar, F., Lencastre, I., Filipe, A. R. (1998) Is murine typhus re-emerging in Portugal? *Eurosurveillance* **3**(2): 18–20.

Bacellar, F., Nuncio, M. S., Alves, M. J., Filipe, A. R. (1995b) *Rickettsia slovaca*: an agent of the group of exanthematous fevers in Portugal. *Enferme. Infecc. Microbiol. Clin.* (Barcelona) **13**(4):218–23.

Bacellar, F., Nuncio, M. S., Rehacek, J., Filipe, A. R. (1991) Rickettsiae and rickettsioses in Portugal. *Eur. J. Epidemiol.* **7**(3):291–3.

Bajer, A., Behnke, J. M., Pawelczyk, A., Sinski, E. (1999) First evidence of *Ehrlichia* sp. in wild *Microtus arvalis* from Poland. *Acta Parasitol.* **44**(3):204–5.

Bakken, J. S., Krueth, J., Tilden, R. L., Dumler, J. S., Kristiansen, B. E. (1996) Serological evidence of human granulocytic ehrlichiosis in Norway. *Eur. J. Clin. Microbiol. Infect. Dis.* **15**(10):829–32.

Balayeva, N. M., Demkin, V. V., Rydkina, E. B. (1993) Genotypic and biological characteristics of non-identified strain of spotted fever group rickettsiae isolated in Crimea. *Acta Virol.* **37**(6):475–83.

Baldari, M., Tamburro, A., Sabatinelli, G. *et al.* (1998) Malaria in Maremma, Italy. *Lancet* **351**(9111):1246–7.

Banerjee, D., Stanley, P. J. (2001) Malaria chemoprophylaxis in UK general practitioners traveling to South Asia. *J. Travel Med.* **8**:173–5.

Banet-Noach, C., Simanov, L., Malkinson, M. (2003) Direct (non-vector) transmission of West Nile virus in geese. *Avian Pathol.* **32**(5):489–94.

Baptista, S., Quaresma, A., Aires, T., Kurtenbach, K. *et al.* (2004) Lyme borreliosis spirochetes in questing ticks from mainland Portugal. *Int. J. Med. Microbiol.* **293**(suppl. 37):109–16.

Baranova, A. M., Strukova, E. V., Suleimanov, G. D., Sabgaida, T. P., Darchenkova, N. N. (1998) Malaria in Russia and the CIS. *Med. Parazit. (Mosk.)* 3(July–Sept.):58–60.

Barbour, A. G. (1998) Fall and rise of Lyme disease and other *Ixodes* tick-borne infections in North America and Europe. *Br. Med. Bull.* **54**(3):647–58.

Barbour, A. G., Maupin, G. O., Teltow, G. J., Carter, C. J., Piesman, J. (1996) Identification of an uncultivable *Borrelia* species in the hard tick *Amblyomma americanum*: possible agent of a Lyme disease-like illness. *J. Infect. Dis.* **173**(2):403–9.

Bardos, V. (1976) The ecology and medical importance of the Tahyna virus. *MMW Munch. Med. Wochenschr.* **118**(49):1617–20.

Bardos, V., Adamcová, J., Dedei, S., Gjini, N., Rosický, B., Simková, A. (1959) Neutralizing antibodies against some neurotropic viruses determined in human sera in Albania. *J. Hyg. Epidemiol. Microbiol. Immunol.* 3:277–82.

Bardos, V., Danielova, V. (1959) The Tahyna virus – a virus isolated from mosquitoes in Czechoslovakia. *J. Hyg. Epidemiol. Microbiol. Immunol.* 3:264–76.

Bardos, V., Danielova, V. (1959) The Tahyna virus – a virus isolated from mosquitoes in Czechoslovakia. *J. Hyg. Epidemiol. Microbiol. Immunol.* 3:264–76.

Bardos, V., Ryba, J., Hubalek, Z., Olejnicek, J. (1978) Virological examination of mosquito larvae from southern Moravia. *Folia Parasitol.* (Praha) 25(1):75–8.

Barral, M., Garcia Perez, A. L., Juste, R. A. *et al.* (2002) Distribution of *Borrelia burgdorferi* sensu lato in *Ixodes ricinus* (Acari: Ixodidae) ticks from the Basque Country, Spain. *J. Med. Entomol.* 39(1):177–84.

Bartolome Regue, M., Balanzo Fernandez, X., Roca Saumell, C. *et al.* (2002) Imported paludism: an emerging illness. *Med. Clin.* (Barc.) 119(10):372–4.

Basta, J., Plch, J., Hulinska, D., Daniel, M. (1999) Incidence of *Borrelia garinii* and *Borrelia afzelii* in *Ixodes ricinus* ticks in an urban environment, Prague, Czech Republic, between 1995 and 1998. *Eur. J. Clin. Microbiol. Infect. Dis.* 18(7):515–17.

Baumberger, P., Krech, T., Frauchiger, B. (1996) Development of early-summer meningoencephalitis (FSME) in the Thurgau region 1990–1995 – a new endemic area? *Schweiz Med. Wochenschr.* 126(48):2072–7.

Baumgarten, B. U., Rollinghoff, M., Bogdan, C. (1999) Prevalence of *Borrelia burgdorferi* and granulocytic and monocytic ehrlichiae in *Ixodes ricinus* ticks from southern Germany. *J. Clin. Microbiol.* 37(11):3448–51.

Bazlikova, M., Kaaserer, B., Brezina, R., Kovacova, E., Kaaserer, G. (1977) Isolations of a Rickettsia of the spotted-fever group (SF group) from ticks of *Dermacentor marginatus* from Tirol, Austria. *Immun. Infekt.* 5(4):167.

Beati, L., Finidori, J. P., Gilot, B., Raoult, D. (1992) Comparison of serologic typing, sodium dodecyl sulfate-polyacrylamide gel electrophoresis protein analysis, and genetic restriction fragment length polymorphism analysis for identification of rickettsiae: characterization of two new rickettsial strains. *J. Clin. Microbiol.* 30(8):1922–30.

Beati, L., Finidori, J. P., Raoult, D. (1993) First isolation of *Rickettsia slovaca* from *Dermacentor marginatus* in France. *Am. J. Trop. Med. Hyg.* 48(2):257–68.

Beati, L., Humair, P. F., Aeschlimann, A., Raoult, D. (1994) Identification of spotted fever group rickettsiae isolated from *Dermacentor marginatus* and *Ixodes ricinus* ticks in Switzerland. *Am. J. Trop. Med. Hyg.* 51(2):138–48.

Beati, L., Meskini, M., Thiers, B, Raoult, D. (1997) *Rickettsia aeschlimannii* sp. nov., a new spotted fever group rickettsia associated with *Hyalomma marginatum* ticks. *Int. J. Syst. Bacteriol.* 47(2):548–54.

Beati, L., Raoult, D. (1993) *Rickettsia massiliae* sp. nov. a new spotted fever group rickettia. *Int. J. Syst. Bacteriol.* 43(4):839–40.

Beati, L., Roux, V., Ortuño, A. *et al.* (1996) Phenotypic and genotypic characterization of spotted fever group Rickettsiae isolated from Catalan *Rhipicephalus sanguineus* ticks. *J. Clin. Microbiol.* 34(11):2688–94.

Beckendorf, R., Klotz, S. A., Hinkle, N., Bartholomew, W. (2002) Nasal myiasis in an intensive care unit linked to hospital-wide mouse infestation. *Arch. Intern. Med.* **162**(6):638–40.

Behrens, R. H., Roberts, J. A. (1994) Is travel prophylaxis worthwhile? Economic appraisal of prophylactic measures against malaria, hepatitis A, and typhoid in travellers. *Br. Med. J.* **309**(6959):918–22.

Beninati, T., Lo, N., Noda, H. *et al.* (2002) First detection of spotted fever group Rickettsiae in *Ixodes ricinus* from Italy. *Emerg. Infect. Dis.* **8**(9):983–6.

Beran, J. (2004) Tickborne encephalitis in the Czech Republic. *Eurosurveillance Weekly* **8**(26).

Berdyev, A., Pchelkina, A. A., Zhmaeva, Z. M. (1975) A serological survey of foci of West Nile fever among man, farm animals and wild animals in the Turkmenskaya SSR. *Ecology Viruses* (collected papers) 3:117–21.

Bernabeu-Wittel, M., Pachon, J., Alarcon, A. *et al.* (1999) Murine typhus as a common cause of fever of intermediate duration: a 17-year study in the south of Spain. *Arch. Intern. Med.* **159**(8):872–6.

Bernasconi, M. V., Casati, S., Peter, O., Piffaretti, J. C. (2002) *Rhipicephalus* ticks infected with *Rickettsia* and *Coxiella* in Southern Switzerland (Canton Ticino). *Infect. Genet. Evol.* **2**(2):111–20.

Berry, R. L., Parsons, M. A., LaLonde-Weigert, B. J. *et al.* (1986) *Aedes canadensis*, a vector of La Crosse virus (California serogroup) in Ohio. *J. Am. Mosq. Control Assoc.* **2**(1):73–8.

Bettini, S., Gradoni, L., Cocchi, M., Tamburro, A. (1978) Rice culture and *Anopheles labranchiae* in Central Italy. WHO unpublished document. WHO/MAL 78.897, WHO/VBC 78.686. Geneva: WHO.

Bilbie, I., Cristescu, A., Enescu, A. *et al.* (1978) Up-to-date entomological aspects in the previously endemic areas of malaria in the Danube Plain and Dobrudja. *Arch. Roum. Pathol. Exp. Microbiol.* **37**(3–4):389–97.

Bills, G. T. (1965) The occurrence of *Periplaneta brunnea* (Burm.) (Dictyoptera, Blattidae) in an international airport in Britain. *J. Stored Prod. Res.* **1**(2):203–4.

Birnbaum, J., Orlando, J. P., Charpin, D., Vervloet, D. (1995) Cockroaches and mites share the same beds. *J. Allergy Clin. Immunol.* **96**(4):561–2.

Bjoersdorff, A., Berglund, J., Kristiansen, B. E., Soderstrom, C., Eliasson, I. (1999a) Human granulocytic ehrlichiosis: 12 Scandinavian case reports of the new tick-borne zoonosis. *Svensk Veterinartidning* **51**(suppl. 15):29–34.

Bjoersdorff, A., Bergstrom, S., Massung, R. F., Haemig, P. D., Olsen, B. (2001) *Ehrlichia*-infected ticks on migrating birds. *Emerg. Infect. Dis.* **7**(5):877–9.

Bjoersdorff, A., Brouqui, P., Elisasson, I. *et al.* (1999b) Serological evidence of *Ehrlichia* infection in Swedish Lyme borreliosis patients. *Scand. J. Infect. Dis.* **31**(1):51–5.

Bloland, P., Colmenares, J., Gartner, G., Schwartz, I. K., Lobel, H. (1995) Cost and appropriateness of treating *Plasmodium falciparum* infections in the United States. *J. Travel Med.* **2**(1):16–21.

Blythe, M. E., Williams, J. D., Smith, J. M. (1974) Distribution of pyroglyphid mites in Birmingham with particular reference to *Euroglyphus maynei*. *Clin. Allergy* **4**:25–33.

Bonnin, J. L., Delattre, P., Artois, M. *et al.* (1986) Intermediate hosts of *Echinococcus multilocularis* in northeastern France. Description of lesions found in 3 naturally infested rodent species. *Ann. Parasitol. Hum. Comp.* **61**(2):235–43.

Boon, S., den, Schellekens, J. F., Schouls, L. M. *et al.* (2004) Doubling of the number of cases of tick bites and Lyme borreliosis seen by general practitioners in the Netherlands. *Ned. Tijdschr. Geneeskd.* **148**(14):665–70.

Boostrom, A., Beier, M. S., Macaluso, J. A. *et al.* (2002) Geographic association of *Rickettsia felis*-infected opossums with human murine typhus, Texas. *Emerg. Infect. Dis.* **8**(6):549–54.

Borcic, B., Kaic, B., Kralj, V. (1999) Some epidemiological data on TBE and Lyme borreliosis in Croatia. *Zentralbl. Bakteriol.* **289**(5–7):540–7.

Borcic, B., Punda, V. (1987) Sandfly fever epidemiology in Croatia. *Acta Med. Jugosl.* **41**(2):89–97.

Bormane, A., Lucenko, I., Duks, A. *et al.* (2004) Vectors of tick-borne diseases and epidemiological situation in Latvia in 1993–2002. *Int. J. Med. Microbiol.* **293**(suppl. 37):36–47.

Bowen, G. S., McLean, R. G., Shriner, R. B. *et al.* (1981) The ecology of Colorado tick fever in Rocky Mountain National Park in 1974. II. Infection in small mammals. *Am. J. Trop. Med. Hyg.* **30**(2):490–6.

Bradley, D., Lawrence, J., Hart, J. (2003) Consequences of failure to use malaria prophylaxis in The Gambia: an example from the United Kingdom. *Eurosurveillance Weekly* **7**:49.

Braito, A., Corbisiero, R., Corradini, S., Fiorentini, C., Ciufolini, M. G. (1998) Toscana virus infections of the central nervous system in children: a report of 14 cases. *J. Pediatr.* **132**(1):144–8.

Bram, R. A., George, J. E. (2000) Introduction of nonindigenous arthropod pests of animals. *J. Med. Entomol.* **37**(1):1–8.

Briese, T., Rambaut, A., Pathmajeyan, M. *et al.* (2002) Phylogenetic analysis of a human isolate from the 2000 Israel West Nile virus epidemic. *Emerg. Infect. Dis.* **8**(5):528–31.

Brouqui, P., Toga, B., Raoult, D. (1988) La fièvre boutonneuse Méditerranéenne en 1988. *Méd. Malad. Infect.* **6–7**: 323–8.

Brummer-Korvenkontio, M., Saikku, P. (1975) Mosquito-borne viruses in Finland. *Med. Biol.* **53**(5):279–81.

Brummer-Korvenkontio, M., Vapalahti, O., Kuusisto, P. *et al.* (2002) Epidemiology of Sindbis virus infections in Finland 1981–96: possible factors explaining a peculiar disease pattern. *Epidemiol. Infect.* **129**(2):335–45.

Bryan, C. S., Moss, S. W., Kahn, R. J. (2004) Yellow fever in the Americas. *Infect. Dis. Clin. North Am.* **18**(2):275–92.

Bucklar, H., Scheu, U., Mossi, R., Deplazes, P. (1998) Is dirofilariasis in dogs spreading in south Switzerland? *Schweiz. Arch. Tierheilkd.* **140**(6):255–60.

Buckley, A., Dawson, A., Moss, S. R. *et al.* (2003) Serological evidence of West Nile virus, Usutu virus and Sindbis virus infection of birds in the UK. *J. Gen. Virol.* **84**(10):2807–17.

Buffet, M., Dupin, N. (2003) Current treatments for scabies. *Fundam. Clin. Pharmacol.* **17**(2):217–25.

Bukowska, K. (2002) Occurrence of genomic species of *Borrelia burgdorferi* sensu lato in populations of *Ixodes ricinus* ticks from West Pomerania. *Ann. Acad. Med. Stetin* **48**:395–405.

Bunikis, J., Tsao, J., Luke, C. J. *et al.* (2004) *Borrelia burgdorferi* infection in a natural population of *Peromyscus leucopus* mice: a longitudinal study in an area where Lyme borreliosis is highly endemic. *J. Infect. Dis.* **189**(8):1515–23.

Buonavoglia, D., Sagazio, P., Gravino, E. A. *et al.* (1995) Serological evidence of *Ehrlichia canis* in dogs in southern Italy. *New Microbiol.* **18**(1):83–6.

Burek, V., Misic-Mayerus, L., Maretic, T. (1992) Antibodies to *Borrelia burgdorferi* in various population groups in Croatia. *Scand. J. Infect. Dis.* **24**(5):683–4.

Burgdorfer, W., Aeschlimann, A., Peter, O., Hayes, S. F., Philip, R. N. (1979) *Ixodes ricinus*: vector of a hitherto undescribed spotted fever group agent in Switzerland. *Acta Tropica* **36**(4):357–67.

Burgdorfer, W., Barbour, A. G., Hayes, S. F. *et al.* (1982) Lyme disease – a tick-borne spirochetosis? *Science* **216**(4552):1317–19.

Burgdorfer, W., Barbour, A. G., Hayes, S. F., Peter, O., Aeschlimann, A. (1983) Erythema chronicum migrans – a tickborne spirochetosis. *Acta Trop.* **40**(1):79–83.

Burgdorfer, W., Cooney, J. C., Mavros, A. J., Jellison, W. L., Maser, C. (1980) The role of cottontail rabbits (*Sylvilagus* spp.) in the ecology of *Rickettsia rickettsii* in the United States. *Am. J. Trop. Med. Hyg.* **29**(4):686–90.

Burkhart, C. G. (2004) Relationship of treatment-resistant head lice to the safety and efficacy of pediculicides. *Mayo Clin. Proc.* **79**(5):661–6.

Burkhart, C. G., Burkhart, C. N., Burkhart, K. M. (2000) An epidemiologic and therapeutic reassessment of scabies. *Cutis* **65**(4):233–40.

Burkot, T. R., Schneider, B. S., Pieniazek, N. J. *et al.* (2000) *Babesia microti* and *Borrelia bissettii* transmission by *Ixodes spinipalpis* ticks among prairie voles, *Microtus ochrogaster*, in Colorado. *Parasitology* **121**: 595–9.

Butenko, A. M., Demikhov, V. G., Nedialkova, M. S., Lavrova, N. A. (1995) Serodiagnosis and epidemiology of a California encephalitis group of infections in the Ryazan region. *Vopr. Virusol.* **40**(1)17–21.

Butenko, A. M., Galkina, I. V., Kuznetsov, A. A. *et al.* (1990) Serological evidence of the distribution of California serogroup virus in the USSR. *Arch. Virol.* (Suppl. 1):235–41.

Cabral, M., McNerney, R., Gomes, S. *et al.* (1993) Demonstration of natural *Leishmania* infection in asymptomatic dogs in the absence of specific humoral immunity. *Arch. Inst. Pasteur Tunis* **70**(3–4):473–9.

Cabral, M., O'Grady, J. E., Gomes, S. *et al.* (1998) The immunology of canine leishmaniasis: strong evidence for a developing disease spectrum from asymptomatic dogs. *Vet. Parasitol.* **76**(3):173–80.

Cacciapuoti, B., Rivosecchi, L., Stella, E., Ciceroni, L., Khoury, C. (1985) Preliminary studies on the occurrence of Rickettsiae of the spotted fever group in

Rhipicephalus sanguineus captured in suburban areas. *Boll. Ist. Sieroterapico* **64**(1):77–81.

Cachia, E. A., Fenech, F. F. (1964) A review of kala-azar in Malta from 1947 to 1962. *Trans. Roy. Soc. Trop. Med. Hyg.* **58**(3):234–41.

Caflisch, U., Tonz, O., Schaad, U. B., Aeschlimann, A., Burgdorfer, W. (1984) Tick-borne meningoradiculitis – a form of spirochetosis. *Schweiz Med. Wochenschr.* **114**(18):630–4.

Calisher, C. H. (1994) Medically important arboviruses of the United States and Canada. *Clin. Microbiol. Rev.* **7**(1):89–116.

Calisher, C. H., Craven, R. B. (1998) Colorado tick fever: Current approaches to diagnosis and treatment. *Infect. Med.* **15**(8):524–5, 528–30, 532–3.

Calisher, C. H., McLean, R. G., Smith, G. C. *et al.* (1977) Rio Grande – a new phlebotomus fever group virus from south Texas. *Am. J. Trop. Med. Hyg.* **26**(5 Pt 1):997–1002.

Calisher, C. H., Monath, T. P., Muth, D. J. *et al.* (1980) Characterization of Fort Morgan virus, an alphavirus of the western equine encephalitis virus complex in an unusual ecosystem. *Am. J. Trop. Med. Hyg.* **29**(6):1428–40.

Calisher, C. H., Tzianabos, T., Lord, R. D., Coleman, P. H. (1970) Tamiami virus, a new member of the TaCaribe group. *Am. J. Trop. Med. Hyg.* **19**(3):520–6.

Campbell, G. L., Ceianu, C. S., Savage, H. M. (2001) Epidemic West Nile encephalitis in Romania. Waiting for history to repeat itself. *Ann. N. Y. Acad. Sci.* **951**:94–101.

Campbell, G. L., Paul, W. S., Schriefer, M. E. *et al.* (1995) Epidemiological and diagnostic studies of patients with suspected early Lyme disease, Missouri, 1990–1993. *J. Infect. Dis.* **172**(2):470–80.

Campbell, G. L., Reeves, W. C., Hardy, J. L., Eldridge, B. F. (1990) Distribution of neutralizing antibodies to California and Bunyamwera serogroup viruses in horses and rodents in California. *Am. J. Trop. Med. Hyg.* **42**(3):282–90.

Canada Communicable Disease Report (2000) Hantavirus pulmonary syndrome in Canada, 1989–1999. *CCDR* **26**(8):65–9.

Cantile, C., Di Guardo, G., Eleni, C., Arispici, M. (2000) Clinical and neuropathological features of West Nile virus equine encephalomyelitis in Italy. *Equine Vet. J.* **32**(1):31–5.

Capelli, G., Baldelli, R., Ferroglio, E. *et al.* (2004) Monitoring of canine leishmaniasis in northern Italy: an update from a scientific network. *Parassitologia* **46**(1–2):193–7.

Caruso, G. (2003) TBE (Tick-borne encephalitis) in Italy. Abstract from VII International Potsdam Symposium on Tick-borne Diseases (IPS-VII) 2003.

Cascio, A., Colomba, C. (2002) Childhood Mediterranean visceral leishmaniasis. *Infez. Med.* **11**(1):5–10.

Casher, L., Lane, R., Barrett, R., Eisen, L. (2002) Relative importance of lizards and mammals as hosts for ixodid ticks in northern California. *Exp. Appl. Acarol.* **26**(1–2):127–43.

Castel, Y., Chastel, C., Le Moigne, C., Simitzis, A. M., Le Fur, J. M. (1982) Autochthonous kala-azar. *Arch. Fr. Pediatr.* **39**(2):97–8.

Castelli, F., Caligaris, S., Matteelli, A. *et al.* (1993) 'Baggage malaria' in Italy: cryptic malaria explained? *Trans. Roy. Soc. Trop. Med. Hyg.* **87**(4):394.

Caten, J. L., Kartman, L. (1968) Human plague in the United States during 1966, case reports. *Southwestern Med.* **49**(6):102–8.

Cave, W., Pandey, P., Osrin, D., Shlim, D. R. (2003) Chemoprophylaxis use and the risk of malaria in travelers to Nepal. *J. Travel Med.* **10**(2):100–5.

CDC (2002) Lyme disease – United States, 2000. *MMWR* **51**(2):29–31.

CDC (2004d) Malaria Facts. Malaria in the United States 4 May. MMWR **53**: 1195–8.

Ceianu, C. S., Ungureanu, A., Nicolescu, G. *et al.* (2001) West Nile virus surveillance in Romania: 1997–2000. *Viral Immunol.* **14**(3):251–62.

Centre for Disease Control (1984) Epidemic typhus – Georgia. *MMWR* **33**(43):618–19.

Centre for Disease Control (1990) Imported bubonic plague – District of Columbia. *MMWR* **39**(49):895–901.

Centre for Disease Control (1993) Update: Hantavirus disease – United States 1993. *MMWR* **43**(3):45–8.

Centre for Disease Control (1994a) Human plague – United States, 1993–1994. *MMWR* **43**(13):242–246.

Centre for Disease Control (1994b) Hantavirus – pulmonary syndrome – United States, 1994. *MMWR* **43**(30):548–56.

Centre for Disease Control (1994c) Epidemiologic notes and reports – Rat-bite fever in a college student – California. *MMWR* **33**(22):318–20.

Centre for Disease Control (1996) Tick paralysis Washington 1995. *MMWR* **45**(16):325–6.

Centre for Disease Control (1999) Outbreak of West Nile-like viral encephalitis – New York, 1999. *MMWR* **48**(38):845–9.

Centre for Disease Control (2000a) Update: Surveillance for West Nile Virus in overwintering mosquitoes – New York, 2000. *MMWR* **49**(9):178–9.

Centre for Disease Control (2000b) Update: West Nile virus activity – Eastern United States, 2000. *MMWR* **49**(46):1044–7.

Centre for Disease Control (2000c) Fatal illnesses associated with a New World Arenavirus – California, 1999–2000. *MMWR* **49**(31):709–11.

Centre for Disease Control (2000d) Provisional surveillance summary of the West Nile virus epidemic – United States, January – November 2002. *MMWR* **51**(50):1129–33.

Centre for Disease Control (2001) Outbreak of Powassan encephalitis – Maine and Vermont, 1999–2001. *MMWR* **50**(35):761–4.

Centre for Disease Control (2002a) Cat-scratch disease in children – Texas, September 2000–August 2001. *MMWR* **51**(10):212–14.

Centre for Disease Control (2002b) Sexually transmitted diseases treatment guidelines 2002. *MMWR* **51**(RR–O6):1–80.

Centre for Disease Control (2003) Imported plague, New York City, 2002. *MMWR* **52**(31):725–8.

Centre for Disease Control (2004a) Lyme disease – United States, 2001–2002. *MMWR* **53**(17):365–9.

Centre for Disease Control (2004b) Fatal cases of Rocky Mountain spotted fever in family clusters – three states, 2003. *MMWR* **53**(19):407–10.

Centre for Disease Control (2004c) Malaria surveillance – United States, 2002. *MMWR* **53**(1):21–34.

Centre for Disease Control (2005) Salmonellosis, pet rodents 2003–2004. *MMWR* **54**(17):429–33.

Cernescu, C., Ruta, S. M., Tardei, G. *et al.* (1997). A high number of severe neurologic clinical forms during an epidemic of West Nile virus infection. *Rom. J. Virol.* **48**(1–4):13–25.

Chan-Yeung, M., Becker A., Lam, J. *et al.* (1995) House dust mite allergen levels in two cities in Canada: effects of season, humidity, city and home characteristics. *Clin. Exp. Allergy* **25**(3):240–6.

Chaniotis, B., Psarulaki, A., Chaliotis, G. *et al.* (1994) Transmission cycle of murine typhus in Greece. *Ann. Trop. Med. Parasitol.* **88**(6):645–7.

Chaniotis, B., Tselentis, Y. (1994) Leishmaniasis, sandfly fever and phlebotomine sandflies in Greece: An annotated bibliography. WHO unpublished document, WHO/LEISH/94.34. Geneva: WHO.

Chapel, T. A., Katta, T., Kuszmar, T., DeGiusti, D. (1979) Pediculosis pubis in a clinic for treatment of sexually transmitted diseases. *Sex. Transm. Dis.* **6**(4):257–60.

Chapman, L. E., Khabbaz, R. F. (1994) Etiology and epidemiology of the Four Corners hantavirus outbreak. *Infect. Agents Dis.* **3**(5):234–44.

Charpin, D., Kleisbauer, J. P., Lanteaume, A. *et al.* (1988) Asthma and allergy to house-dust mites in populations living in high altitudes. *Chest* **93**(4):758–61.

Chastel, C. (1998) Erve and Eyach: two viruses isolated in France, neuropathogenic for man and widely distributed in western Europe. *Bull. Acad. Natl. Med.* **182**(4):801–9.

Chastel, C. (1999) The "plague" of Barcelona. Yellow fever epidemic of 1821. *Bull. Soc. Pathol. Exot.* **92**(5 pt 2):405–7.

Chastel, C., Launay, H., Rogues, G., Beaucournu, J. C. (1980) Infections a arbovirus au Espagne: enquete serologique chez les petits mammiferes. *Bull. Soc. Pathol. Exot. Filiales* **78**:384–90.

Chastel, C., Main, A. J., Couatarmanac'h, H. A. *et al.* (1984) Isolation of Eyach virus (Reoviridae, Colorado tick fever group) from *Ixodes ricinus* and *I. ventalloi* ticks in France. *Arch. Virol.* **82**(3–4):161–71.

Chernin, E. (1987) The disappearance of bancroftian filariasis from Charlestown, South Carolina. *Am. J. Trop. Med. Hyg.* **37**(1):111–14.

Chicharro, C., Jimenez, M. I., Alvar, J. (2003) Iso-enzymatic variability of *Leishmania infantum* in Spain. *Ann. Trop. Med. Parasitol.* **97**(suppl. 1):57–64.

Childs, J. E., Glass, G. E., Korch, G. W., Ksiazek, T. G., Leduc, J. W. (1992) Lymphocytic choriomeningitis virus infection and house mouse (*Mus musculus*) distribution in urban Baltimore. *Am. J. Trop. Med. Hyg.* **47**(1):27–34.

Childs, J. E., Glass, G. E., Ksiazek, T. G. *et al.* (1991) Human rodent contact and infection with lymphocytic choriomeningitis and Seoul viruses in an inner-city population. *Am. J. Trop. Med. Hyg.* **44**(2):117–21.

Childs, J. E., Ksiazek, T. G., Spiropoulou, C. F. *et al.* 1994. Serologic and genetic identification of *Peromyscus maniculatus* as the primary rodent reservoir for a new hantavirus in the southwestern United States. *J. Infect. Dis.* **169**(6):1271–80.

Chmielewska-Badora, J. (1998) Seroepidemiologic study on Lyme borreliosis in the Lublin region (eastern Poland). *Ann. Agric. Environ. Med.* **5**(2):183–6.

Chodynicka, B., Flisiak, I. (1998) Epidemiology of erythema migrans in north-eastern Poland. *Rocz Akad. Med. Bialymst* **43**:271–7.

Choi, C. M., Lerner, E. A. (2001) Leishmaniasis as an emerging infection. *J. Investig. Dermatol. Symp. Proc.* **6**(3):175–82.

Chomel, B. B. (2000) Cat-scratch disease. *Rev. Sci. Tech.* **19**(1):136–50.

Christensen, K. L., Ronn, A. M., Hansen, P. S., Aarup, M., Buhl, M. R. (1996) Contributing causes of malaria among Danish travellers. *Ugeskr Laeger* **158**(51):7411–14.

Christenson, B. (1984) An outbreak of tularemia in the northern part of central Sweden. *Scand. J. Infect. Dis.* **16**(3):285–90.

Christian, F., Rayert, P., Patey, O., Lafaix, C. (1996) Epidemiology of Lyme disease in France: Lyme borreliosis in the region of Berry sud: A six year retrospective. *Eur. J. Epidemiol.* **12**(4):479–83.

Christmann, D., Staub-Schmidt, T., Gut, J. P. *et al.* (1995) Situation actuelle en France de l'encéphalite a tique. *Méd. Mal. Infect.* **25**(special avril):660–4.

Christova, I. S., Dumler, J. S. (1999) Human granulocytic ehrlichiosis in Bulgaria. *Am. J. Trop. Med. Hyg.* **60**(1):58–61.

Christova, I., Komitova, R. (2004) Clinical and epidemiological features of Lyme borreliosis in Bulgaria. *Wien Klin. Wochenschr.* **116**(1–2):42–6.

Christova, I., Schouls, L., van De Pol, I. *et al.* (2001) High prevalence of granulocytic Ehrlichiae and *Borrelia burgdorferi* sensu lato in *Ixodes ricinus* ticks from Bulgaria. *J. Clin. Microbiol.* **39**(11):4172–4.

Chumakov, M. P., Bashkirtsev, V. N., Golger, E. I. *et al.* (1974) Isolation and identification of Crimean haemorrhagic fever and West Nile fever viruses from ticks collected in Moldavia. *Tr. Inst. Polio Virus Entsef. Akad. Med. Nauk.* **22**(2):45–9.

Chumakov, M. P., Belyaeva, A. P. Y., Butenko, A. M., Martyanova, L. I. (1968) Isolation of West Nile virus from *Hyalomma plumbeum plumbeum* Panz. ticks. *Tr. Inst. Polio Virus Entsef. Akad. Med. Nauk.* **13**:365–73.

Ciceroni, L., Stepan, E., Pinto, A. *et al.* (2000) Epidemiological trend of human leptospirosis in Italy between 1994 and 1996. *Eur. J. Epidemiol.* **16**(1):79–86.

Cimmino, M. A., Fumarola, D., Sambri, V., Accardo, S. (1992) The epidemiology of Lyme borreliosis in Italy. *Microbiologia* **15**(4):419–24.

Cinco, M., Balanzin, D., Benussi, P., Trevisan, G. (1993) Seroprevalence and incidence of Lyme borreliosis in forestry workers in Friuli Venezia Giulia (Northern Italy). *Alpe Adria. Microbiol. J.* **2**(2):91–8.

Cinco, M., Banfi, E., Trevisan, G., Stanek, G. (1989) Characterization of the first tick isolate of *Borrelia burgdorferi* from Italy. *Acta Parasitol. Microbiol. Immunol. Scand.* **97**(4):381–2.

Cinco, M., Murgia, R., Costantini, C. (1995) Prevalence of IgG reactivity in Lyme borreliosis patients versus *Borrelia garinii* and *Borrelia afzelii* in a restricted area of Northern Italy. *FEMS Immunol. Med. Microbiol.* **12**(3–4):217–22.

Cinco, M., Padovan, D., Murgia, R. *et al.* (1998) Rate of infection of *Ixodes ricinus* ticks with *Borrelia burgdorferi* sensu stricto, *Borrelia garinii, Borrelia afzelii* and group VS116 in an endemic focus of Lyme disease in Italy. *Eur. J. Clin. Microbiol. Infect. Dis.* **17**(2):90–4.

Ciufolini, M. G., Nicoletti, L. (1997) Dengue: an emerging health problem. *Giorn. Ital. Med. Trop.* **2**(1–4): 1–8.

Cizman, M., Avsic-Zupanc, T., Petrovec, M., Ruzic-Sabljic, E., Pokorn, M. (2000) Seroprevalence of ehrlichiosis, Lyme borreliosis and tick-borne encephalitis infections in children and young adults in Slovenia. *Wien. Klin. Wochenschr.* **112**(19):842–5.

Clark, G. G., Crabbs, C. L., Bailey, C. L., Calisher, C. H., Craig, G. B. Jr. (1986) Identification of *Aedes campestris* from New Mexico: with notes on the isolation of western equine encephalitis and other arboviruses. *J. Am. Mosq. Control Assoc.* **2**(4):529–34.

Clark, J., Friesen, D. L., Williams, W. A. (1992) Management of an outbreak of Norwegian scabies. *Am. J. Infect. Control* **20**(4):217–20.

Cleary, V. A., Figueroa, J. I., Heathcock, R., Warren, L. (2003) Improving malaria surveillance in inner city London: is there a need for targeted intervention? *Commun. Dis. Publ. Hlth.* **6**(4):300–4.

Clement, J., Heyman, P., McKenna, P., Colson, P., Avsic-Zupanc, T. (1997) The hantaviruses of Europe: from the bedside to the bench. *Emerg. Infect. Dis.* **3**(2):205–11.

Clore, E. R., Longyear, L. A. (1990) Comprehensive pediculosis screening programs for elementary schools. *J. Sch. Hlth.* **60**(5):212–14.

Coleman, P. H. (1969) Tensaw virus, a new member of the Bunyamwera arbovirus group from the Southern United States. *Am. J. Trop.Med. Hyg.* **18**(1):81–91.

Coleman, W. (1984) Epidemiological method in the 1860s: yellow fever at Saint Nazaire. *Bull. Hist. Med.* **58**(2):145–63.

Collard, M.,Gut, J. P., Christmann, D. *et al.* (1993) Tick-borne encephalitis in Alsace. *Rev. Neurol. (Paris)*, **149**(3):198–201.

Collares-Pereira, M., Couceiro, S., Franca, I. *et al.* (2004) First isolation of *Borrelia lusitaniae* from a human patient. *J. Clin. Microbiol.* **42**(3):1316–18.

Colloff, M. J. (1998). Taxonomy and identification of dust mites. *Allergy* **53**(suppl. 48):7–12.

Comer, J. A., Flynn, C., Regnery, R. L., Vlahov, D., Childs, J. E. (1996) Antibodies to *Bartonella* species in inner-city intravenous drug users in Baltimore, MD. *Arch. Intern. Med.* **156**(21):2491–5.

Comer, J. A., Tzianabos, T., Flynn, C., Vlahov, D., Childs, J. E. (1999) Serologic evidence of rickettsial pox (*Rickettsia akari*) infection among intravenous drug users in inner-city Baltimore, Maryland. *Am. J. Trop. Med. Hyg.* **60**(6):894–8.

Comer, K. M. (1991) Rocky Mountain spotted fever. *Vet. Clin. North Am. Small Anim. Pract.* **21**(1):27–44.

Comiskey, N., Wesson, D. M. (1995) *Dirofilaria* (Filarioidea:Onchocercidae) infection in *Aedes albopictus* (Diptera: Culicidae) collected in Louisiana. *J. Med. Entomol.* **32**(5):734–7.

Conrad, M. E. (1989) Review: *Ehrlichia canis*: a tick-borne rickettsial-like infection in humans living in the southeastern United States. *Am. J. Med. Sci.* **297**(1):35–7.

Couatrmanac'h, A., Chastel, C., Chastel, O., Beaucournu, J. C. (1989) Tiques d'Importation observees en Bretagne. *Bull. Soc. Fr. Parasitol.* **7**(1):127–32.

Courtiade, C., Labrèze, C., Fontan, I., Taieb, A., Maleville, J. (1993) La pédiculose du cuir chevelu: enquête par questionnaire dans quatre groupes scolaires de l'Académie de Bordeaux en 1990–1991. *Ann. Dermatol. Venérol.* **120**(5):363–8.

Craven, R. B., Eliason, D. A., Francy, D. B. *et al.* (1988) Importation of *Aedes albopictus* and other exotic mosquito species into the United States in used tires from Asia. *J. Am. Mosq. Control Assoc.* **4**(2):138–42.

Craven, R. B., Maupin, G. O., Beard, M. L., Quan, T. J., Barnes, A. M. (1993) Reported cases of human plague infections in the United States, 1970–1991. *J. Med. Entomol.* **30**(4):758–61.

Cruz, I., Morales, M. A., Noguer, I., Rodriguez, A., Alvar, J. (2003) *Leishmania* in discarded syringes from intravenous drug users. *Lancet* **359**(9312):1124–5.

Cuadros, J., Calvente, M. J., Benito, A. *et al.* (2002) *Plasmodium ovale* malaria acquired in central Spain. *Emerg. Infect. Dis.* **8**(12):1506–8.

Currie, B. J., Harumal, P., McKinnon, M., Walton, S. F. (2004) First documentation of in vivo and in vitro ivermectin resistance in *Sarcoptes scabiei*. *Clin. Infect. Dis.* **39**(1):8–12.

Custovic, A., Green, R., Smith, A., Chapman, M. D., Woodcok, A. (1996) New mattresses: how fast do they become a significant source of exposure to house dust mite allergens? *Clin. Exper. Allergy* **26**(11):1243–5.

Dabrowski, J., Schönberg, A., Wegner, Z., Stanczak, J., Kruminis-Lozowska, W. (1994) The first isolation of *Borrelia burgdorferi* from *Ixodes ricinus* ticks in Poland. *Bull. Inst. Marit. Trop. Med.* (*Gdynia*) (1993–1994) **44–45**(1–4):61–4.

Dalton, M. J., Clarke, M. J., Holman, R. C. *et al.* (1995) National surveillance for Rocky mountain spotted fever, 1981–1992: epidemiologic summary and evaluation of risk factors for fatal outcome. *Am. J. Trop. Med. Hyg.* **52**(5):405–13.

Daniel, M., Danielova, V., Kriz, B., Kott, I. (2004) An attempt to elucidate the increased incidence of tick-borne encephalitis and its spread to higher altitudes in the Czech Republic. *Int. J. Med. Microbiol.* **293**(suppl. 37):55–62.

Daniel, M., Sramova, H., Zablabska, E. (1994) *Lucilia sericata* (Diptera: Calliphoridae) causing hospital-acquired myiasis of a traumatic wound. *J. Hosp. Infect.* **28**(2):149–52.

Daniel, S. A., Manika, K., Arvanmdou, M., Antoniadis, A. (2002) Prevalence of *Rickettsia conorii* and *Rickettsia typhi* infections in the population of northern Greece. *Am. J. Trop. Med. Hyg.* **66**(1):76–9.

Danielova, V. (1990) Dissemination of arboviruses transmitted by mosquitoes in Czechoslovakia and some epidemiologic consequences. *Cske. Epidemiol. Mikrobiol. Immunol.* **39**(6):353–8.

Danielova, V. (2002) Natural foci of tick-borne encephalitis and prerequisites for their existence. *Int. J. Med. Microbiol.* **291**(suppl. 33):183–6.

Danielova, V., Holubova, J. (1977) Two more mosquito species proved as vectors of Tahyna virus in Czechoslovakia. *Folia Parasitol.* (*Praha*) **24**(2):187–9.

Danielova, V., Malkova, D., Minar, J. *et al.* (1978) Arbovirus isolations from mosquitoes in South Slovakia. *Folia Parasitol.* (*Praha*) **25**(2):187–90.

Danielova, V., Malkova, D., Minar, J., Ryba, J. (1976) Dynamics of the natural focus of Tahyna virus in southern Moravia and species succession of its vectors, the mosquitoes of the genus *Aedes*. *Folia Parasitol.* (*Praha*) **23**(3):243–9.

Danis, M., Legros, F., Thellier, M., Caumes, E. (2002) Current data on malaria in metropolitan France. *Méd. Trop.* **62**(3):214–18.

Danis, M., Mouchet, M., Giacomini, T. *et al.* (1996) Paludisme autochtone et introduit en Europe. *Méd. Mal. Infect.* **26**(special):393–6.

Dashkova, N. G., Stepenko, A. A., Glushkova, M. R., Yarotskii, M. S. (1978) Characteristics of malaria introduced into Moscow from abroad (1974–1976). *Med. Parazit* (*Mosk.*). **47**(1):105–9.

Dautel, H., Kahl, O., Knulle, W. (1991) The soft tick *Argas reflexus* (F.) (Acari, Argasidae) in urban environments and its medical significance in Berlin (West). *J. Appl. Entomol.* **111**(4):380–90.

Dawson, J. E., Anderson, B. E., Fishbein, D. B. *et al.* (1991) Isolation and characterization of an *Ehrlichia* sp. from a patient diagnosed with human ehrlichiosis. *J. Clin. Microbiol.* **29**(12):2741–5.

de Mik, E. L., van Pelt, W., Docters-van Leeuwen, B. D. *et al.* (1997) The geographical distribution of tick bites and erythema migrans in general practice in The Netherlands. *Int. J. Epidemiol.* **26**(2):451–7.

De Zulueta, J. (1994) Malaria and ecosystems: from prehistory to posteradication. *Parassitologia* **36**(1–2):7–15.

Deblock, S., Petavy, A. F. (1983) Hepatic larvae of cestode parasites of the vole rat *Arvicola terrestris* in Auvergne (France). *Ann. Parasitol. Hum. Comp.* **58**(5):423–37.

Dekonenko, E. J., Steere, A. C., Berardi, V. P., Kravchuk, L. N. (1988) Lyme borrelioses in the Soviet Union: a cooperative US–USSR report. *J. Infect. Dis.* **158**(4):748–53.

Del Giudice, P., Mary-Krause, M., Pradier, C. *et al.* (2002) Impact of highly active antiretroviral therapy on the incidence of visceral leishmaniasis in a French cohort of patients infected with human immunodeficiency virus. *J. Infect. Dis.* **186**(9):1366–70.

Del Giudice, P., Schuffenecker, I., Vandenbos, F., Evelyne, C., Zeller, H. (2004) Human West Nile virus, France. [letter] *Emerg. Infect. Dis.* **10**(10):1885–6.

Delmont, J., Brouqui, P., Poullin, P., Bourgeaade, A. (1994) Harbour-acquired *Plasmodium falciparum* malaria. *Lancet* **344**(8918):330–1.

Demikhov, V. G. (1995) Outcomes and prognosis of diseases caused by Inkoo and Tahyna viruses. *Vopr. Virusol.* **40**(2):72–4.

Demikhov, V. G., Chaitsev, V. G. (1995) Neurologic characteristics of diseases caused by Inkoo and Tahyna viruses. *Vopr. Virusol.* **40**(1):21–5.

Derdakova, M., Beati, L., Pet'ko, B., Stanko, M., Fish, D. (2003) Genetic variability within *Borrelia burgdorferi* sensu lato genospecies established by PCR-single-strand conformation polymorphism analysis of the rrfA-rrlB intergenic spacer in *Ixodes ricinus* ticks from the Czech Republic. *Appl. Environ. Microbiol.* **69**(1):509–16.

Desjeux, P., Alvar, J. (2003) Leishmania/HIV co-infections: epidemiology in Europe. *Ann. Trop. Med. Parasitol.* **97**(suppl. 1):3–15.

Desowitz, R. S. (1991) *The Malaria Capers. More Tales of Parasites and People, Research and Reality*. New York: W. W. Norton & Company.

Develoux, M. (2004) Ivermectin. *Ann. Dermatol. Venereol.* **131**(6–7):561–570.

Di Luca, M., Toma, L., Severini, F., D'Ancona, F., Romi, R. (2001) *Aedes albopictus* in Rome: monitoring in the 3-year period of 1998–2000. *Ann. Ist. Super. Sanita* **37**(2):249–54.

Di Martino, L., Gramiccia, M., Occorsio, P. *et al.* (2004) Infantile visceral leishmaniasis in the Campania region, Italy: experience from a paediatric reference centre. *Parassitologia* **46**(1–2):221–3.

Dietz, V. J., Gubler, D. J., Ortiz, S. *et al.* (1996) The 1986 dengue and dengue hemorrhagic fever epidemic in Puerto Rico: epidemiologic and clinical observations. *P. R. Hlth Sci. J.* **15**(3):201–10.

Diglisic, G., Rossi, C. A., Doti, A., Walshe, D. K. (1999) Seroprevalence study of Hantavirus infection in the community based population. *Md. Med. J.* **48**(6):303–6.

Dirckx, J. H. (1980) The pied Piper of Hamelin. A medical–historical interpretation. *Am. J. Dermatopathol.* **2**(1):39–45.

Dobec, M., Hrabar, A. (1990) Murine typhus on the northern Dalmatian Islands, Yugoslavia. *J. Hyg. Epidemiol. Microbiol. Immunol.* **34**(2):175–181.

Dobler, G. (1996) Arboviruses causing neurological disorders in the central nervous system. *Arch. Virol.* (Vienna) *Suppl.* **11**:33–40.

Dobson, M. J. (1994) Malaria in England: a geographical and historical perspective. *Parassitologia* **36**(1–2):35–60.

Doby, J. M., Anderson, J. F., Couatarmanac'h, A., Magnarelli, L. A., Martin, A. (1987) Lyme disease in Canada with possible transmission by an insect. *Zentralbl. Bakteriol. Mikrobiol. Hyg.* [A] **263**(3):488–90.

Doby, J. M., Couatarmanac'h, A., Aznar, C. (1986) Filarioses canines par *Dirofilaria immitis*, Ledyi, 1856 et *Dirofilaria repens* (Raillet et Henry, 1911) dans l' Ouest de la France. *Bull. Soc. Fr. Parasitol.* **2**:229–33.

Doby, J. M., Deunff, J., Couatarmanac'h, A., Guiguen, C. (1985) Human hypodermyiasis in France in 1984: 266 cases recorded to date. Distribution according to known geographic origin. *Bull. Soc. Pathol. Exot. Filiales.* **78**(2):205–15.

dos Santos, C. C., Kain, K. C. (1999) Two tick-borne diseases in one: a case report of concurrent babesiosis and Lyme disease in Ontario. *Canad. Med. Assoc. J.* **160**:1851–3.

Draganescu, N., Duca, M., Girjabu, E. *et al.* (1977) Epidemic outbreak caused by West Nile virus in the crew of a Roumanian cargo ship passing the Suez Canal and the Red Sea on route to Yokohama. *Virologie* **28**(4):259–62.

Draganescu, N., Girjabu, E. (1979) Investigations on the presence of antibodies to Tahyna virus in Romania. *Virologie* **30**(2):91–3.

Drancourt, M., Aboudharam, G., Signoli, M., Dutour, O., Raoult, D. (1998) Detection of 400-year-old *Yersinia pestis* DNA in human dental pulp: an approach to the diagnosis of ancient septicemia. *Proc. Natl. Acad. Sci. USA* **95**(21):12637–40.

Drancourt, M., Mainardi, J. L., Brouqui, P. *et al.* (1995) *Bartonella* (*Rochalimaea*) *quintana* endocarditis in three homeless men. *N. Engl. J. Med.* **332**(7):419–23.

Drebot, M. A., Lindsay, R., Barker, I. K., Artsob, H. (2001) Characterization of a human granulocytic ehrlichiosis-like agent from *Ixodes scapularis*, Ontario, Canada. *Emerg. Infect. Dis.* **7**(3):479–80.

Drevova, H., Hulinska, D., Kurzova, Z., Plch, J., Janovska, D. (2003) Study of awareness of tick-borne diseases among children and young people in the Czech Republic. *Cent. Eur. J. Publ. Hlth.* **11**(3):138–41.

Drraganescu, N., Iftimovici, R., Iacobescu, V. *et al.* (1975) Investigations on the presence of antibodies to several flaviviruses in humans and some domestic animals in a biotope with a high frequency of migratory birds. *Virologie* **26**(2):103–8.

Dubey, J. P., Frenkel, J. K. (1998) Toxoplasmosis of rats: a review, with considerations of their value as an animal model and their possible role in epidemiology. *Vet. Parasitol.* **77**(1):1–32.

Dubus, J. C., Guerra, M. T., Bodiou, A. C. (2001). Cockroach allergy and asthma. *Allergy* **56**(4):351–2.

Duh, D., Petrovec, M., Avsic Zupanc, T. (2001) Diversity of *Babesia* infecting European sheep ticks (*Ixodes ricinus*). *J. Clin. Microbiol.* **39**(9):3395–7.

Dumpis, U., Crook, D., Oksi, J. (1999) Tick-borne encephalitis. *Clin. Infect. Dis.* **28**(4):882–90.

Dworkin, M. S., Anderson, D. E. Jr., Schwan, T. G. *et al.* (1998) Tick-borne relapsing fever in the northwestern United States and southwestern Canada. *Clin. Infect. Dis.* **26**(1):122–31.

Echevarria, J. M., de Ory, F., Guisasola, M. E. *et al.* (2003) Acute meningitis due to Toscana virus infection among patients from both the Spanish Mediterranean region and the region of Madrid. *J. Clin. Virol.* **26**(1):79–84.

Edwards, J. F., Hendricks, K. (1997) Lack of serologic evidence for an association between Cache Valley virus infection and anencephaly and other neural tube defects in Texas. *Emerg. Infect. Dis.* **3**(2):195–7.

Effler, P. V., Pang, L., Kitsutani, P. *et al.* (2005) Dengue fever, Hawaii, 2001–2002. *Emerg. Infect. Dis.* **11**(5):742–9.

Ehrenkranz, N. J., Ventura, A. K., Cuadrado, R. R., Pond, W. L., Poreter, J. E. (1971) Pandemic dengue in Caribbean countries and the southern United States – past present, and potential problems. *N. Engl. J. Med.* **285**(26):1460–9.

Eitrem, R., Niklasson, B., Weiland, O. (1991b) Sandfly fever among Swedish tourists. *Scand. J. Infect. Dis.* **23**(4):451–7.

Eitrem, R., Stylianou, M., Niklasson, B. (1991a) High prevalence rates of antibody to three sandfly fever viruses (Sicilian, Naples, and Toscana) among Cypriots. *Epidemiol. Infect.* **107**(3):685–91.

Eleckova, E., Labuda, M., Rajani, J. (2003) Arboviruses in Slovakia. *Acta Med. Martiniana* **3**(1):19–28.

Eliasson, H., Lindback, J., Nuorti, J. P. *et al.* (2002) The 2000 tularemia outbreak: a case-control study of risk factors in disease-endemic and emergent areas, Sweden. *Emerg. Infect. Dis.* **8**(9):956–60.

Ellert-Zygadlowska, J., Radowska, D., Orlowski, M. *et al.* (1996) Borreliosis – Lyme disease – a growing clinical problem. *Przegl. Lek.* **53**(8):587–91.

Elliot, L. B., Fournier, P. V., Teltow, G. J. (1990) Rickettsia in Texas. *Ann. N. Y. Acad. Sci.* **590**(1):221–6.

Eloy, O., Bruneel, F., Diebold, C. *et al.* (2003) Pediatric imported malaria. – Experience of the hospital center of Versailles (1997–2001). *Ann. Biol. Clin. (Paris)* **61**(4):449–53.

Eltari, E., Gina, A., Bitri, T., Sharofi, F. (1993) Some data on arboviruses, especially tick-borne encephalitis in Albania. *Giorn. Malattie Infett. Parassit.* **45**(5):404–11.

Emek, E., Kozuch, O., Nosek, J., Hudec, K., Folk, C. (1975) Virus neutralizing antibodies to arboviruses in birds of the order Anseriformes in Czechoslovakia. *Acta Virol.* **19**(4):349–53.

Emek, E., Kozuch, O., Nosek, J., Teplan, J., Folk, C. (1977) Arboviruses in birds captured in Slovakia. *J. Hyg. Epidemiol. Microbiol. Immunol.* **21**(3):353–9.

Ena, P., Spano, G., Leigheb, G. (1999) Treatment of scabies: recurrent problems. Reasons for failure and new proposals. *Giorn. Ital. Dermatol. Venereol.* **1349**(2):115–22.

Encinas Grandes, A., Gomez-Bautista, M., Martin Novo, M., Simon Martin, F. (1988) Leishmaniasis in the province of Salamanca, Spain. Prevalence in dogs and seasonal dynamics of vectors. *Ann.Parasitol. Hum. Comp.* **63**(6):387–97.

Engel, A. (1949) Haverhill fever (in connection with a case observed in Sweden). *Acta Med. Scand.* **132**:562–71.

Eriksson, N. E., Ryden, B., Jonsson, P. (1989) Hypersensitivity to larvae of chironomids (non-biting midges). Cross-sensitization with crustaceans. *Allergy* **44**(5):305–13.

Escudero, R., Barral, M., Perez, A. *et al.* (2000) Molecular and pathogenic characterization of *Borrelia burgdorferi* sensu lato isolates from Spain. *J. Clin. Microbiol.* **38**(11):4026–33.

Eskow, E. S., Krause, P. J., Spielman, A., Freeman, K., Aslanzadeh, J. (1999) Southern extension of the range of human babesiosis in the eastern United States. *J. Clin. Microbiol.* **37**(6):2051–2.

Espejo Arenas, E., Font Creus, B., Alegre Segura, M. D., Segura Porta, F., Bella Cueto, F. (1990) Seroepidemiological survey of Mediterranean spotted fever in an endemic area ("Valles Ocidental", Barcelona, Spain). *Trop. Geograph. Med.* **43**(2): 212–16.

Espy, P. D., Jolly, W. R. (1976) Norwegian scabies. Occurrence in a patient undergoing imunosuppression. *Arch. Dermatol.* **112**(3):193–6.

Eurosurveillance Weekly (2001) Malaria in Austria, 1990–9. *Eurosurveillance Weekly* **5**(1).

Eurosurveillance Weekly (2004) Tickborne encephalitis in Europe: basic information, country by country. **8**(29).

Euzeby, J. P., Raffi, A. (1988). Mise en evidence d'anticorps anti-*Borrelia burgdorferi* chez le chien: sondage epidemiologique en region midi-Pyrenees. *Rev. Med. Vet.* **139**(6):589–93.

Evstafev, I. I. (2001) Results of the 20-year study of tick-borne encephalitis in Crimea. *Zh. Mikrobiol. Epidemigiol. Immunol.* **2**:111–14.

Fabbi, M., De Giuli, L., Tranquillo, M. *et al.* (2004) Prevalence of *Bartonella henselae* in Italian stray cats: evaluation of serology to assess the risk of transmission of *Bartonella* to humans. *J. Clin. Microbiol.* **42**(1):264–8.

Fagbo, S. F. (2002) The evolving transmission pattern of Rift Valley fever in the Arabian Peninsula. *Ann. N. Y. Acad. Sci.* **969**:201–4.

Faulde, M., Sobe, D., Kimmig, P., Scharninghausen, J. (2000) Renal failure and hantavirus infection in Europe. *Nephrol. Dial. Transplant* **15**(6):751–3.

Federico, G., Damiano, F., Caldarola, G., Tartaglione, R., Fantoni, M. (1989) Indagine sieroepiemiologica sulla diffusione dell'infezione da *Rickettsia conori* nel Lazio: risultati preliminary. *Giorn Malattie Infett. Parassit.* **40**(11):1196–7.

Feldman, K. A., Enscore, R. E., Lathrop, S. L. *et al.* (2001) An outbreak of primary pneumonic tularemia on Martha's Vineyard. *N. Engl. J. Med.* **345**(22):1601–6.

Fenech, F. F. (1997) Leishmaniasis in Malta and in the Mediterranean basin. *Ann. Trop. Med. Parasitol.* **91**(7):747–53.

Ferté, H., Postic, D., Baranton, G. *et al.* (1994) Premier isolement en France (Marne) de *Borrelia afzeli* a partir d'*Ixodes ricinus*. *Bull. Soc. Pathol. Exot.* **87**(4):226–7.

Filipe, A. R. (1967) Antibodies against virus transmitted by arbovirus of the B group arthropods in animals of the south of Portugal. Preliminary blood analysis with the West Nile virus, Egypt 101 species. *An. Esc. Nacl. Saude Publ. Med. Trop.* (*Lisb*). **1**(1):197–204.

Filipe, A. R. (1972) Isolation in Portugal of West Nile virus from *Anopheles maculipennis* mosquitoes. *Acta Virol.* **16**(4):361.

Filipe, A. R., Calisher, C. H., Lazuick, J. (1985) Antibodies to Congo–Crimean Haemorrhagic fever, Dhori, Thogoto and Bhanja viruses in Southern Portugal. *Acta Virol.* **29**(4):324–8.

Filipe, A. R., Casals, J. (1979) Isolation of Dhori virus from *Hyalomma marginatum* ticks in Portugal. *Intervirology* **11**(2):124–7.

Filipe, A. R., Pinto, M. R. (1969) Surveys for antibodies to arbovirus in serum of animals from southern Portugal. *Am. J. Trop. Med. Hyg.* **18**(3):423–6.

Filipe, A. R., Sobral, M., Campanico, F. C. (1973) Encefalomielite equina por arbovirus. A proposito de uma epizootia presuntiva causada pelo virus West Nile. *Rev. Port. Cienc. Vet.* **68**(426):90–101.

Fingerle, V., Goodman, J. L., Johnson, R. C. *et al.* (1997) Human granulocytic ehrlichiosis in southern Germany: increased seroprevalence in high-risk groups. *J. Clin. Microbiol.* **35**(12):3244–7.

Fingerle, V., Munderloh, U., Liegl, G., Wilske, B. (1999) Coexistence of ehrlichiae of the phagocytophila group with *Borrelia burgdorferi* in *Ixodes ricinus* from southern Germany. *Med. Microbiol. Immunol. (Berl.)* **188**(3):145–9.

Finlay, G. A., Brown, J. S., Marcus, L.C, Pollack, R. J. (1999) Nasal myiasis – a "Noso"comial infection. *J. Infect. Dis. Clin. Pract.* **8**(4):218–20.

Fisa, R., Gallego, M., Castillejo, S. *et al.* (1999) Epidemiology of canine leishmaniasis in Catalonia (Spain); the example of the Priorat focus. *Vet. Parasitol.* **83**(2):87–97.

Flacio, E., Lüthy, P., Patocchi, N. *et al.* (2004) Primo ritrovamento di *Aedes albopictus* in Svizzera. *Boll. Soc. Ticinese Sci. Naturali.* **92**(1–2):141–2.

Flanigan, T. P., Schwan, T. G., Armstrong, C., van Voris, L. P., Salata, R. A. (1991) Relapsing fever in the US Virgin Islands: a previously unrecognized focus of infection. *J. Infect. Dis.* **163**(6):1391–2.

Foley, J. E., Chomel, B., Kikuchi, Y., Yamamoto, K., Pedersen, N. C. (1998) Seroprevalence of *Bartonella henselae* in cattery cats: association with cattery hygiene and flea infestation. *Vet. Q.* **20**(1):1–5.

Foppa, I. M., Krause, P. J., Spielman, A. *et al.* (2002) Entomologic and serologic evidence of zoonotic transmission of *Babesia microti*, eastern Switzerland. *Emerg. Infect. Dis.* **8**(7):722–6.

Foucault, C., Barrau, K., Brouqui, P., Raoult, D. (2002) *Bartonella* bacteremia among homeless people. *Clin. Infect. Dis.* **35**(6):684–9.

Fournier, P. E., Grunnenberger, F., Jaulhac, B., Gastinger, G., Raoult, D. (2000) Evidence of *Rickettsia helvetica* infection in humans, eastern France. *Emerg. Infect. Dis.* **6**(4):389–92.

Fournier, S., Seegers, H., Jolivet, F., L'Hostis, M. (1998) Assessment of the risk of infestation of pastures by *Ixodes ricinus* due to their phyto-ecological characteristics. *Vet. Res.* **29**(5):487–96.

Fraenkel, C. J., Garpmo, U., Berglund, J. (2002) Determination of novel *Borrelia* genospecies in Swedish *Ixodes ricinus* ticks. *J. Clin. Microbiol.* **40**(9):3308–12.

Frampton, J. W., Lanser, S., Nichols, C. R. (1995) Sin Nombre virus infection in 1959 [Letter]. *Lancet* **346**(8977):781–2.

Francisci, D., Papili, R., Camanni, G. *et al.* (2003) Evidence of Toscana virus circulation in Umbria: first report. *Eur. J. Epidemiol.* **18**(5):457–9.

Francy, D. B., Jaenson, T. G., Lundstrom, J. O. *et al.* (1989) Ecologic studies of mosquitoes and birds as hosts of ockelbo virus in Sweden and isolation of inkoo and batai viruses from mosquitoes. *Am. J. Trop. Med. Hyg.* **41**(3):355–63.

Francy, D. B., Karabatsos, N., Wesson, D. M. *et al.* (1990) A new arbovirus from *Aedes albopictus*, an Asian mosquito established in the United States. *Science* **250**:1738–40.

Frandsen, F., Bresciani, J., Hansen, H. G. (1995) Prevalence of antibodies to *Borrelia burgdorferi* in Danish rodents. *Acta Parasitol. Microbiol. Immunol. Scand.* **103**(4):247–53.

Frank, C., Schöneberg, I., Krause, G. et al. (2004) Increase in imported dengue, Germany, 2001–2002. *Emerg. Infect. Dis.* **10**(5):903–6.

Frese, M., Weeber, M., Weber, F., Speth, V., Haller, O. (1997) Mx1 sensitivity: Batken virus is an orthomyxovirus closely related to Dhori virus. *J. Gen. Virol.* **78**(10):2453–8.

Friedman, C. R., Malcolm, G., Rigau-Perez, J. G., Arambulo, P. 3rd, Tauxe, R. V. (1996) Public health risk from *Salmonella*-based rodenticides. *Lancet* **347**(9016):1705–6.

Fulhorst, C. F., Bowen, M. D., Ksiazek, T. G. *et al.* (1996) Isolation and characterization of Whitewater Arroyo virus, a novel North American arenavirus. *Virology* **224**(1):114–20.

Fulhorst, C. F., Charrel, R. N., Weaver, S. C. *et al.* (2001) Geographic distribution and genetic diversity of Whitewater Arroyo virus in the southwestern United States. *Emerg. Infect. Dis.* **7**(3):403–7.

Gage, K. L., Dennis, D. T., Orloski, K. A. *et al.* (2000) Cases of Cat-Associated Human Plague in the Western US, 1977–1998. *Clin. Infect. Dis.* **30**(6):893–900.

Galinovic-Weisglass, M., Vesenjak-Hirjan, J. (1986) Tick-borne encephalitis in Croatia. *Mikrobiologija (Beograd)* **23**(2):163–8.

Gardner, S. L., Holman, R. C., Krebs, J. W., Berkelman, R., Childs, J. E. (2003) National surveillance for the human ehrlichioses in the United States, 1997–2001, and proposed methods for evaluation of data quality. *Ann. N. Y. Acad. Sci.* **990**:80–9.

Gargan, T. P. 2nd, Clark, G. G., Dohm, D. J., Turell, M. J., Bailey, C. L. (1988) Vector potential of selected North American mosquito species for Rift Valley fever virus. *Am. J. Trop. Med. Hyg.* **38**(2):440–6.

Gelfand, M. S. (2003) West Nile virus infection. What you need to know about this emerging threat. *Postgrad. Med.* **114**(1):31–8.

Gelli, G. P., Peralta, M. (1966) On 3 cases of encephalitis caused by arbovirus of the B group (West Nile). *Minerva Pediatrica* **18**(37):2185–90.

Gentilini, M., Danis, M. (1981) Le paludisme autochtone. *Méd. Mal. Infect.* **11**(6):356–62.

George, J. C., Chastel, C. (2002) Tick-borne diseases and changes in the ecosystem in Lorraine. *Bull. Soc. Pathol. Exot.* **95**(2):95–9.

Georgieva, G., Manev, Kh., Georgieva, V., Matev, G. (1993) Ixodid ticks infected with *Borrelia* in two different eco-geographical sites. *Infectology* **30**(5):38–40.

Gerhardt, R. R., Gottfried, K. L., Apperson, C. S. *et al.* (2001) First isolation of La Crosse virus from naturally infected *Aedes albopictus*. *Emerg. Infect. Dis.* **7**(5):807–11.

Giammanco, G. M., Mansueto, S., Ammatuna, P., Vitale, G. (2003) Israeli spotted fever *Rickettsia* in Sicilian *Rhipicephalus sanguineus* ticks. *Emerg. Infect. Dis.* **9**(7):892–3.

Gilot, B., Degeilh, B., Pichot, J. *et al.* (1996) Prevalence of *Borrelia burgdorferi* (sensu lato) in *Ixodes ricinus* populations in France according to a phytoecological zoning of the territory. *Eur. J. Epidemiol.* **12**(4):395–401.

Gjorup, I. E. (2001) Malaria. An evaluation of 127 patients treated in Rigshospitalet in 1999 and 2000. *Ugeskr Laeger* **163**(35):4732–5.

Gligic, A., Adamovic, Z. R. (1976) Isolation of Tahyna virus from *Aedes vexans* mosquitoes in Serbia. *Mikrobiologija (Beograd)* **13**(2):119–29.

Glinskikh, N. P., Fedotova, T. T., Pereskokova, I. G., Mel'nikov, V. G., Volkova, L. I. (1994) The potentials for the comprehensive diagnosis of viral encephalitis in Sverdlovsk Province. *Vopr. Virusol.* **39**(4):190–1.

Goddard, J. (1998) Arthropod transmission of tularemia. *Infect. Med.* **15**(5):306–8.

Goethert, H. K., Telford, S. R. IIIrd. (2003) Enzootic transmission of *Babesia divergens* among cottontail rabbits on Nantucket Island, Massachusetts. *Am. J. Trop. Med. Hyg.* **69**(5):455–60.

Goldblum, N., Sterk, V. V., Paderski, B. (1954) West Nile fever. The clinical features of the disease and the isolation of West Nile virus from the blood of nine human cases. *Am. J. Hyg.* **59**(1):89–103.

Gonzalez, M. T., Filipe, A. R. (1977) Antibodies to arboviruses in northwestern Spain. *Am. J. Trop. Med. Hyg.* **26**(4):792–7.

Gordon, J. R., McLaughlin, B. G., Nitiuthai, S. (1983) Tularaemia transmitted by ticks (*Dermacentor andersoni*) in Saskatchewan. *Can. J. Comp. Med.* **47**(4):408–11.

Gorelova, N. B., Korenberg, E. I., Postic, D., Iunicheva, Iu. V., Riabova, T. E. (2001) The first isolation of *Borrelia burgdorferi* sensu stricto in Russia. *Zh. Mikrobiol. Epidemiol. Immunobiol.* **4**:10–12.

Gothe, R. (1999) *Rhipicephalus sanguineus* (Ixodidae): frequency of infestation and ehrlichial infections transmitted by this tick species in dogs in Germany; an epidemiological study and consideration. *Wien. Tierarztliche Monatsschrift.* **86**(2):49–56.

Gotzsche, P. C., Hammarquist, C., Burr, M. (1998) House dust mite control measures in the management of asthma: meta-analysis. *Br. Med. J.* **317**:1105–10.

Goubau, P. F. (1984) Relapsing fevers. A review. *Ann. Soc. Belge Med. Trop.* **4**:335–64.

Gourgouli, K., Bethimouti, A., Raftopoulou, O., Papadaki, E., Fotiou, K. (1992) Epidemiological research on Mediterranean spotted fever. *Acta Microbiol. Hellen.* **37**(2):241–6.

Gradoni, L., Gramiccia, M. (1990) Studies on the control of leishmaniasis in Malta. *Parassitologia* **32**(suppl.):137–8.

Gradoni, L., Gramiccia, M., Léger, N. *et al.* (1991) Isoenzyme characterization of leishmaniasis from man, dog and sandflies in the Maltese islands. *Trans. Roy. Soc. Trop. Med. Hyg.* **85**(2):217–19.

Gradoni, L., Gramiccia, M., Mancianti, F., Pieri, S. (1988) Studies on canine leishmaniasis control. 2. Effectiveness of control measures against canine leishmaniasis in the Isle of Elba, Italy. *Trans. Roy. Soc. Trop. Med. Hyg.* **82**(4):568–71.

Gradoni, L., Gramiccia, M., Pozio, E. (1984) Status of the taxonomy of *Leishmania* from the Mediterranean basin. *Parassitologia* **26**(3):289–97.

Gratz, N. G. (1988) Rodents and human disease: A global appreciation. In *Rodent Pest Management* (ed. I. Prakesh), pp. 101–65. Boca Raton, FL: CRC Press.

Gratz, N. G. (1997) Human lice – their prevalence, control and resistance to insecticides. WHO/CTD/WHOPES/978. Geneva: WHO.

Gratz, N. G. (1999a) Rodent reservoirs and flea vectors of natural foci of plague. In *Plague Manual*. WHO/CDS/CSR/EDC/99.2, pp. 63–96. Geneva: WHO.

Gratz, N. G. (1999b) Emerging and resurging vector-borne diseases. *Ann. Rev. Entomol.* **44**:51–75.

Gratz, N. G. (2004) Critical review of the vector status of *Aedes albopictus*. *Med. Vet. Entomol.* **18**(3):215–27.

Gratz, N. G., Steffen, R., Cocksedge, W. (2000) Why aircraft disinsection? *Bull. Wld. Hlth. Org.* **78**(8):995–1004.

Gray, J., von Stedingk, L. V., Gurtelschmid, M., Granstrom, M. (2002) Transmission studies of *Babesia microti* in *Ixodes ricinus* ticks and gerbils. *J. Clin. Microbiol.* **40**(4):1259–63.

Grazioli, D. (1996) Tick-borne diseases in the province of Belluno. *Alpe Adria. Microbiol. J.* **5**(3):215 (correspondence).

Grech, V., Mizzi, J., Mangion, M., Vella, C. (2000) Visceral leishmaniasis in Malta – an 18 year paediatric, population based study. *Arch. Dis. Child.* **82**(5):381–5.

Grimm, F., Gessler, M., Jenni, L. (1993) Aspects of sandfly biology in southern Switzerland. *Med. Vet. Entomol.* **7**(2):170–6.

Grimstad, P. R., Calisher, C. H., Harroff, R. N., Wentworth, B. B. (1986) Jamestown Canyon virus (California serogroup) is the etiologic agent of widespread infection in Michigan humans. *Am. J. Trop. Med. Hyg.* **35**:376–86.

Grimstead, P. R., Shabino, C. L., Calisher, C. H., Waldman, R. J. (1982) A case of encephalitis in a human associated with a serologic rise to Jamestown Canyon virus. *Am. J. Trop. Med. Hyg.* **31**(6):1238–44.

Grist, N. R., Burgess, N. R. H. (1994) *Aedes* and dengue. *Lancet* **343**(8895): 477.

Gromashevsky, V. L., Lvov, D. K., Sidorova, G. A. *et al.* (1973) A complex natural focus of arboviruses on Glinyanyi Island, Baku Archipelago, Azerbaidzhan S. S. R. *Acta Virol.* **17**(2):155–8.

Gryczynska, A., Zgodka, A., Ploski, R., Siemiatkowski, M. (2004) *Borrelia burgdorferi* sensu lato infection in passerine birds from the Mazurian Lake region (Northeastern Poland). *Avian Pathol.* **33**(1):67–75.

Grzeszczuk, A., Stanczak, J., Kubica Biernat, B. (2002) Serological and molecular evidence of human granulocytic ehrlichiosis focus in the Bialowieza Primeval Forest (Puszcza Bialowieska), northeastern Poland. *Eur. J. Clin. Microbiol. Infect. Dis.* **21**(1):6–11.

Guerra, M., Walker, E., Jones, C. *et al.* (2002) Predicting the risk of Lyme disease: habitat suitability for *Ixodes scapularis* in the North Central United States. *Emerg. Infect. Dis.* **8**(3):289–97.

Guillaume, B., Heyman, P., Lafontaine, S. *et al.* (2002) Seroprevalence of human granulocytic ehrlichiosis infection in Belgium. *Eur. J. Clin. Microbiol. Infect. Dis.* **21**(5):397–400.

Guptill, L., Wu, C. C., HogenEsch, H. *et al.* (2004) Prevalence, risk factors, and genetic diversity of *Bartonella henselae* infections in pet cats in four regions of the United States. *J. Clin. Microbiol.* **42**(2):652–9.

Gurycova, D., Kocianova, E., Vyrostekova, V., Rehacek, J. (1995) Prevalence of ticks infected with *Francisella tularensis* in natural foci of tularemia in western Slovakia. *Eur. J. Epidemiol.* **11**:469–74.

Gustafson, R. (1994) Epidemiological studies on Lyme borreliosis and tick-borne encephalitis. *Scand. J. Infect. Dis.* (suppl. 92):1–63.

Gustafson, R., Artursson, K. (1999) Ehrlichiosis is among animals but also occurs in humans. *Lakartidningen* **96**(37):3884–7.

Gustafson, R., Jaenson, T. G. T., Gardulf, A., Mejlon, H., Svenungsson, B. (1995) Prevalence of *Borrelia burgdorferi* sensu lato infection in *Ixodes ricinus* in Sweden. *Scand. J. Infect. Dis.* **27**(6):597–601.

Hagen, I. J., Aandahl, E., Hasseltvedt, V. (2005) Five case histories of tularaemia infection in Oppland and Hedmark Counties, Norway. *Eurosurveillance Weekly* **10**(3).

Haglund, M., Gunther, G. (2003) Tick-borne encephalitis – pathogenesis, clinical course and long-term follow-up. *Vaccine* **21**(suppl. 1):S11–18.

Hagstrom, T., Ljungberg, H. (1999) The Surinam cockroach *Pyenoscelus surinamensis* (L.) (Blattodea:Panchloridae) established in Sweden. *Entomol. Tidskrift.* **120**(3):113–15.

Halouzka, J., Pejcoch, M., Hubalek, Z., Knoz, J. (1991) Isolation of Tahyna virus from biting midges (Diptera, Ceratopogonidae) in Czechoslovakia. *Acta Virol.* **35**(3):247–51.

Halstead, S. B. (1980) Dengue haemorrhagic fever – a public health problem and a field for research. *Bull. Wld. Hlth. Org.* **58**(1):1–21.

Halstead, S. B., Papaevangelou, G. (1980) Transmission of dengue 1 and 2 viruses in Greece in 1928. *Am. J. Trop. Med. Hyg.* **29**(4):635–7.

Hamrick, H. J., Bowdre, J. H., Church, S. M. (1990) Rat tapeworm (*Hymenolepis diminuta*) infection in a child. *Pediatr. Infect. Dis. J.* **9**(3):216–19.

Han, L. L., Popovici, F., Alexander, J. P. Jr. *et al.* (1999) Risk factors for West Nile virus infection and meningoencephalitis, Romania, 1996. *J. Infect. Dis.* **179**(1):230–3.

Han, X., Aho, M., Vene, S. *et al.* (2001) Prevalence of tick-borne encephalitis virus in *Ixodes ricinus* ticks in Finland. *J. Med. Virol.* **64**(1):21–8.

Hannoun, C., Chatelain, J., Krams, S., Guillon, J. C. (1971) Isolement en Alsace du virus de l'encephalite a tiques (Arbovirus group B). *C. R. Acad. Sci. Sér. D Paris* **272**(5):766–8.

Hanscheid, T., Melo-Cristino, J., Grobusch, M. P., Pinto, B. G. (2003) Avoiding misdiagnosis of imported malaria: screening of emergency department samples with thrombocytopenia detects clinically unsuspected cases. *J. Travel Med.* **10**(3):155–9.

Hansen, K. (1994) Lyme neuroborreliosis: improvements of the laboratory diagnosis and a survey of the epidemiological and clinical features in Denmark 1985–1990. *Acta Neurol. Scand. Suppl.* **89**(151):1–44.

Hansen, R. C., O'Haver, J. (2004) Economic considerations associated with *Pediculus humanus capitis* infestation. *Clin. Pediatr.* **43**(6):523–7.

Hanson, M. S., Edelman, R. (2003) Progress and controversy surrounding vaccines against Lyme disease. *Expert Rev. Vaccines* **2**(5):683–703.

Harden, F. W., Hepburn, H. R., Ethridge, B. J. (1967) A history of mosquitoes and mosquito-borne diseases in Mississippi 1699–1965. *Mosq. News* **27**(1):60–6.

Harling, R., Crook, P., Lewthwaite, P. *et al.* (2004) Burden and cost of imported infections admitted to infectious diseases units in England and Wales in 1998 and 1999. *J. Infect.* **48**(2):139–44.

Harms, G., Dorner, F., Bienzle, U., Stark, K. (2002) Infections and diseases after travelling. *Dtsch. Med. Wochenschr.* **127**(34–35):1748–53.

Harms, G., Schoenian, G., Feldmeier, H. (2003) Leishmaniasis in Germany. *Emerg. Infect. Dis.* **9**(7):872–5.

Harris, J., Crawshaw, J. G., Millership, S. (2003) Incidence and prevalence of head lice in a district health authority area. *Commun. Dis. Publ. Hlth.* **6**(3):246–9.

Hatcher, J. C., Greenberg, P. D., Antique, J., Jimenez-Lucho, V. E. (2001) Severe babesiosis in Long Island: review of 34 cases and their complications. *Clin. Infect. Dis.* **32**(8):1117–25.

Headington, C. E., Barbara, C. H., Lambson, B. E., Hart, D. T., Barker, D. C. (2002) Diagnosis of leishmaniasis in Maltese dogs with the aid of the polymerase chain reaction. *Trans. Roy. Soc. Trop. Med. Hyg.* **96**(suppl. 1):S195–7.

Healing, T. D. (1991) Salmonella in rodents: a risk to man? *CDR (Lond Engl Rev)* **1**(10):R114–16.

Heinz, F. X., Kunz, C. (2004) Tick-borne encephalitis and the impact of vaccination. *Arch. Virol. Suppl.* (suppl. 18): 201–5.

Hellenbrand, W., Breuer, T., Petersen, L. (2001) Changing epidemiology of Q fever in Germany, 1947–1999. *Emerg. Infect. Dis.* **7**(5):789–96.

Henke, S. E., Pence, D. B., Demarais, S., Johnson, J. R. (1990) Serologic survey of selected zoonotic disease agents in black-tailed jack rabbits from western Texas. *J. Wildl. Dis.* **26**(1):107–11.

Hernandez-Cabrera, M., Angel-Moreno, A., Santana, E. *et al.* (2004) Murine typhus with renal involvement in Canary Islands, Spain. *Emerg. Infect. Dis.* **10**(4):740–3.

Herrero, C., Pelaz, C., Alvar, J. *et al.* (1992) Evidence of the presence of spotted fever rickettsiae in dogs and dog ticks of the central provinces of Spain. *Eur. J. Epidemiol.* **8**(4):575–9.

Herwaldt, B. L., Grijalva, M. J., Newsome, A. L. *et al.* (2000) Use of polymerase chain reaction to diagnose the fifth reported US case of autochthonous transmission of *Trypanosoma cruzi*, in Tennessee, 98. *J. Infect. Dis.* **181**(1):395–9.

Herwaldt, B. L., Persing, D. H., Precigout, E. A. *et al.* (1996) A fatal case of babesiosis in Missouri: identification of another piroplasm that infects humans. *Ann. Intern. Med.* **124**(7):643–50.

Higa, H. H., Fujinaka, I. T. (1976) Prevalence of rodent and mongoose leptospirosis on the Island of Oahu. *Publ. Hlth. Rept.* **91**(2):171–7.

Hildebrandt, A., Schmidt, K. H., Wilske, B. *et al.* (2003) Prevalence of four species of *Borrelia burgdorferi* sensu lato and coinfection with *Anaplasma phagocytophila* in *Ixodes ricinus* ticks in central Germany. *Eur. J. Clin. Microbiol. Infect. Dis.* **22**(6):364–7.

Hilton, A. C., Willis, R. J., Hickie, S. J. (2002) Isolation of *Salmonella* from urban wild brown rats (*Rattus norvegicus*) in the West Midlands, UK. *Int. J. Environ. Hlth. Res.* **12**(2):163–8.

Hinman, A. R., Fraser, D. W., Douglas, R. G. *et al.* (1975) Outbreak of lymphocytic choriomeningitis virus infections in medical center personnel. *Am. J. Epidemiol.* **101**(2):103–10.

Hinson, E. R., Shone, S. M., Zink, M. C., Glass, G. E., Klein, S. L. (2004) Wounding: the primary mode of Seoul virus transmission among male Norway rats. *Am. J. Trop. Med. Hyg.* **70**(3):310–17.

Hirsch, T., Stappenbeck, C., Neumeister, V. *et al.* (2000) Exposure and allergic sensitization to cockroach allergen in East Germany. *Clin. Exp. Allergy* **30**:529–37.

Hoey, J. (2000) West Nile fever in New York City. *Canad. Med. Assoc. J.* **162**(7):1036.

Hofer, S., Gloor, S., Muller, U. *et al.* (2000) High prevalence of *Echinococcus multilocularis* in urban red foxes (*Vulpes vulpes*) and voles (*Arvicola terrestris*) in the city of Zurich, Switzerland. *Parasitology* **120**(2):135–42.

Hohenschild, S. (1999) Babesiosis – a dangerous infection for splenectomized children and adults. *Klin. Padiatr.* **211**(3):137–40.

Holbach, M., Oehme, R. (2002) Tick-borne encephalitis and Lyme borreliosis. Spread of pathogens and risk of illness in a tick-borne encephalitis region. *Fortschr. Med. Orig.* **120**(4):113–18.

Holick, J., Kyle, A., Ferraro, W., Delaney, R. R., Iwaseczko, M. (2002) Discovery of *Aedes albopictus* infected with West Nile virus in southeastern Pennsylvania. *J. Am. Mosq. Control Assoc.* **18**(2):131.

Holk, K., Nielsen, S. V., Ronne, T. (2000) Human leptospirosis in Denmark 1970–1996: an epidemiological and clinical study. *Scand. J. Infect. Dis.* **32**(5):533–8.

Homer, M. J., Aguilar-Delfin, I., Telford, S. R. III, Krause, P. J., Persing, D. H. (2000) Babesiosis. *Clin. Microbiol. Rev.* **13**(3):451–69.

Hoogstraal, H. (1973) Viruses and ticks. In *Viruses and Invertebrates* (ed. A. J. Gibbs), pp. 349–417. Amsterdam: North-Holland Publ. Co.

Hovmark, A., Jaenson, T. G., Asbrink, E., Forsman, A., Jansson, E. (1988) First isolations of *Borrelia burgdorferi* from rodents collected in northern Europe. *APMIS* **96**(10):917–20.

Hribar, L. J., Vlach, J. J., Demay, D. J. *et al.* (2003) Mosquitoes infected with West Nile virus in the Florida Keys, Monroe County, Florida, USA. *J. Med. Entomol.* **40**(3):361–3.

Hubalek, Z. (1987) Geographic distribution of Bhanja virus. *Folia Parasitol.* (*Praha*). **34**(1):77–86.

Hubalek, Z. (2000) European experience with the West Nile virus ecology and epidemiology: could it be relevant for the New World? *Viral Immunol.* **13**(4):415–26.

Hubalek, Z., Halouzka, J. (1996) Arthropod-borne viruses of vertebrates in Europe. *Acta Sci. Nat. Acad. Sci. Bohemicae* (*Brno*) **30**(4–5): 1–95.

Hubalek, Z., Halouzka, J. (1999) West Nile fever – a reemerging mosquito-borne viral disease in Europe. *Emerg. Infect. Dis.* **5**(9):643–50.

Hubalek, Z., Halouzka, J., Juricova, Z. (1999) West Nile Fever in Czechland. *Emerg. Infect. Dis.* **5**(4):594–5

Hubalek, Z., Halouzka, J., Juricova, Z. (2003) Longitudinal surveillance of the tick *Ixodes ricinus* for borreliae. *Med. Vet. Entomol.* **17**(1):46–51.

Hubalek, Z., Halouzka, J., Juricova, Z. *et al.* (1999) Surveillance of mosquito-borne viruses in the Breclav area (Czech Republic) after the 1997 flood. *Epidemiol. Mikrobiol. Immunol.* **48**(3):91–6.

Hubalek, Z., Halouzka, J., Juricova, Z., Sebesta, O. (1998) First isolation of the mosquito-borne West Nile Virus in the Czech Republic. *Acta Virol.* **42**(2):119–20.

Hubalek, Z., Juricova, Z., Halouzka, J., Pelantova, J., Hudec, K. (1989) Arboviruses associated with birds in southern Moravia, Czechoslovakia. *Pirodvedne Prace Ustavi Cesk Akad. Ved v Brne* **23**(7):3–50.

Hubalek, Z., Juricova, Z., Svobodova, S., Halouzka, J. (1993) A serologic survey for some bacterial and viral zoonoses in game animals in the Czech Republic. *J. Wildl. Dis.* **29**(4):604–7.

Hubalek, Z., Mittermayer, T., Halouzka, J. (1988) Bhanja virus (Bunyaviridae) isolated from *Dermacentor marginatus* ticks in Czechoslovakia. *Acta Virol.* **32**(6):526.

Hubalek, Z., Savage, H. M., Halouzka, J. *et al.* (2000) West Nile virus investigations in South Moravia, Czechland. *Viral Immunol.* **13**(4):427–33.

Huebner, R. J., Stamps, P., Armstrong, C. (1946) Rickettsialpox. A newly recognized rickettsial disease. 1. Isolation of the etiological agent. *Publ. Hlth. Rept.* **61**:1605–14.

Hughes, J. H. (1961) Mosquito interceptions and related problems in aerial traffic arriving in the United States. *Mosq. News* **21**(2):93–100.

Hulinska, D., Kurzova, D., Drevova, H., Votypka, J. (2001) First detection of Ehrlichiosis detected serologically and with the polymerase chain reaction in patients with borreliosis in the Czech Republic. *Cas. Lek. Cske.* **140**(6):181–4.

Hulinska, D., Votypka, J., Plch, J. *et al.* (2002) Molecular and microscopical evidence of *Ehrlichia* spp. and *Borrelia burgdorferi* sensu lato in patients, animals and ticks in the Czech Republic. *New Microbiol.* **25**(4):437–48.

Humair, P. F., Turrian, N., Aeschlimann, A., Gern, L. (1993) *Borrelia burgdorferi* in a focus of Lyme borreliosis: epizootiologic contribution of small mammals. *Folia Parasitol. (Praha)* **40**(1):65–70.

Hunfeld, K. P., Brade, V. (1999) Prevalence of antibodies against the human granulocytic ehrlichiosis agent in Lyme borreliosis patients from Germany. *Eur. J. Clin. Microbiol. Infect. Dis.* **18**(3):221–4.

Ivovic, V., Ivovic, M., Miscevic, Z. (2003) Sandflies (Diptera: Psychodidae) in the Bar area of Montenegro (Yugoslavia). *Ann. Trop. Med. Parasitol.* **97**(2):193–7.

Izri, M. A., Fauran, P., Le Fichoux, Y., Rousset, J. J. (1994) *Phlebotomus perfiliewi* Parrot, 1930 (Diptera, Psychodidae) dans le sud-est de la France. *Parasite* **1**(3):286.

Izri, M. A., Marty, P., Rahal, A. *et al*. (1992) *Phlebotomus perniciosus* Newstead, 1911 naturally infected by promastigotes in the region of Nice (France). *Bull. Soc. Pathol. Exot* **85**(5):385–7.

Jack, M. (1993) Scabies outbreak in an extended care unit – a positive outcome. *Canad. J. Infect. Control* **8**(1):11–13.

Jackson, L. A., Spach, D. H., Kippen, D. A. *et al*. (1996) Seroprevalence to *Bartonella* among patients at a community clinic in downtown Seattle. *J. Infect. Dis.* **173**(4):1023–6.

Jaenson, T. G. T., Talleklint, L., Lundqvist, L. *et al*. (1994) Geographical distribution, host association and vector roles of ticks (Acari: Ixodidae & Argasidae) in Sweden. *J. Med. Entomol.* **31**(2):240–56.

Jameson, P., Greene, C., Regnery, R. *et al*. (1995) Prevalence of *Bartonella henselae* antibodies in pet cats throughout regions of North America. *J. Infect. Dis.* **172**(4):1145–9.

Janouskovcova, E., Zakovska, A., Halouzka, J., Dendis, M. (2004) Occurrence of *Borrelia afzelii* and *Borrelia garinii* in *Ixodes ricinus* ticks from Southern Moravia, Czech Republic. *Vector-Borne Zoon. Dis.* **4**(1):43–52.

Jansen, A., Schöneberg, I., Frank, C. *et al*. (2005) Leptospirosis in Germany, 1962–2003. *Emerg. Infect. Dis.* **11**(7):1048–54.

Januska, J., Bruj, J., Farnik, J. (1990) To the prevalence of certain zoonoses with natural focality in West-Bohemian Region. Part II. Antibodies to Tick-borne encephalitis, Tahyna and Tribec viruses in the local population. *Cske. Epidemiol. Mikrobiol. Immunol.* **39**(3):134–8.

Jenkins, A., Handeland, K., Stuen, S. *et al*. (1998) Ehrlichiosis in a moose calf in Norway. *J. Wildlife Dis.* **37**(1):201–3.

Jensen, P. M., Frandsen, F. (2000) Temporal risk assessment for Lyme borreliosis in Denmark. *Scand. J. Infect. Dis.* **32**(5):539–44.

Jirous, J. (1987) Lyme borreliosis, a new clinical phenomenon in Czechoslovakia. *Bull. Cske. Spolecnosti Mikrobiol. Pri. CSAV* **28**(5):12–14.

Johnson, R. T., Irani, D. N. (2002) West nile virus encephalitis in the United States. *Curr. Neurol. Neurosci. Rep.* **2**(6):496–500.

Jones, J., Lopez, A., Wilson, M. (2003) Congenital toxoplasmosis. *Am. Fam. Phys.* **67**(10):2131–8.

Joss, A. W., Davidson, M. M., Ho-Yen, D. O., Ludbrook, A. (2003) Lyme disease – what is the cost for Scotland? *Publ. Hlth.* **117**(4):264–73.

Joubert, L. (1975) L'arbovirose West Nile, zoonose du midi mediterraneen de la France. *Bull. Acad. Natl. Med.* **159**(6):499–503.

Jouda, F. F., Perret, J.- L., Gern, L. (2004) Density of questing *Ixodes ricinus* nymphs and adults infected by *Borrelia burgdorferi* sensu lato in Switzerland: Spatio-temporal pattern at a regional scale. *Vector-Borne Zoon. Dis.* **4**, 23–32.

Jucowics, P., Ramon, M. E., Don, P. C., Stone, R. K., Bamji, M. (1989) Norwegian scabies in an infant with acquired immunodeficiency syndrome. *Arch. Dermatol.* **125**(12):1670–1

Jupp, P. G., Kemp, A., Grobbelaar, A. *et al*. (2002) The 2000 epidemic of Rift Valley fever in Saudi Arabia: mosquito vector studies. *Med. Vet. Entomol.* **16**(3):245–52.

Juricova, Z. (1988) Arbovirus antibodies in birds caught in the Krkonose mountains. *Biologia (Bratislava)* **43**(3):259–64.

Juricova, Z., Hubalek, Z. (1993) Serological examination of domestic ducks (*Anas platyrhynchos f. domestica*) in southern Moravia for antibodies against arboviruses of groups A, B, California and Bunyamwera. *Biologia* **48**(5):481–4.

Juricova, Z., Hubalek, Z., Halouzka, J., Machacek, P. (1993) Virologic detection of arboviruses in greater cormorants. *Vet. Med. (Praha)* **38**(6):375–9.

Juricova, Z., Literak, I., Pinowski, J. (2000) Antibodies to arboviruses in house sparrows (*Passer domesticus*) in the Czech Republic. *Acta Vet. Brno* **69**:213–15.

Juricova, Z., Mitterpak, J., Prokopic, J., Hubalek, Z. (1986) Circulation of mosquito-borne viruses in large-scale sheep farms in eastern Slovakia. *Folia Parasitol. (Praha)* **33**(3):285–8.

Juricova, Z., Pinowski, J., Literak, I., Hahm, K. H., Romanowski, J. (1998) Antibodies to alphavirus, flavivirus, and bunyavirus arboviruses in house sparrows (*Passer domesticus*) and tree sparrows (*P. montanus*) in Poland. *Avian Dis.* **42**(1):182–5.

Kahl, O., Schmidt, K., Schönberg, A. *et al.* (1989) Prevalence of *Borrelia burgdoferi* in *Ixodes ricinus* ticks in Berlin (West). *Zentralbl. Bakteriol. Mikrobiol. Hyg.* [A] **270**(3):434–40.

Kain, K. C., MacPherson, D. W., Kelton, T. *et al.* (2001) Malaria deaths in visitors to Canada and in Canadian travellers: a case series. *Can. Med. Assoc. J.* **164**(5):654–9.

Kaiser, M. N., Hoogstraal, H., Watson, G. E. (1974) Ticks (Ixodoidea) on migrating birds in Cyprus, fall 1967 and spring 1968 and epidemiological considerations. *Bull. Ent. Res.* **64**(1):97–110.

Kaiser, R., Seitz, A., Strub, O. (2002) Prevalence of *Borrelia burgdorferi* sensu lato in the nightingale (*Luscinia megarhynchos*) and other passerine birds. *Int. J. Med. Microbiol.* **291**(suppl. 33):75–9.

Kamarinchev, B., Kovacheva, T., Christova, T., Goergieva, G., Zlatanova, V. (1991) Studies on mosquitoes and ticks as carriers of alpha-, flavi-, and bunyaviruses. In *Modern Acarology*, Vol II (eds F. Dusbabek and V. Bukva), pp. 89–92. Prague: Proc. 8th Int. Congr. Acarol.

Kampen, H., Maltezos, E., Pagonaki, M. *et al.* (2002) Individual cases of autochthonous malaria in Evros Province, northern Greece: serological aspects. *Parasitol. Res.* **88**(3):261–6.

Kampen, H., Proft, J., Etti, S. *et al.* (2003) Individual cases of autochthonous malaria in Evros Province, northern Greece: entomological aspects. *Parasitol. Res.* **89**(4):252–8.

Kampen, H., Rotzel, D. C., Kurtenbach, K., Maier, W. A., Seitz, H. M. (2004) Substantial rise in the prevalence of Lyme borreliosis spirochetes in a region of western Germany over a 10-year period. *Appl. Environ. Microbiol.* **70**(3):1576–82.

Kang, B. C., Sulit, N. (1978) A comparative study of prevalence of skin hypersensitivity to cockroach and house dust antigens. *Ann. Allergy* **41**(6):333–6.

Karabatsos, N., Lewis, A. L., Calisher, C. H., Hunt, A. R., Roehrig, J. T. (1988) Identification of Highlands J virus from a Florida horse. *Am. J. Trop. Med. Hyg.* **39**(6):603–6.

Karadzhov, I., Greshikova, M., Ignatov, G. (1982) Haemagglutination-inhibiting antibodies against arboviruses in bovine sera in Bulgaria (Sindbis, West Nile and Bhanja viruses). *Veterinnamo-meditsinki Nauki.* **19**(2):25–8.

Karch, S., Dellile, M. F., Guillet, P., Mouchet, J. (2001) African malaria vectors in European aircraft. *Lancet* **357**(9251):235.

Kasbohrer, A., Schonberg, A. (1990) Serologic studies of the occurrence of *Borrelia burgdorferi* in domestic animals in Berlin (West). *Berl. Münch. Tierartzl. Wochenschr.* **103**(11):374–8.

Keefe, T. J., Holland, C. J., Salyer, P. E., Ristic, M. (1982) Distribution of *Ehrlichia canis* among military working dogs in the world and selected civilian dogs in the United States. *J. Am. Vet. Med. Assoc.* **181**(3):236–8.

Keirans, J. E., Durden, L. A. (2001) Invasion: exotic ticks (Acari: Argasidae, Ixodidae) imported into the United States. A review and new records. *J. Med. Entomol.* **38**(6):850–61.

Keirans, J. E., Hutcheson, H. J., Durden, L. A., Klompen, J. S. (1996) *Ixodes (Ixodes) scapularis*: redescription of all active stages, distribution, hosts, geographical variation and medical and veterinary importance. *J. Med. Entomol.* **33**(3):297–318.

Kern, P., Bardonnet, K., Renner, E. *et al.* (2003) European echinococcosis registry: human alveolar echinococcosis, Europe, 1982–2000. *Emerg. Infect. Dis.* **9**(3):343–9.

Kero, A., Xinxo, A. (1998) Epidemiological characteristics of leishmaniasis in Albania in the period 1984–1996. *Giorn. Ital. Med. Trop.* **3**(3–4):55–7.

Kerr, S. F., McHugh, C. P., Dronen, N. O. Jr. (1995) Leishmaniasis in Texas: prevalence and seasonal transmission of *Leishmania mexicana* in *Neotoma micropus. Am. J. Trop. Med. Hyg.* **53**(1):73–7.

Kerr, S. F., McHugh, C. P., Merkelz, R. (1999) Short report: a focus of *Leishmania mexicana* near Tucson, Arizona. *Am. J. Trop. Med. Hyg.* **61**(3):378–9.

Khan, A., Khan, A. S. (2003) Hantaviruses: a tale of two hemispheres. *Panminerva Med.* **45**(1):43–51.

Khanakahg, G., Kocianova, E., Vyrostekova, V. *et al.* (2003) Geographical and seasonal variation in detecting *Borrelia burgdorferi* sensu lato in rodents of eastern Austria. VII International Potsdam Symposium on Tick-borne Diseases (IPS-VII) 2003.

Kirchhoff, L. V. (1993) American trypanosomiasis, (Chagas' disease) – A tropical disease now in the United States. *N. Engl. J. Med.* **329**(9):639–44.

Kirkland, K. B., Klimko, T. B., Meriwether, R. A. *et al.* (1997) Erythema migrans-like rash illness at a camp in North Carolina: a new tick-borne disease? *Arch. Intern. Med.* **157**(22):2635–41.

Kitch, B. T., Chew, G., Burge, H. A. *et al.* (2000) Socioeconomic predictors of high allergen levels in homes in the greater Boston area. *Environ. Hlth. Perspect.* **108**(4):301–7.

Kjemtrup, A. M., Conrad, P. A. (2000) Human babesiosis: an emerging tick-borne disease. *Int. J. Parasitol.* **30**(12–13):1323–37.

Knuth, T., Liebermann, H., Urbaneck, D. (1990) New types of virus infections of domestic animals in the German Democratic Republic. 4. Tahyna virus infections – a review. *Arch. Exp. Vet.* **44**(2):265–77.

Kocarek, P., Holusa, J., Vidlicka, L. (1999) Check-list of Blattaria, Mantodea, Orthoptera and Dermaptera of the Czech and Slovak Republics. *Articulata* **14**(2):177–84.

Kockaerts, Y., Vanhees, S., Knockaert, D. C. *et al.* (2001) Imported malaria in the 1990s: a review of 101 patients. *Eur. J. Emerg. Med.* **8**(4):287–90.

Koehler, J. E., Quinn, F. D., Berger, T. G., LeBoit, P. E., Tappero, J. W. (1992) Isolation of *Rochalimaea* species from cutaneous and osseous lesions of bacillary angiomatosis. *N. Engl. J. Med.* **327**(23):1625–31.

Koehler, K., Stechele, M., Hetzel, U. *et al.* (2002) Cutaneous leishmaniasis in a horse in southern Germany caused by *Leishmania infantum*. *Vet. Parasitol.* **109**(1–2):9–17.

Kofoed, K., Petersen, E. (2003) The efficacy of chemoprophylaxis against malaria with chloroquine plus proguanil, mefloquine, and atovaquone plus proguanil in travelers from Denmark. *J. Travel Med.* **10**(3):150–4.

Kolar, K. A., Rapini, R. P. (1991) Crusted (Norwegian) scabies. *Am. Fam. Phys.* **44**(4):1317–21.

Kolman, J. M., Kopecky, K., Rac, O. (1979) Serologic examination of human population in South Moravia (Czechoslovakia) on the presence of antibodies to arboviruses of the Alfavirus, Flavivirus, Turlock groups and Bunyamwera supergroup. *Folia Parasitol.* (*Praha*) **26**(1):55–60.

Kolobukhina, L. V., L'vov, D. K., Butenko, A. M., Kuznetsov, A. A., Galkina, I. V. (1989) The clinico-laboratory characteristics of cases of diseases connected with viruses of the California encephalitis complex in the inhabitants of Moscow. *Zh Mikrobiol. Epidemiol. Immunobiol.* **10**:68–73.

Komar, N. (2003) West Nile virus: epidemiology and ecology in North America. *Adv. Virus Res.* **61**:185–234.

Komar, N., Langevin, S., Hinten, S. *et al.* (2003) Experimental infection of North American birds with the New York 1999 strain of West Nile virus. *Emerg. Infect. Dis.* **9**(3):311–22.

Koptopoulos, G., Papadopoulos, O. (1980) A serological survey for tick-borne encephalitis and West Nile viruses in Greece. In *Arboviruses in the Mediterranean Countries: 6. FEMS Symposium* (ed. J. Vesenjak-Hirjan). *Zentralbl. Bakteriol. Microbiol. Hyg. 1 Abt.* Suppl. 9:185–8.

Korenberg, E. I. (1994) Comparative ecology and epidemiology of Lyme disease and tick-borne encephalitis in the former Soviet Union. *Parasitol. Today* **4**(1):157–60.

Korenberg, E. I. (1997) Tick-borne encephalitis (TBE) morbidity in Russia. *Vect. Ecol. Newsl.* **28**(4):3–4.

Korenberg, E. I. (1998) Ixodid tick-borne borreliosies (infections of the Lyme disease group) morbidity in Russia. *Vect. Ecol. Newsl.* **29**(1):4–5.

Korzan, A. I., Samoylova, T. I., Protas, I. I. *et al.* (1996) Tick-borne encephalitis (TBE) epidemiology in the Brest Province of the Republic of Belarus. *Rocz. Akad. Med. Bialymst* **41**(1):28–34.

Kosoy, M. Y., Elliott, L. H., Ksiazek, T. G. *et al.* (1996) Prevalence of antibodies to arenaviruses in rodents from the southern and western United States: evidence for an arenavirus associated with the genus *Neotoma. Am. J. Trop. Med. Hyg.* **54**(6):570–6.

Kovalevskii, Y. V., Korenberg, E. I. (1995) Differences in *Borrelia* infections in adult *Ixodes persulcatus* and *Ixodes ricinus* ticks (Acari:Ixodidae) in populations of north-western Russia. *Exp. Appl. Acarol.* **19**(1):19–29.

Kovats, S., Haines, A., Stanwell-Smith, R. *et al.* (1999) Climate change and human health in Europe. *Br. Med. J.* **318**(7199):1682–5.

Kozuch, O., Nosek, J., Gresikova, M., Ciampor, F., Chelma, J. (1978) Isolation of Tettnang virus from *Ixodes ricinus* ticks in Czechoslovakia. *Acta Virol.* **22**(1):74–6.

Krause, P. J., McKay, K., Gadbaw, J. *et al.* (2003) Increasing health burden of human babesiosis in endemic sites. *Am. J. Trop. Med. Hyg.* **68**(4):431–6.

Krause, P. J., Telford, S. R. 3rd, Spielman, A. *et al.* (1996) Concurrent Lyme disease and babesiosis: Evidence for increased severity and duration of illness. *J. Am. Med. Assoc.* **275**(21):1657–60.

Krech, T. (2002) TBE foci in Switzerland. *Int. J. Med. Microbiol.* **291**(suppl. 33):30–3.

Krishna, G., Chitkara, R. K. (2003) Pneumonic plague. *Semin. Respir. Infect.* **18**(3):159–67.

Kruger, A., Rech, A., Su, X. Z., Tannich, E. (2001) Two cases of autochthonous *Plasmodium falciparum* malaria in Germany with evidence for local transmission by indigenous *Anopheles plumbeus. Trop. Med. Int. Hlth.* **6**(12):983–5.

Krusell, A., Comer, J. A., Sexton, D. J. (2002) Rickettsialpox in North Carolina: a case report. *Emerg. Infect. Dis.* **8**(7):727–8.

Kuenzi, A. J., Morrison, M. L., Swann, D. E., Hardy, P. C., Downard, G. T. (1999) A longitudinal study of Sin Nombre virus prevalence in rodents, southeastern Arizona. *Emerg. Infect. Dis.* **5**(1):113–17.

Kuhn, K. G. (1999) Global warming and leishmaniasis in Italy. *Bull. Trop. Med. Int. Hlth.* **7**(2):1–2.

Kuhn, K. G., Campbell-Lendrum, D. H., Armstrong, B., Davies, C. R. (2003) Malaria in Britain: past, present, and future. *Proc. Natl. Acad. Sci. USA* **100**(17):9997–10001.

Kuiper, H., de Jongh, B. M., Nauta, A. P. *et al.* (1991) Lyme borreliosis in Dutch forestry workers. *J. Infect.* **23**(3):279–86.

Kurtenbach, K., Peacey, M., Rijpkema, S. G. *et al.* (1998) Differential transmission of the genospecies of *Borrelia burgdorferi* sensu lato by game birds and small rodents in England. *Appl. Environ. Microbiol.* **64**(4):1169–74.

Kuster, P. A. (1996) Reducing risk of house dust mite and cockroach allergen exposure in inner-city children with asthma. *Pediatr. Nurs.* **22**(4):297–303.

L'vov, D. K. (1973) Results of 3 years of field work by the Department of Ecology of Viruses. *Akad. Med. Nauk.* 5–12.

L'vov, D. K., Butenko, A. M., Gaidamovich, S.Ia. *et al.* (2000a) Epidemic outbreak of meningitis and meningoencephalitis, caused by West Nile virus, in the Krasnodar territory and Volgograd region (preliminary report). *Vopr. Virusol.* **45**(1):37–8.

L'vov, D. K., Butenko, A. M., Gromashevsky, V. L. *et al.* (2004) West Nile virus and other zoonotic viruses in Russia: examples of emerging-reemerging situations. *Arch. Virol.* **18**:85–96.

L'vov, D. K., Butenko, A. M., Vyshemirskii, O. I. *et al.* (2000b) Isolation of the West Nile fever virus from human patients during an epidemic outbreak in the Volgograd and Astrakhan regions. *Vopr. Virusol.* **45**(3):9–12.

L'vov, D. K., Dzharkenov, A. F., L'vov, D. N. *et al.* (2002a) Isolation of the West Nile fever virus from the great cormorant *Phalacrocorax carbo*, the crow *Corvus corone*, and *Hyalomma marginatum* ticks associated with them in natural and synanthroic biocenosis in the Volga delta (Astrakhan region, 2001). *Vopr. Virusol.* **47**(5):7–12.

L'vov, D. K., Gromashevskii, V. L., Skvortsova, T. M. *et al.* (1998) Circulation of viruses of the California serocomplex (Bunyaviridae, Bunyavirus) in the central and southern parts of the Russian plain. *Vopr. Virusol.* **43**(1):10–14.

L'vov, D. K., Vladimirtseva, E. A., Butenko, A. M. *et al.* (1988) Identity of Karelian fever and Ockelbo viruses determined by serum dilution-plaque reduction neutralization tests and oligonucleotide mapping. *Am. J. Trop. Med. Hyg.* **39**(6):607–10.

L'vov, D. N., Dzharkenov, A. F., Aristova, V. A. *et al.* (2002b) The isolation of Dhori viruses (Orthomyxoviridae, Thogotovirus) and Crimean-Congo hemorrhagic fever virus (Bunyaviridae, Nairovirus) from the hare (*Lepus europaeus*) and its ticks *Hyalomma marginatum* in the middle zone of the Volga delta, Astrakhan region, 2001. *Vopr. Virusol.* **47**(4):32–6.

La Scola, B., Davoust, B., Boni, M., Raoult, D. (2002) Lack of correlation between *Bartonella* DNA detection within fleas, serological results, and results of blood culture in a *Bartonella*-infected stray cat population. *Clin. Microbiol. Infect.* **8**(6):345–51.

La Scola, B., Fournier, P. E., Brouqui, P., Raoult, D. (2001) Detection and culture of *Bartonella quintana, Serratia marcescens*, and *Acinetobacter* spp. from decontaminated human body lice. *J. Clin. Microbiol.* **39**(5):1707–9.

Laan, J. R. van der, Smit, R. B. (1996) Back again: the clothes louse (*Pediculus humanus* var. *corporis*). *Ned. Tijdschr. Geneeskd.* **140**(38):1912–15.

Labuda, M., Kozuch, O., Gresikova, M. (1974) Isolation of West Nile virus from *Aedes cantans* mosquitoes in west Slovakia. *Acta Virol.* **18**(5):429–33.

Lademann, M., Wild, B., Reisinger, E. C. (2003) Tick-borne encephalitis (FSME) – how great is the danger really? *MMW Fortschr. Med.* **145**(15):45, 47–9.

Ladhani, S., El Bashir, H., Patel, V. S., Shingadia, D. (2003) Childhood malaria in East London. *Pediatr. Infect. Dis. J.* **22**(9):814–19.

Laiskonis, A., Mickiene, A. (2002) Tick-borne encephalitis in Eastern Europe. *Méd. Mal. Infect.* **32**(5):203–11.

Lakos, A. (2002) Tick-borne lymphadenopathy (TIBOLA). *Wiener Klinische Wochenschrift* **114**(13–14):648–54.

Lakos, A., Ferenczi, E., Ferencz, A., Toth, E. (1996) Tick-borne encephalitis. *Parasitol. Hung.* **29–30**:5–16.

Landbo, A. S., Flong, P. T. (1992) *Borrelia burgdorferi* in *Ixodes ricinus* from habitats in Denmark. *Med. Vet. Entomol.* **6**(2):165–7.

Lay-Rogues, G. le, Valle, M., Chastel, C., Beaucournu, J. C. (1983) Petits mammifères sauvages et arbovirus en Italie. *Bull. Soc. Pathol. Exot.* **76**(4):333–45.

Le Duc, J. W., Smith, G. A., Bagley, L. R., Hasty, S. E., Johnson, K. M. (1982) Letter to editor, ["Preliminary evidence that Hantaan or a closely related virus is enzootic in domestic rodents"]. *N. Engl. J. Med.* **307**(10):624–5.

Lebech, A. M., Hansen, K., Pancholi, P. *et al.* (1998) Immunoserologic evidence of Human Granulocytic Ehrlichiosis in Danish patients with Lyme neuroborreliosis. *Scand. J. Infect. Dis.* **30**(2):173–6.

Lednicky, J. A. (2003) Hantaviruses. a short review. *Arch. Pathol. Lab. Med.* **127**(1):30–5.

LeDuc, J. W., Smith, G. A., Johnson, K. M. (1984) Hantaan-like viruses from domestic rats captured in the United States. *Am. J. Trop. Med. Hyg.* **35**(3):992–8.

Lefebvre, J. P. (1981) A seasonal neurologic disease: paralysis following tick-bites. *Sem. Hop.* **57**(43–44):1809–14.

Lefebvre, J. P., Pouget-Abadie, J. F., Bontoux, D., Sudre, Y., Gil, R. (1976) Meningoradiculitis after tick bites. Apropos of 9 cases. *Sem. Hop.* **52**(11):687–94.

Leger, N., Marchais, R., Madulo-Leblond, G. *et al.* (1991) Sandflies implicated in the transmission of leishmaniasis on the island of Gozo, Malta. *Ann. Parasitol. Hum. Comp.* **66**(1):33–41.

Legros, F., Gay, F., Belkaid, M., Danis, M. (1998) Imported malaria in continental France, 1996. *Eurosurveillance Weekly* **3**(4):37–8.

Leiby, D. A., Duffy, C. H., Murrell, K. D., Schad, G. A. (1990) *Trichinella spiralis* in an agricultural ecosystem: transmission in the rat population. *J. Parasitol.* **76**(3):360–4.

Libikova, H., Heinz, F., Ujhazyova, D., Stunzner, D. (1978) Orbiviruses of the Kemerovo complex and neurological diseases. *Med. Microbiol. Immunol. (Berl.)* **166**(1–4):255–63.

Lindgren, E. (1998) Climate change, tick-borne encephalitis and vaccination needs in Sweden – a prediction model. *Ecol. Model.* **110**(1):55–63.

Lindgren, E., Gustafson, R. (2001) Tick-borne encephalitis in Sweden and climate change. *Lancet* **358**(9275):16–18.

Lindgren, E., Talleklint, L., Polfeldt, T. (2000) Impact of climatic change on the northern latitude limit and population density of the disease-transmitting European tick, *Ixodes ricinus*. *Environ. Hlth. Perspect.* **108**(2):119–23.

Linnemann, C. C. Jr., Barber, L. C., Dine, M. S., Body, A. E. (1978) Tick-borne relapsing fever in the Eastern United States. *Am. J. Dis. Child.* **132**(1):40–2.

Linthicum, K. J., Kramer, V. L., Madon, M. B., Fujioka, K. (2003) Introduction and potential establishment of *Aedes albopictus* in California in 2001. *J. Am. Mosq. Control Assoc.* **19**(4):301–8.

Lipsker, D., Hansmann, Y., Limbach, F. *et al.* (2001) Disease expression of Lyme borreliosis in northeastern France. *Eur. J. Clin. Microbiol. Infect. Dis.* **20**:225–30.

Liz, J. S., Anderes, L., Sumner, J. W. *et al.* (2000) PCR detection of granulocytic ehrlichiae in *Ixodes ricinus* ticks and wild small mammals in western Switzerland. *J. Clin. Microbiol.* **38**(3):1002–7.

Lledo, L., Gegundez, M. I., Serrano, J. L., Saz, J. V., Beltran, M. (2003) A sero-epidemiological study of *Rickettsia typhi* infection in dogs from Soria province, central Spain. *Ann. Trop. Med. Parasitol.* **97**(8):861–4.

Lopez-Garcia, J. S., Garcia-Lozano, I., Martinez-Garchitorena, J. (2003) Phthiriasis palpebrarum: diagnosis and treatment. *Arch. Soc. Esp. Oftalmol.* **78**(7):365–74.

Lopez-Vélez, R. (2003) The impact of highly active antiretroviral therapy (HAART) on visceral leishmaniasis in Spanish patients who are co-infected with HIV. *Ann. Trop. Med. Parasitol.* **97**(suppl. 1):143–7.

Lopez-Vélez, R., Pérez-Casas, C., Vorndam, A. V., Rigau, J. (1996) Dengue in Spanish travellers returning from the tropics. *Eur. J. Clin. Microbiol. Infect. Dis.* **15**(10):823–6.

Lopez-Vélez, R., Viana, A., Pérez-Casas, C. *et al.* (1999) Clinicoepidemiological study of imported malaria in travellers and immigrants to Madrid. *J. Travel Med.* **6**(2):81–6.

Lotric-Furlan, S., Petrovec, M., Asvic-Zupanc, T. *et al.* (2001) Prospective assessment of the etiology of acute febrile illness after a tick bite in Slovenia. *Clin. Infect. Dis.* **33**(4):503–10.

Lotric-Furlan, S., Petrovec, M., Zupanc, T. A. *et al.* (1998) Human granulocytic ehrlichiosis in Europe: clinical and laboratory findings for four patients from Slovenia. *Clin. Infect. Dis.* **27**(3):424–8.

Lozano, A., Filipe, A. R. (1998) Anticuerpos frente a virus West Nile y otros virus transmitidos por artropodos en la poblacion del Delta de Ebro. *Rev. Esp. Salud Publ.* **72**(3):245–50.

Lozyns'kyi, I. M., Vynohrad, I. A. (1998) Arboviruses and arbovirus infections in the forest steppe zone of Ukraine. *Mikrobiol. Z.* **60**(2):49–60.

Lucenko, I., Jansone, I., Velicko, I., Pujate, E. (2004) Tickborne encephalitis in Latvia. *Eurosurveillance Weekly* **8**(26).

Lundstrom, J. O. (1994) Vector competence of Western European mosquitoes for arboviruses: a review of field and experimental studies. *Bull. Soc. Vector Ecol.* **19**(1):23–6.

Lundstrom, J. O. (1999) Mosquito-borne viruses in Western Europe: a review. *J. Vector Ecol.* **24**(1):1–39.

Lundstrom, J. O., Turell, M. J., Niklasson, B. (1992) Antibodies to Ockelbo virus in three orders of birds (Anseriformes, Galliformes and Passeriformes) in Sweden. *J. Wildl. Dis.* **28**(1):144–7.

Lundstrom, J. O., Vene, S., Espmark, A., Engvall, M., Niklasson, B. (1991) Geographical and temporal distribution of Ockelbo disease in Sweden. *Epidemiol. Infect.* **106**(3):567–74.

Macan, J., Plavec, D., Kanceljak, B., Milkovic-Kraus, S. (2003) Exposure levels and skin reactivity to German Cockroach (*Blattella germanica*) in Croatia. *Croat. Med. J.* **44**(6):756–60.

Machurot, P. Y., Fumal, A., Sadzot, B. (2001) Neuroborreliosis. *Rev. Med. Liege* **56**(1):11–16.

MacLean, J. D., Demers, A.-M., Ndao, M. *et al.* (2004) Malaria epidemics and surveillance systems in Canada. *Emerg. Infect. Dis.* **10**(7):1195–201.

MacLean, J. D.,Ward, B. J. (1999) The return of swamp fever: malaria in Canadians. *Can. Med. Assoc. J.* **160**(2):211–12.

Madic, J., Huber, D., Lugovic, B. (1993) Serologic survey for selected viral and rickettsial agents of brown bears (*Ursus arctos*) in Croatia. *J. Wildl. Dis.* **29**(4):572–6.

Maeda, K., Markowitz, N., Hawley, R. C. *et al.* (1987) Human infection with *Ehrlichia canis*, a leukocytic rickettsia. *N. Engl. J. Med.* **316**(14):853–6.

Maes, E., Lecomte, P., Ray, N. (1998) A cost-of-illness study of Lyme disease in the United States. *Clin. Ther.* **20**(5):993–1008.

Mailles, A., Abu Sin, M., Ducoffre, G. *et al.* (2005) Larger than usual increase in cases of hantavirus infections in France, Belgium and Germany, June 2005. *Eurosurveillance Weekly* **10**(8).

Makdoembaks, A. M., Kager, P. A. (2000) Increase of malaria among migrants in Amsterdam-Zuidoost. *Ned. Tijdschr. Geneeskd.* **144**(2):83–5.

Makhnev, M. V. (2002) Local cases of three-day malaria in the Moscow area. *Ter. Arkh.* **74**(11):31–3.

Malkinson, M., Banet, C. (2002) The role of birds in the ecology of West Nile virus in Europe and Africa. *Curr. Top. Microbiol. Immunol.* **267**:309–22.

Malkova, D., Holubova, J., Kolman, J. M. *et al.* (1980) Isolation of Tettnang coronavirus from man? *Acta Virol.* **24**(5):363–6.

Mann, J. M., Schmid, G. P., Stoesz, P. A., Skinner, M. D., Kaufmann, A. F. (1982) Peripatetic plague. *J. Am. Med. Assoc.* **247**(1):47–8.

Manor, E., Ighbarieh, J., Sarov, B., Kassis, I., Regnery, R. (1992) Human and tick spotted fever group *Rickettsia* isolates from Israel: a genotypic analysis. *J. Clin. Microbiol.* **30**(10):2653–6.

Mansueto, S., Domingo, S., Vitale, G. *et al.* (1984) Indagini siero-epidemiologiche sull febbre Bottonosa in Scilia occidentale. V. Presenza di anticorpi anti-*R. conorii* in sieri di cani del trapense. *Riv. Parassitologia* **1**(45):197–200.

Mantel, C. F., Klose, C., Scheurer, S. *et al.* (1998) *Plasmodium falciparum* malaria acquired in Berlin, Germany. *Lancet* **346**(8970):320–1.

Marfin, A. A., Petersen, L. R., Eidson, M. *et al.* and the ArboNET Cooperative Surveillance Group (2001) Widespread West Nile virus activity, Eastern United States, 2000. *Emerg. Infect. Dis.* **7**:730–6.

Markeshin, Sla., Smirnova, S. E., Evstaf'ev, I. I. (1992) Assessment of the status of natural foci of Crimean–Congo haemorrhagic fever in the Crimea. *Zh. Mikrobiol. Epidemiol. Immunol.* **4**:28–31.

Maroli, M., Khoury, C. (1998) Leishmaniasis vectors in Italy. *Giorn. Ital. Med. Trop.* **3**(3/4):69–75.

Maroli, M., Mizzoni, V., Siragusa, C., D'Orazi, A., Gradoni, L. (2002) Evidence for an impact on the incidence of canine leishmaniasis by the mass use of deltamethrin-impregnated dog collars in southern Italy. *J. Med. Entomol.* **15**(4):358–63.

Maroli, M., Pampiglione, S., Tosti, A. (1988) Cutaneous leishmaniasis in western Sicily (Italy) and preliminary survey of phlebotomine sandflies. *Parassitologia* **30**(2–3):211–17.

Marquez, F. J., Muniain, M. A., Perez, J. M., Pacho, J. (2002) Presence of *Rickettsia felis* in the cat flea from southwestern Europe. *Emerg. Infect. Dis.* **8**(1):89–91.

Martin-Sanchez, J., Gramiccia, M., Di Muccio, T., Ludovisi, A., Morillas-Marquez, F. (2004) Isoenzymatic polymorphism of *Leishmania infantum* in southern Spain. *Trans. Roy. Soc. Trop. Med. Hyg.* **98**(4):228–32.

Martinez-Ortega, E. (1984) *Phlebotomus perniciosus* possible vector de la leishmaniasis cutanea. *Rev. Iberica Parasitol.* **44**(1):59–64.

Marty, P., Le Fichoux, Y. (1988) Epidemiologie de la leishmaniose dans le sud de la France. *Prat. Med. Chirur. Anim. Comp.* **23**(suppl. 5):11–15.

Marty, P., Le Fichoux, Y., Izri, M. A. *et al.* (1992) Autochthonous *Plasmodium falciparum* malaria in southern France. *Trans. Roy. Soc. Trop. Med. Hyg.* **86**(5):478.

Marty, P., Leger, I., Albertini, M. *et al.* (1994) Infantile visceral leishmaniasis in the Alpes-Maritimes from 1985 to 1992. *Bull. Soc. Pathol. Exot.* **87**(2):105–9.

Masoero, L., Dadone, R., Guercio, A. (1991) *Rickettsia conorii*: indagine sierologia sulla popolazione canina si un comune del Cuneese in seguito an un focolaio di febbre bottosa. *Nuovo Progr. Vet.* **46**(19):716–18.

Massei, F., Messina, F., Gori, L., Macchia, P., Maggiore, G. (2004) High prevalence of antibodies to *Bartonella henselae* among Italian children without evidence of cat scratch disease. *Clin. Infect. Dis.* **38**(1):145–8.

Matteelli, A., Volonterio, A., Gulletta, M. *et al.* (2001) Malaria in illegal Chinese immigrants, Italy. *Emerg. Infect. Dis.* **7**(6):1055–8.

Matuschka, F. R., Endepols, S., Richter, D., Spielman, A. (1997) Competence of urban rats as reservoir hosts for Lyme disease spirochetes. *J. Med. Entomol.* **34**(4):489–93.

Matuschka, F. R., Klug, B., Schinkel, T. W., Spielman, A., Richter, D. (1998) Diversity of European Lyme disease spirochetes at the southern margin of their range. *Appl. Environ. Microbiol.* **64**(5):1980–2.

Matuschka, F. R., Lange, R., Spielman, A., Richter, D., Fischer, P. (1990) Subadult *Ixodes ricinus* on rodents in Berlin West Germany. *J. Med. Entomol.* **27**(3):385–90.

Matuschka, F. R., Ohlenbusch, A., Eiffert, H., Richter, D., Spielman A. (1996) Characteristics of Lyme disease spirochetes in archived European ticks. *J. Infect. Dis.* **174**(2):424–6.

McDade, J. E. (1987) Flying squirrels and their ectoparasites: disseminators of epidemic typhus. *Parasitol. Today* **3**(3):85–7.

McDade, J. E., Shepard, C. C., Redus, M. A., Newhouse, V. F., Smith, J. D. (1980) Evidence of *Rickettsia prowazekii* infections in the United States. *Am. J. Trop. Med. Hyg.* **29**(2):277–84.

McHugh, C. P., Grogl, M., Kreutzer, R. D. (1993) Isolation of *Leishmania mexicana* (Kinetoplastida: Trypanosomatidae) from *Lutzomyia anthophora* (Diptera: Psychodidae) collected in Texas. *J. Med. Entomol.* **30**(3):631–3.

McJunkin, J. E., de los Reyes, E. C., Irazuzta, J. E. *et al.* (2001) La Crosse encephalitis in children. *N. Engl. J. Med.* **344**(11):801–7.

McLean, D. M. (1975) Mosquito-borne arboviruses in arctic America. *Med. Biol.* **53**(5):264–70.

McLean, D. M., Donohue, W. L. (1959). Powassan virus: isolation of virus from a fatal case of encephalitis. *Can. Med. Assoc. J.* **80**:708–11.

McQuiston, J. H., Childs, J. E. (2001) Q Fever in humans and animals in the United States. *Vector-Borne Zoon. Dis.* **2**(3):179–91.

McQuiston, J. H., Paddock, C. D., Holman, R. C., Childs, J. E. (1999) The Human Ehrlichioses in the United States. *Emerg. Infect. Dis.* **5**(5):635–42.

Meador, C. N. (1915) Five cases of relapsing fever originating in Colorado, with positive blood findings in two. *Colo. Med.* **12**:365–9.

Meek, J. I., Roberts, C. L., Smith, E. V. Jr., Cartter, M. L. (1996) Underreporting of Lyme disease by Connecticut physicians, 1992. *J. Publ. Hlth. Manag. Pract.* **2**(4):61–5.

Meites, E., Jay, M. T., Deresinski, S. *et al.* (2004) Reemerging leptospirosis, California. *Emerg. Infect. Dis.* **10**(3):406–12.

Mellor, P. S., Boorman, J., Baylis, M. (2000) *Culicoides* biting midges: Their role as arbovirus vectors. *Ann. Rev. Entomol.* **45**:307–40.

Melnick, J. L., Paul, J. R., Riordan, J. T. *et al.* (1950) Isolation from human sera in Egypt of a virus apparently identical to West Nile Virus (18884). *Proc. Soc. Exp. Biol. Med.* **77**(4):661–5.

Meltzer, M. I., Rigau-Perez, J. G., Clark, G. G., Reiter, P., Gubler, D. J. (1998) Using disability-adjusted life years to assess the economic impact of dengue in Puerto Rico: 1984–1994. *Am. J. Trop. Med. Hyg.* **59**(2):265–71.

Mendoza-Montero, J., Gamez-Rueda, M. I., Navarro-Mari, J. M., de la Rosa-Fraile, M., Oyonarte-Gomez, S. (1998) Infections due to sandfly fever virus serotype Toscana in Spain. *Clin. Infect. Dis.* **27**(3):434–6.

Mickiene, A., Laiskonis, A., Gunther, G. *et al.* (2002) Tickborne encephalitis in an area of high endemicity in Lithuania: disease severity and long-term prognosis. *Clin. Infect. Dis.* **35**(6):650–8.

Mielke, U. (1997) Nosocomial myiasis. *J. Hosp. Infect.* **37**(1):1–5.

Milutinovic, M., Pavlovic, I., Miscevic, Z., Bisevac, L. (1989) Dynamics of sandfly (Diptera, Phlebotomidae) and tick (Acarina, Ixodidae) populations in the endemic foci of visceral leishmaniasis in Yugoslavia. *Acta Vet.* (*Beograd*) **39**(2–3):143–54.

Minar, J., Herold, J., Eliskova, J. (1995) Nosocomial myiasis in Central Europe. *Epidemiol. Mikrobiol. Imunol.* **44**(2):81–3.

Misonne, M.-C., Van Impe, G., Hoet, P. P. (1998) Genetic heterogeneity of *Borrelia burgdorferi* sensu lato in *Ixodes ricinus* ticks collected in Belgium. *J. Clin. Microbiol.* **36**(11):3352–4.

Mitchell, C. J., L'vov, S. D., Savage, H. M. *et al.* (1993) Vector and host relationships of California serogroup viruses in Western Siberia. *Am. J. Trop. Med. Hyg.* **49**(1):53–62.

Mohd-Jaafar, F., Attoui, H., De Micco, P., De Lamballerie, X. (2004) Recombinant VP6-based enzyme-linked immunosorbent assay for detection of immunoglobulin G antibodies to Eyach virus (genus Coltivirus). *J. Clin. Virol.* **30**(3):248–53.

Molina, R., Gradoni, L., Alvar, J. (2003) HIV and the transmission of Leishmania. *Ann. Trop. Med. Parasitol.* **97**(suppl. 1):29–45.

Molle, I., Christensen, K. L., Hansen, P. S. *et al.* (2000) Use of medical chemoprophylaxis and antimosquito precautions in Danish malaria patients and their traveling companions. *J. Travel Med.* **7**(5):253–8.

Molnar, E. (1982) Occurrence of tick-borne encephalitis and other arboviruses in Hungary. *Geogr. Med.* **12**:78–120.

Molnar, E., Gulyas, M. S., Kubinyi, L. *et al.* (1976) Studies on the occurrence of tick-borne encephalitis in Hungary. *Acta. Vet. Acad. Sci. Hung.* **26**(4):419–37.

Monath, T. P. (1980) (ed.) *St. Louis Encephalitis.* Washington, DC: Am. Publ. Hlth Assoc.

Moore, C. G., Mitchell, C. J. (1997) *Aedes albopictus* in the United States: Ten-year presence and public health implications. *Emerg. Infect. Dis.* **3**(3):329–34.

Moore, D. A., Grant, A. D., Armstrong, M., Stumpfle, R., Behrens, R. H. (2004) Risk factors for malaria in UK travellers. *Trans. Roy. Soc. Trop. Med. Hyg.* **98**(1):55–63.

Moore, V. A. 4th, Varela, A. S., Yabsley, M. J., Davidson, W. R., Little, S. E. (2003) Detection of *Borrelia lonestari,* putative agent of southern tick-associated rash illness, in white-tailed deer (*Odocoileus virginianus*) from the southeastern United States. *J. Clin. Microbiol.* **41**(1):424–7.

Morais, J. D., Dawson, J. E., Greene, C. *et al.* (1991) First European cases of Ehrlichiosis (Correspondence). *Lancet* **338**(8767):633–4.

Moral, L., Rubio, E. M., Moya, M. (2002) A leishmanin skin test survey in the human population of l' Alacanti region (Spain): implications for the epidemiology of *Leishmania infantum* infection in southern Europe. *Trans. Roy. Soc. Trop. Med. Hyg.* **96**(2):129–32.

Morbidity and Mortality Weekly Records (MMWR) (2000) **49**:25–28.

Morillas, F., Sanchez-Rabasco, F.,Ocaña, J. *et al.* (1996) Leishmaniasis in the focus of the Axarquía region, Malaga province, southern Spain: a survey of the human, dog, and vector. *Parasitol. Res.* **82**(6):569–70.

Mosimann, B., Peitrequin, R., Blanc, C., Pecoud, A. (1992) Allergy to cockroaches in a Swiss population with asthma and chronic rhinitis. *Schweiz. Med. Wochenschr.* **122**(34):1245–8.

Muckenfuss, R. S., Armstrong, C., McCordock, H. A. (1933) Encephalitis: studies on experimental transmission. *Publ. Hlth Rept.* **48**(1):1341–3.

Muentener, P., Schlagenhauf, P., Steffen, R. (1999) Imported malaria (1985–95): trends and perspectives. *Bull. Wld. Hlth. Org.* **77**(7):560–6.

Muhlberger, N., Jelinek, T., Behrens, R. H. *et al.* (2003) Age as a risk factor for severe manifestations and fatal outcome of falciparum malaria in European patients: observations from TropNetEurop and SIMPID surveillance data. *Clin. Infect. Dis.* **36**(8):990–5.

Muhlemann, M. F., Wright, D. J. (1987) Emerging pattern of Lyme disease in the United Kingdom and Irish Republic. *Lancet* **1**(8527):260–2.

Musken, H., Fernandez-Caldas, E., Maranon, F. *et al.* (2002) In vivo and in vitro sensitization to domestic mites in German urban and rural allergic patients. *J. Investig. Allerg. Clin. Immunol.* **12**(3):177–81.

Naucke, T. J., Schmitt, C. (2004) Leishmaniasis becoming endemic in Germany? *Int. J. Med. Microbiol.* **293**(suppl. 37):179–81.

Navin, T. R., Roberto, R. R., Juranek, D. D. *et al.* (1985) Human and sylvatic *Trypanosoma cruzi* infection in California. *Am. J. Publ. Hlth.* **75**(4):366–9.

Nelson, H. S., Fernandez-Caldas, E. (1995) Prevalence of house dust mites in the Rocky Mountain states. *Ann. Allergy Asthma Immunol.* **75**(4):337–9.

Nichol, S. T., Spiropoulou, C. F., Morzunov, S. *et al.* (1993) Genetic identification of a novel hantavirus associated with an outbreak of acute respiratory illness in the southwestern United States. *Science* **262**:914–17.

Nicoletti, L., Ciufolini, M. G., Verani, P. (1996) Sandfly fever viruses in Italy. *Arch. Virol.* suppl. 11:41–7.

Nielsen, H., Fournier, P. E., Pedersen, I. S. *et al.* (2004) Serological and molecular evidence of *Rickettsia helvetica* in Denmark. *Scand. J. Infect. Dis.* **36**(8):559–63.

Niklasson, B., Espmark, A., LeDuc, J. W. *et al.* (1984) Association of a Sindbis-like virus with Ockelbo disease in Sweden. *Am. J. Trop. Med. Hyg.* **33**(6):1212–17.

Niklasson, B., Espmark, A., Lundstrom, J. (1988) Occurrence of arthralgia and specific IgM antibodies three to four years after Ockelbo disease. *J. Infect. Dis.* **157**(4):832–5.

Niklasson, B., Vene, S. (1996) Vector-borne viral diseases in Sweden – a short review. *Arch. Virol.* Suppl. 11:29–55.

Nilsson, K., Jaenson, T. G. T., Uhnoo, I. *et al.* (1997) Characterization of a spotted fever group rickettsia from *Ixodes ricinus* ticks in Sweden. *J. Clin. Microbiol.* **35**(1):243–7.

Nilsson, K., Lindquist, O., Liu, A. J. *et al.* (1999a) *Rickettsia helvetica* in *Ixodes ricinus* ticks in Sweden. *J. Clin. Microbiol.* **37**(2):400–3.

Nilsson, K., Lindquist, O., Pahlson, C. (1999b) Association of *Rickettsia helvetica* with chronic perimyocarditis in sudden cardiac death. *Lancet* **354**(9185):1169–73.

Niscigorska, J. (1999) Epidemiologic-clinical aspects of tick borne borreliosis in the Szczecin Province. *Ann. Acad. Med. Stetin.* **45**:157–73.

Nissen, M. D., Walker, J. C. (2002) Dirofilariasis. Ebay web site http://www.emedicine.com/med/topic3446.htm

Nohlmans, M. K., de Boer, R., van den Bogaard, A. E., Blaauw, A. A., van Boven, C. P. (1990) Occurrence of *Borrelia burgdorferi* in *Ixodes ricinus* ticks in the Netherlands. *Ned. Tijdschr. Geneeskd.* **134**(27):1300–3.

Novakovic, S., Avsic Zupanc, T., Hren Venceli, H. (1991) Mediterranean spotted fever in Slovenia. *Zdravstveni Vestnik* **59**(11):509–11.

Nuti, M., Amaddeo, D., Crovatto, M. *et al.* (1993) Infections in an alpine environment: antibodies to hantaviruses, leptospira, rickettsiae and *Borrelia burgdorferi* in defined Italian populations. *Am. J. Trop. Med. Hyg.* **48**(1):20–5.

Nuti, M., Serafini, D. A., Bassetti, D. *et al.* (1998) Ehrlichia infection in Italy. *Emerg. Infect. Dis.* **4**(4):663–5.

Ogden, N. H., Bown, K., Horrocks, B. K., Woldehiwet, Z., Bennett, M. (1998) Granulocytic *Ehrlichia* infection in ixodid ticks and mammals in woodlands and uplands of the U. K. *Med. Vet. Entomol.* **12**(4):423–9.

Oliveira, J., Corte Real, R. (1999) Rickettsia infections in Portugal. *Acta. Med. Port.* **12**(12):313–21.

Oliver, J. H. Jr., Owsley, M. R., Hutcheson, H. J. *et al.* (1993) Conspecificity of the ticks *Ixodes scapularis* and *I. dammini* (Acari: Ixodidae). *J. Med. Entomol.* **30**(1):54–63.

Opaneye, A. A., Jayaweera, D. T., Walzman, M., Wade, A. A. (1993) *Pediculosis pubis*: a surrogate marker for sexually transmitted diseases. *J. Roy. Soc. Hlth.* **113**(1):6–7.

Ostergard, N. H. (2000) Borreliosis and ehrlichiosis in hunting dogs in Vendsyssel. *Dansk Vet.* **83**(14):6–9.

Ostfeld, R. S., Roy, P., Haumaier, W. *et al.* (2004) Sand fly (*Lutzomyia vexator*) (Diptera: Psychodidae) populations in upstate New York: abundance, microhabitat, and phenology. *J. Med. Entomol.* **41**(4):774–8.

Oteo, J. A., Blanco, J. R., de Artola, V. M., Ibarra, V. (2000) First report of human granulocytic ehrlichiosis from southern Europe (Spain). *Emerg. Infect. Dis.* **6**(4):430–2.

Oteo, J. A., Gil, H., Barral, M. *et al.* (2001) Presence of granulocytic ehrlichia in ticks and serological evidence of human infection in La Rioja, Spain. *Epidemiol. Infect.* **127**(2):353–8.

Ozdemir, F. A., Rosenow, F., Slenczka, W., Kleine, T. O., Oertel, W. H. (1999) Early summer meningoencephalitis. Extension of the endemic area to mid-Hessia. *Nervenarzt* **70**(2):119–22.

Paddock, C. D., Holman, R. C., Krebs, J. W., Childs, J. E. (2002) Assessing the magnitude of fatal Rocky Mountain spotted fever in the United States: comparison of two national data sources. *Am. J. Trop. Med. Hyg.* **67**(4):349–54.

Paddock, C. D., Sumner, J. W., Comer, J. A. *et al.* (2004) *Rickettsia parkeri*: a newly recognized cause of spotted fever rickettsiosis in the United States. *Clin. Infect. Dis.* **38**(6):805–11.

Paddock, C. D., Zaki, S. R., Koss, T. *et al.* (2003) Rickettsialpox in New York City: a persistent urban zoonosis. *Ann. N. Y. Acad. Sci.* **990**:36–44.

Painter, J. A., Molbak, K., Sonne Hansen, J. *et al.* (2004) *Salmonella*-based rodenticides and public health. *Emerg. Infect. Dis.* **10**(6):985–7.

Pampiglione, S., Rivasi, F., Angeli, G. *et al.* (2001) Dirofilariasis due to *Dirofilaria repens* in Italy, an emergent zoonosis: report of 60 new cases. *Histopathology* **38**(4):344–54.

Pancewicz, S. A., Olszewska, B., Hermanowska-Szpakowicz, T. *et al.* (2001) Epidemiologic aspect of lyme borreliosis among the inhabitants of Podlasie Province. *Przegl. Epidemiol.* **55**(suppl. 3):187–94.

Panthier, R. (1964) Épidémiologie du virus West Nile. Étude d'un foyer en Camargue. *Ann. Inst. Pasteur* **114**(4):518–20.

Panthier, R., Hannoun, C., Beytout, D., Mouchet, J. (1968) Épidémiologie du virus West Nile. Étude d'un foyer en Camargue: III Les maladies humanines. *Ann. Inst. Pasteur* **115**(3):435–45.

Papa, A., Bino, S., Llagami, A. *et al.* (2002) Crimean–Congo hemorrhagic fever in Albania, 2001. *Eur. J. Clin. Microbiol. Infect. Dis.* **21**(8):603–6.

Papa, A., Christova, I., Papadimitriou, E., Antoniadis, A. (2004) Crimean–Congo hemorrhagic fever in Bulgaria. *Emerg. Infect. Dis.* **10**(8):1465–7.

Papadopoulos, O., Koptopoulos, G. (1978) Isolation of Crimean–Congo haemorrhagic fever (CCHF) virus from *Rhipicephalus bursa* ticks in Greece. *Acta Microbiol. Hellen.* **23**(1): 20–8.

Papapanagiotuo, J., Kyriazopouluo, V., Antoniadis, A. *et al.* (1974) Haemagglutionation–inhibition antibodies to arboviruses in a human population in Greece. *Zentralbl. Bakteriol. Hyg.* (Orig. A) **228**(4):443–6.

Paredes, R., Munoz, J., Diaz, I. *et al.* (2003) Leishmaniasis in HIV infection. *J. Postgrad. Med.* **49**(1):39–49.

Parola, P., Beati, L., Cambon, M., Brouqui, P., Raoult, D. (1998a) Ehrlichial DNA amplified from *Ixodes ricinus* (Acari: Ixodidae) in France. *J. Med. Entomol.* **35**(2):180–3.

Parola, P., Raoult, D. (2001) Tick-borne bacterial diseases emerging in Europe. *Clin. Microbiol. Infect.* **7**(2):80–3.

Patz, J. A., Reisen, W. K. (2001) Immunology, climate change and vector-borne diseases. *Trends Immunol.* **22**(4):171–2.

Pauli, G. (2004) West Nile virus. Prevalence and significance as a zoonotic pathogen. *Bundesgesundheitsblatt Gesundheitsforschung Gesundheitsschutz.* **47**(7):653–60.

Pavlatos, M., Gordon-Smith, C. E. (1964) Antibodies to arthropod-borne viruses in Greece. *Trans. Roy. Soc. Trop. Med. Hyg.* **58**(3):422–4.

Pawelczyk, A., Sinski, E. (2000) Prevalence of IgG antibodies response to *Borrelia burgdorferi* s.l. in populations of wild rodents from Mazury lakes district region of Poland. *Ann. Agric. Environ. Med.* **7**(2):79–83.

Pazdiora, P., Benesova, J., Böhm, K. *et al.* (1994) Incidence of Lyme borreliosis in the West Bohemian region (1988–1992). *Epidemiol. Mikrobiol. Imunol.* **43**(2):71–4.

Pazdiora, P., Struncova, V., Mlada, L. (1991) Lyme borreliosis in the west Bohemian region – 2 year's experience. *Cas. Lek. Cske.* **9**(36):1125–1128.

Peavy, C. A., Lane, R. S., Kleinjan, J. E. (1997) Role of small mammals in the ecology of *Borrelia burgdorferi* in a peri-urban park in north coastal California. *Exp. Appl. Acarol.* **21**(8):569–84.

Peleman, R., Benoit, D., Goossens, L. *et al.* (2000) Indigenous malaria in a suburb of Ghent, Belgium. *J. Travel Med.* **7**(1):48–9.

Perevoscikovs, J., Jansone, I., Jansone, S. *et al.* (2005) Trichinellosis outbreak in Latvia linked to bacon bought at a market. *Eurosuveillance Weekly* **10**:5.

Perez-Eid, C., Hannoun, C., Rodhain, F. (1992) The Alsatian tick-borne encephalitis focus: presence of the virus among ticks and small mammals. *Eur. J. Epidemiol.* **8**(2):178–86.

Persing, D. H., Herwaldt, B. L., Glaser, C. *et al.* (1995) Infection with a babesia-like organism in northern California. *N. Engl. J. Med.* **332**(5):298–303.

Pether, J. V., Jones, W., Lloyd, G., Rutter, D. A., Barry, M. (1994) Fatal murine typhus from Spain. *Lancet* **344**(8926):897–8.

Petric, D., Pajovuic, I., Ignjatovic, C., Zgomba, M. (2001) *Aedes albopictus* (Skuse, 1894), a new mosquito species (Diptera:Culicidae) in the entomofauna of Yugoslavia. *Symposia of Serbian Entomologists' 2001. Ent. Soc.* Abstract Volume, p. 29. Serbia, Goic.

Petruschke, G., Rossi, L., Genchi, C., Pollono, F. (2001) Canine dirofilariasis in the canton of Ticino and in the neighboring areas of northern Italy. *Schweiz. Archiv. Tierheilkd.* **143**(3):141–7.

Peyton, E. L., Campbell, S. R., Candeletti, T. M., Romanowski, M., Crans, W. J. *Aedes (Finlaya) japonicus japonicus* (Theobald), a new introduction into the United States. *J. Am. Mosq. Control Assoc.* **15**(2):238–41.

Pezzella, M., Lillini, E., Sturchio, E. *et al.* (2004) Leptospirosis survey in wild rodents living in urban areas of Rome. *Ann. Ig.* **16**(6):721–6.

Pichon, B., Godfroid, E., Hoyois, B. *et al.* (1995) Simultaneous infection of *Ixodes ricinus* nymphs by two *Borrelia burgdorferi* sensu lato species: Possible implications for clinical manifestations. *Emerg. Infect. Dis.* **1**(3):89–90.

Pichon, B., Mousson, L., Figureau, C., Rodhain, F., Perez-Eid, C. (1999) Density of deer in relation to the prevalence of *Borrelia burgdorferi* s.l. in *Ixodes ricinus* nymphs in Rambouillet forest, France. *Exp. Appl. Acarol.* **23**(3):267–75.

Piemont, Y., Heller, R. (1998) Bartonellosis: I. *Bartonella henselae. Ann. de Biol. Clin. (Paris)* **56**(6):681–92.

Pierer, K., Kock, T., Freidl, W. *et al.* (1993) Prevalence of antibodies to *Borrelia burgdorferi* flagellin in Styrian blood donors. *Zentralbl. Bakteriol.* **279**(2):239–43.

Pierzchalski, J. L., Bretl, D. A., Matson, S. C. (2002) *Phthirus pubis* as a predictor for chlamydia infections in adolescents. *Sex. Transm. Dis.* **29**(6):331–4.

Pilaski, J. (1987) Contributions to the ecology of Tahyna virus in Central Europe. *Bull. Soc. Vector Ecol.* **12**(2):544–53.

Pilaski, J., Mackenstein, H. (1989) Isolation of Tahyna virus from mosquitoes in 2 different European natural foci. *Zentralbl. Bakteriol. Mikrobiol. Hyg. [B]* **180**(4):394–420.

Pineda, J. A., Gallardo, J. A., Macias, J. *et al.* (1998) Prevalence of and factors associated with visceral leishmaniasis in human immunodeficiency virus type 1-infected patients in southern Spain. *J. Clin. Microbiol.* **36**(9):2419–22.

Platonov, A. E., Shipulin, G. A., Shipulina, O. Y. *et al.* (2001) Outbreak of West Nile virus infection, Volgograd region, Russia, 1999. *Emerg. Infect. Dis.* **7**(1):128–32.

Platts-Mills, T. A. (1994) How environment affects patients with allergic disease: indoor allergens and asthma. *Ann. Allergy* **72**(4):381–4.

Podsiadly, E., Sokolowska, E., Tylewska Wierzbanowska, S. (2002) Seroprevalence of *Bartonella henselae* and *Bartonella quintana* infections in Poland in 1998–2001. *Przegl. Epidemiol.* **56**(3):399–407.

Pokorny, P. V. (1990) *Borrelia* sp. in the common tick (*Ixodes ricinus*) in the locality of Prague. *Cske. Epidemiol. Mikrobiol. Immunol.* **39**(1):32–8.

Pokorny, P., Zahradkova, S. (1990) Incidence of *Borrelia* in the common tick (*Ixodes ricinus*) in the area of the city of Brno. *Cske. Epidemiol. Mikrobiol. Immunol.* **39**(3):166–70.

Pollitzer, R. (1954) *Plague*. Geneva: WHO.

Pollono, F., Rossi, L., Cancrini, G. (1998) An investigation of the attraction of Culicidae to dogs used as bait in Piemonte. *Parassitologia* **40**(4):439–45.

Pozio, E. (2001) New patterns of *Trichinella* infection. *Vet. Parasitol.* **98**(1–3):133–148.

Prazniakova, E., Sekeyova, M., Batikova, M. *et al.* (1975) The communications of the Twelth Annual Meeting of the Czechoslovak Society for Microbiology. Abstracts of Communications. Kosice, September 9–11, 1975. *Folia Microbiol. (Praha)* **21**(3):185–256.

Proenca, P., Cabral, T., Carmo, G., do Ferreira, L., Xavier, R. (1997) Le paludisme d'importation a l'hopital de Santa Maria de Lisbonne (1989–1995). *Méd. Mal. Infect.* **27**:691–5.

Pugliese, P., Martini-Wehrlen, S., Roger, P. M. *et al.* (1997) Malaria attacks after returning from endemic areas. Failure or inadequate chemoprophylaxis? *Presse Med.* **26**(29):1378–80.

Punda Polic, V., Bradaric, N., Klismanic Nuber, Z., Mrljak, V., Giljanovic, M. (1995) Antibodies to spotted fever group rickettsiae in dogs in Croatia. *Eur. J. Epidemiol.* **11**(4):389–92.

Punda-Polic, V., Leko-Grbic, J., Radulovic, S. (1995) Prevalence of antibodies to rickettsiae in the north-western part of Bosnia and Herzegovina. *Eur. J. Epidemiol.* **11**(6):697–9.

Purnell, R. E., Brocklesby, D. W. (1978) Isolation of a virulent strain of *Ehrlichia phagocytophila* from the blood of cattle in the Isle of Man. *Vet. Rec.* **102**(25):552–3.

Pusterla, N., Pusterla, J. B., Deplazes, P. *et al.* (1998a) Seroprevalence of *Ehrlichia canis* and of canine granulocytic *Ehrlichia* infection in dogs in Switzerland. *J. Clin. Microbiol.* **36**(12):3460–2.

Pusterla, N., Weber, R., Wolfensberger, C. *et al.* (1998b) Serological evidence of human granulocytic ehrlichiosis in Switzerland. *Eur. J. Clin. Microbiol. Infect. Dis.* **17**(3):207–9.

Pusterla, N., Wolfensberger, C., Lutz, H., Braun, U. (1997) Serologic studies on the occurrence of bovine ehrlichiosis in the cantons Zurich, Schaffhausen, Thurgau, St Gallen and Obwalden. *Schweiz. Archiv Tierheilkd.* **139**(12):543–9.

Quessada, T., Martial-Convert, F., Arnaud, S. *et al.* (2003) Prevalence of *Borrelia burgdorferi* species and identification of *Borrelia valaisiana* in questing *Ixodes ricinus* in the Lyon region of France as determined by polymerase chain reaction-restriction fragment length polymorphism. *Eur. J. Clin. Microbiol. Infect. Dis.* **22**(3):165–73.

Quick, R. E., Herwaldt, B. L., Thomford, J. W. *et al.* (1993) Babesiosis in Washington State – A new species of Babesia. *Ann. Intern. Med.* **119**(4):284–90.

Radulovic, S., Feng, H. M., Morovic, M. *et al.* (1996) Isolation of *Rickettsia akari* from a patient in a region where Mediterranean spotted fever is endemic. *Clin. Infect. Dis.* **22**(2):216–20.

Ralph, E. D. (1999) Powassan encephalitis. *Can. Med. Assoc. J.* **161**(11):1416–17.

Randolph, S. E. (2001) The shifting landscape of tick-borne zoonoses: tick- borne encephalitis and Lyme borreliosis in Europe. *Philos. Trans. Roy. Soc. Lond. B Biol. Sci.* **356**(1411):1045–56.

Raoult, D., Aboudharam, G., Crubezy, E. *et al.* (2000) Molecular identification by "suicide PCR" of *Yersinia pestis* as the agent of medieval black death. *Proc. Natl. Acad. Sci. USA* **97**(23):12800–3.

Raoult, D., Brouqui, P., Roux, V. A. (1996) A new spotted fever group rickettsiosis. *Lancet* **348**:412.

Raoult, D., Dupont, H. T., Chicheportiche, C. *et al.* (1993) Mediterranean spotted fever in Marseille, France: correlations between prevalence of hospitalized patients, seroepidemiology, and prevalence of infected ticks in three different areas. *Am. J. Trop. Med. Hyg.* **48**(2):249–56.

Raoult, D., La Scola, B., Enea, M. *et al.* (2001) A flea-associated *Rickettsia* pathogenic for humans. *Emerg. Infect. Dis.* **7**(1):73–81.

Raoult, D., Lakos, A., Fenollar, F. *et al.* (2002) Spotless rickettsiosis caused by *Rickettsia slovaca* and associated with *Dermacentor* ticks. *Clin. Infect. Dis.* **34**(10):1331–6.

Raoult, D., Nicolas, D., De Micco, Ph., Gallais, H., Casanova, P. (1985) Aspects épidémiologiques de la fièvre boutonneuse Méditerranéenne en Corse du Sud. [Epidemiologic aspects of Mediterranean boutonneuse fever in the south of Corsica]. *Bull. Soc. Pathol. Exot. Filiales* **78**(4):446–451.

Raoult, D., Tissot-Dupont, H., Foucault, C. *et al.* (2000) Q fever 1985–1998. Clinical and epidemiologic features of 1,383 infections. *Medicine* (*Baltimore*) **79**(2):109–23.

Raoult, D., Toga, B., Chaudet, H., Chiche Portiche, C. (1987) Rickettsial antibody in southern France: antibodies to *Rickettsia concorii* and *Coxiella burnetii* among urban, suburban and semi-rural blood donors. *Trans. Roy. Soc. Trop. Med. Hyg.* **81**(1):80–1.

Raoult, D., Weiller, P. J., Chagnon, A. *et al.* (1986) Mediterranean spotted fever: clinical, laboratory and epidemiological features of 199 cases. *Am. J. Trop. Med. Hyg.* **35**(4):845–50.

Rawlings, J. A., Torrez-Martinez, N., Neill, S. U. *et al.* (1996) Cocirculation of multiple hantaviruses in Texas, with characterization of the small (S) genome of a previously undescribed virus of cotton rats (*Sigmodon hispidus*). *Am. J. Trop. Med. Hyg.* **55**(6):672–9.

Reep Van den Bergh, C. M., Leeuwen, W. M., Kessel, R. P., Van Lelijveld, J. L. (1996) Malaria: under-notification and risk assessment for travellers to the tropics. *Ned. Tijdschr. Geneeskd.* **140**(16):878–82.

Reeves, W. C. (2001) Partners: serendipity in arbovirus research. *J. Vector Ecol.* **26**(1):1–6.

Reeves, W. C., Hardy, J. L., Reisen, W. K., Milby, M. M. (1994) Potential effect of global warming on mosquito-borne arboviruses. *J. Med. Entomol.* **31**(3):323–32.

Rehacek, J., Krauss, H., Kocianova, E. *et al.* (1993) Studies of the prevalence of *Coxiella burnetii*, the agent of Q fever, in the foothills of the southern Bavarian Forest Germany. *Zentralbl. Bakteriol.* **278**(1):132–8.

Rehacek, J., Liebisch, A., Urvolgyi, J., Kovacova, E. (1977) Rickettsiae of the spotted fever isolated from *Dermacentor marginatus* ticks in South Germany. *Zentralbl. Bakteriol.* [*A*] **239**(2):275–81.

Rehacek, J., Nosek, J., Urvolgyi, J., Sztankay, M. (1979) Rickettsiae of the spotted fever group in Hungary. *Folia Parasitol.* (*Praha*) **26**(4):367–71.

Rehacek, J., Urvolgyi, J., Kocianova, E. *et al.* (1991) Extensive examination of different tick species for infestation with *Coxiella burnetii* in Slovakia. *Eur. J. Epidemiol.* **7**(3):299–303.

Rehse-Krüpper, B., Casals, J., Rehse, E., Ackermann, R. (1976) Eyach – an arthropod-borne virus related to Colorado tick fever virus in the Federal Republic of Germany. *Acta Virol.* **20**(4):339–42.

Reid, A. J. C., Whitty, C. J. M., Ayles, H. M. *et al.* (1998) Malaria at Christmas: risks of prophylaxis versus risks of malaria. Health professionals need to educate travellers about the dangers of malaria and the importance of prophylaxis. *Br. Med. J.* **317**(7171):1506–8.

Reider, N., Gothe, R. (1993) Ehrlichiosis in dogs in Germany: fauna, biology and ecology of the causative agent, pathogenesis, clinical signs, diagnosis, therapy and prophylaxis. *Kleintierpraxis* **38**(12):775–90.

Reintjes, R., Dedushaj, I., Gjini, A. *et al.* (2002) Tularemia outbreak investigation in Kosovo: case control and environmental studies. *Emerg. Infect. Dis.* **8**(1):69–73.

Reisen, W. K., Chiles, R. E., Martinez, V. M., Fang, Y., Green, E. N. (2003) Experimental infection of California birds with western equine encephalomyelitis and St. Louis encephalitis viruses. *J. Med. Entomol.* **40**(6):968–82.

Richter, J., Fournier, P. E., Petridou, J., Haussinger, D., Raoult, D. (2002) *Rickettsia felis* infection acquired in Europe and documented by polymerase chain reaction. *Emerg. Infect. Dis.* **8**(2):207–8.

Rieke, B., Fleischer, K. (1993) Chronic morbidity after travel in endemic malaria regions. *Versicherungmedizin* **45**(6):197–202.

Rigau-Perez, J. G., Aayala-Lopez, A., Garcia-Rivera, E. J. *et al.* (2002) The reappearance of dengue-3 and a subsequent dengue-4 and dengue-1 epidemic in Puerto Rico in 1998. *Am. J. Trop. Med. Hyg.* **67**(4):355–62.

Rijpkema, S., Golubic, D., Molkenboer, M., Verbeek de Kruif, N., Schellekens, J. (1996) Identification of four genomic groups of *Borrelia burgdorferi* sensu lato in *Ixodes ricinus* ticks collected in a Lyme borreliosis endemic region of northern Croatia. *Exp. Appl. Acarol.* **20**(1):23–30.

Rijssenbeek-Nouwens, L. H., Oosting, A. J., de Bruin-Weller, M. S. *et al.* (2002) Clinical evaluation of the effect of anti-allergic mattress covers in patients with moderate to severe asthma and house dust mite allergy: a randomised double blind placebo controlled study. *Thorax* **57**(9):784–90.

Robert, L., Yelton, J. (2002) Imported furuncular myiasis in Germany. *Milit. Med.* **167**(12):990–3.

Robinson, D., Leo, N., Prociv, P., Barker, S. C. (2003) Potential role of head lice, *Pediculus humanus capitis*, as vectors of *Rickettsia prowazekii*. *Parasitol. Res.* **90**(3):209–11.

Rolain, J. M., Arnoux, D., Parzy, D., Sampol, J., Raoult, D. (2003a) Experimental infection of human erythrocytes from alcoholic patients with *Bartonella quintana*. *Ann. NY Acad. Sci.* **990**:605–11.

Rolain, J. M., Franc, M., Davoust, B., Raoult, D. (2003b) Molecular detection of *Bartonella quintana, B. koehlerae, B. henselae, B. clarridgeiae, Rickettsia felis*, and *Wolbachia pipientis* in cat fleas, France. *Emerg. Infect. Dis.* **9**(3):338–42.

Rollin, P. E., Ksiazek, T. G., Elliott, L. H. *et al.* (1995) Isolation of black creek canal virus, a new hantavirus from *Sigmodon hispidus* in Florida. *J. Med. Virol.* **46**(1):35–9.

Romi, R., Boccolini, D., Majori, G. (2001) Malaria incidence and mortality in Italy in 1999–2000. *Eurosurveillance Weekly* **6**(10):143–7.

Romi, R., Di Luca, M., Majori, G. (1999) Current status of *Aedes albopictus* and *Aedes atropalpus* in Italy. *J. Am. Mosq. Control Assoc.* **15**(3):425–7.

Romi, R., Sabatinelli, G., Savelli, L. G. *et al.* (1997) Identification of a North American mosquito species, *Aedes atropalpus* (Diptera: Culicidae) in Italy. *J. Am. Mosq. Control Assoc.* **13**(3):245–6.

Ronne, T. (2000) Malaria 1999 and suggested prophylaxis. *EPI News* 33.

Roppo, M. R., Lilja, J. L., Maloney, F. A., Sames, W. J. (2004) First occurrence of *Ochlerotatus japonicus* in the state of Washington. *J. Am. Mosq. Control Assoc.* **20**(1):83–4.

Rosa, S. di, Vitale, G., Todaro, R. *et al.* (1987). Antibodies anti *R. conorii* in sera of equines, ovines and cattle. *Giorn. Malattie Infect. Parassit.* **39**(1):110–11.

Rosenstreich, D. L., Eggleston, P., Kattan, M. *et al.* (1997) The role of cockroach allergy and exposure to cockroach allergen in causing morbidity among inner-city children with asthma. *N. Engl. J. Med.* **336**:1356–63.

Rosiu, N., Topciu, V., Munteanu, S. (1980) Presence of antibodies to some flaviviruses in the dog. *Rev. Roum. Virol.* **31**(2):143–4.

Rossi, L., Pollono, F., Meneguz, P. G., Gribaudo, L., Balbo, T. (1996) An epidemiological study of canine filarioses in north-west Italy: what has changed in 25 years? *Vet. Res. Commun.* **20**(4):308–15.

Rosypal, A. C., Zajac, A. M., Lindsay, D. S. (2003) Canine visceral leishmaniasis and its emergence in the United States. *Vet. Clin. North Am. Small Anim. Pract.* **33**(4):921–37.

Rotaeche-Montalvo, V., Hernández Pezzi, G., Mateo-Ontañón, S. de (2001) Epidemiological vigilance of malaria in Spain. 1996–1999. Vigilancia epidemiológica del paludismo en España. 1996–1999.) *Bol. Epidemiol. Semanal.* **9**(3):21–5.

Roux, V., Raoult, D. (1999) Body lice as tools for diagnosis and surveillance of reemerging diseases. *J. Clin. Microbiol.* **37**(3):596–9.

Rowin, K. S., Tanowitz, H. B., Rubinstein, A., Kunkel, M., Wittner, M. (1984) Babesiosis in asplenic hosts. *Trans. Roy. Soc. Trop. Med. Hyg.* **78**(4):442–4.

Ruiz Beltran, R., Herrero Herrero, J. I., Martin Sanchez, A. M., Martin Gonzalez, J. A. (1990) Prevalence of antibodies to *Rickettsia conorii, Coxiella burnetii* and *Rickettsia typhi* in Salamanca Province (Spain). Serosurvey in the human population. *Eur. J. Epidemiol.* **6**(3):293–9.

Russo, R., Nigro, L., Panarello, G., Montineri, A. (2003) Clinical survey of Leishmania/HIV co-infection in Catania, Italy: the impact of highly active antiretroviral therapy (HAART). *Ann. Trop. Med. Parasitol.* **97**(suppl. 1):149–55.

Rutledge, C. R., Day, J. F., Lord, C. C., Stark, L. M., Tabachnick, W. J. (2003) West Nile virus infection rates in *Culex nigripalpus* (Diptera: Culicidae) do not reflect transmission rates in Florida. *J. Med. Entomol.* **40**(3):253–8.

Ruzic-Sabljic, E., Maraspin, V., Lotric-Furlan, S. *et al.* (2002) Characterization of *Borrelia burgdorferi* sensu lato strains isolated from human material in Slovenia. *Wien. Klin. Wochenschr.* **114**(13–14):544–50.

Ryckman, R. E. (1984) The triatominae of North and Central America and the West Indies: a checklist with synonymy (Hemiptera: Reduviidae: Triatominae). *Bull. Soc. Vector Ecol.* **9**(2):71–83.

Rydkina, E. B., Roux, V., Gagua, E. M. *et al.* (1999) *Bartonella* in body lice collected from homeless persons in Russia. *Emerg. Infect. Dis.* **5**(1):176–8.

Sabatinelli, G., Ejov, M., Joergensen, P. (2001) Malaria in the WHO European region (1971–1999). *Eurosurveillance Weekly* **6**(4):61–5.

Sabatinelli, G., Majori, G. (1998) Malaria surveillance in Italy: 1986–1996 analysis and 1997 provisional data. *Eurosurveillance* **3**(4):38–40.

Sabatini, A., Raineri, V., Trovoto, G., Coluzzi, M. (1990) *Aedes albopictus* in Italia e possible diffusione della species nell'area mediterranea. *Parassitologia* **32**(3):301–4.

Safdar, N., Young, D. K., Andes, D. (2003) Autochthonous furuncular myiasis in the United States: case report and literature review. *Clin. Infect. Dis.* **36**(7):73–80.

Sall, A. A., Zanotto, P. M., de A., Vialat, P., Sene, O. K., Bouloy, M. (1998) Origin of 1997–98 Rift Valley fever outbreak in East Africa. *Lancet* **352**(9140): 1596–7.

Samanidou-Voyadjoglou, A., Patsoula, E., Spanakos, G., Vakalis, N. C. (2005) Confirmation of the presence of *Aedes albopictus* (Skuse) (Diptera:Culicidae) in Greece. *Eur. Mosq. Bull.* **19**:10–12.

Samoilova, T. I., Votiakov, V. I., Titov, L. P. (2003) Virologic and serologic investigations of West Nile virus circulation in Belarus. *Cent. Eur. J. Publ. Hlth.* **11**(2):55–62.

Sanders, F. H. Jr., Oliver, J. H. Jr. (1995) Evaluation of *Ixodes scapularis, Amblyomma americanum*, and *Dermacentor variabilis* (Acari: Ixodidae) from Georgia as vectors of a Florida strain of the Lyme disease spirochete, *Borrelia burgdorferi. J. Med. Entomol.* **32**(4): 402–6.

Sanogo, Y. O., Zeaiter, Z., Caruso, G. *et al.* (2003) *Bartonella henselae* in *Ixodes ricinus* ticks (Acari: Ixodida) removed from humans, Belluno Province, Italy. *Emerg. Infect. Dis.* **9**(3):329–32.

Santino, I., Dastoli, F., Lavorino, C. *et al.* (1996) Determination of antibodies to *Borrelia burgdorferi* in the serum of patients living in Calabria, southern Italy. *Panminerva Med.* **38**(3):167–72.

Santino, I., Dastoli, F., Sessa, R., Del Piano, M. (1997) Geographical incidence of infection with *Borrelia burgdorferi* in Europe. *Panminerva Med.* **39**(3):208–14.

Santino, I., Iori, A., Sessa, R. *et al.* (1998) *Borrelia burgdorferi* s.l. and *Ehrlichia chaffeensis* in the National Park of Abruzzo. *FEMS Microbiol. Lett.* **164**(1):1–6.

Santos-Gomes, G., Campino, L., Gomes-Pereira, S., Pires, R., Abranches, P. (1998) Leishmaniasis in Portugal: epidemiological situation and evaluation of diagnostic methods. *Giorn. Ital. Med Trop.* **3**(3–4):89–94.

Sarpong, S. B., Hamilton, R. G., Eggleston, P. A., Adkinson, N. F. Jr. (1996) Socioeconomic status and race as risk factors for cockroach allergen exposure and sensitization in children with asthma. *J. Allergy Clin. Immunol.* **97**(6):1393–401.

Sastre, J., Ibanez, M. D., Lombardero, M., Laso, M. T., Lehrer, S. (1996) Allergy to cockroaches in patients with asthma and rhinitis in an urban area (Madrid). *Allergy* **51**(8):582–6.

Satz, N., Knoblauch, M. (1989) Diagnostic possibilities and limitations in Lyme borreliosis. *Schweiz. Med. Wochenschr.* **119**(52):1883–93.

Sawchuk, L. A., Burke, S. D. (1998) Gibraltar's 1804 yellow fever scourge: the search for scapegoats. *J. Hist. Med. Allied. Sci.* **53**(1):3–42.

Schaarschmidt, D., Oehme, R., Kimmig, P., Hesch, R. D., Englisch, S. (2001) Detection and molecular typing of *Borrelia burgdorferi* sensu lato in *Ixodes ricinus* ticks and in different patient samples from southwest Germany. *Eur. J. Epidemiol.* **17**(12):1067–74.

Schaffner, F., Bouletreau, B., Guillet, B., Guilloteau, J., Karch, S. (2001) *Aedes albopictus* (Skuse, 1894) established in metropolitan France. *Eur. Mosq. Bull.* **9**:1–3.

Schaffner, F., Chouin, S., Guilloteau, J. (2003) First record of *Ochlerotatus* (*Finlaya*) *japonicus japonicus* (Theobald, 1901) in metropolitan France. *J. Am. Mosq. Control Assoc.* **19**(1):1–5.

Schaffner, F., Van Bortel, W., Coosemans, M. (2004) First record of *Aedes* (*Stegomyia*) *albopictus* in Belgium. *J. Am. Mosq. Control Assoc.* **20**(2):201–3.

Scharninghausen, J. J., Meyer, H., Pfeffer, M., Davis, D. S., Honeycutt, R. L. (1999) Genetic evidence of Dobrava virus in *Apodemus agrarius* in Hungary. *Emerg. Infect. Dis.* **5**(3):468–70.

Schlagenhauf, P., Steffan, R., Loutan, L. (2003) Migrants as a major risk group for imported malaria in European countries. *J. Travel Med.* **10**(2):106–7.

Schlagenhauf, P., Steffen, R., Tschopp, A. *et al.* (1995) Behavioural aspects of travellers in their use of malaria presumptive treatment. *Bull. Wld. Hlth. Org.* **73**(2):215–21.

Schmidt, R., Ackermann, R. (1985) Meningopolyneuritis (Garin-Bujadoux, Bannwarth) erythema chronicum migrans disease of the nervous system transmitted by ticks. *Fortschr. Neurol. Psychiatr.* **53**(5):145–53.

Schmidt, R., Kabatzki, J., Hartung, S., Ackermann, R. (1987) Erythema chronicum migrans disease in the Federal Republic of Germany. *Zentralbl. Bakteriol. Mikrobiol. Hyg. [A]* **263**(3):435–41.

Schoffel, I., Schein, E., Wittstadt, U., Hentsche, J. (1991) Parasite fauna of red foxes in Berlin (West). *Berl. Münch. Tierarztl Wochenschr.* **104**(5):153–7.

Schoneberg, I., Krause, G., Ammon, A., Strobel, H., Stark, K. (2003) Malaria surveillance in Germany 2000/2001 – results and experience with a new reporting system. *Gesundheitswesen* **65**(4):263–9.

Schouls, L. M., Van De Pol, I., Rijpkema, S. G., Schot, C. S. (1999) Detection and identification of *Ehrlichia*, *Borrelia burgdorferi* sensu lato, and *Bartonella* species in Dutch *Ixodes ricinus* ticks. *J. Clin. Microbiol.* **37**(7):2215–22.

Schriefer, M. E., Sacci, J. B. Jr., Dumler, J. S., Bullen, A. G., Azad, A. F. (1994a) Identification of a novel rickettsial infection in a patient diagnosed with murine typhus. *J. Clin. Microbiol.* **32**(4):949–54.

Schriefer, M. E., Sacci, J. B. Jr., Taylor, J. P., Higgins, J. A., Azad, A. F. (1994b) Murine typhus: Updated roles of multiple urban components and a second typhus like rickettsia. *J. Med. Entomol.* **31**(5):681–5.

Schuhmacher, H., Hoen, B., Baty, V. *et al.* (1999) Seroprevalence of Central European tick-borne encephalitis in the Lorraine region. *Presse Méd.* **28**(5):221–4.

Schutt, M. G., Meisel, H., Kruger, D. H. *et al.* (2004) Life-threatening Dobrava hantavirus infection with unusually extended pulmonary involvement. *Clin. Nephrol.* **62**(1):54–7.

Schwab, P. M. (1968) Economic cost of St. Louis encephalitis epidemic in Dallas, Texas, 1966. *Publ. Hlth. Rept.* **83**(10):860–6.

Schwanda, M., Oertli, S., Frauchiger, B., Krause, M. (2000) Tick-borne meningoencephalitis in Thurgau Canton: a clinical and epidemiological analysis. *Schweiz. Med. Wochenschr.* **130**(41):1447–55.

Schwarz, B. (1993) Health economics of early summer meningoencephalitis in Austria. Effects of a vaccination campaign 1981 to 1990. *Wien. Med. Wochenschr.* **143**(21):551–5.

Schwarz, T. F., Jager, G., Gilch, S. *et al.* (1996) Travel-related vector-borne virus infections in Germany. *Arch. Virol.* Suppl. 11: 57–65.

Scolari, C., Tedoldi, S., Casalini, C. *et al.* (2002) Knowledge, attitudes, and practices on malaria preventive measures of migrants attending a public health clinic in northern Italy. *J. Travel Med.* **9**(3):160–2.

Seguin, B., Boucaud-Maitre, Y., Quenin, P., Lorgue, G. (1986) Epidemiologic evaluation of a sample of 91 rats (*Rattus norvegicus*) captured in the sewers of Lyon. *Zentralbl. Bakteriol. Mikrobiol. Hyg.[A]* **261**(4):539–46.

Segura Porta, F., Diestre Ortin, G., Ortuno Romero, A. (1998) Prevalence of antibodies to spotted fever group rickettsiae in human beings and dogs from an endemic area of Mediterranean spotted fever in Catalonia, Spain. *Eur. J. Epidemiol.* **14**(4):395–8.

Sekeyova, Z., Roux, V., Xu, W., Rehacek, J., Raoult, D. (1998) *Rickettsia slovaca* sp. nov., a member of the spotted fever group rickettsiae. *Int. J. Syst. Bacteriol.* **48**(4):1455–62.

Sellers, R. F. (1989) Eastern equine encephalitis in Quebec and Connecticut, 1972: introduction of infected mosquitoes on the wind? *Can. J. Vet. Res.* **53**(1):76–9.

Sellon, R. K., Menard, M. M., Meuten, D. J. *et al.* (1993) Endemic visceral leishmaniasis in a dog from Texas. *J. Vet. Intern. Med.* **7**(1):16–19.

Semiao Santos, S. J., El Harith, A., Ferreira, E. *et al.* (1995) Evora district as a new focus for canine leishmaniasis in Portugal. *Parasitol. Res.* **81**(3):235–9.

Sergiev, V. P., Baranova, A. M., Orlov, V. S. *et al.* (1993) The importation of malaria to the USSR from Afghanistan 1981–1989. *Bull. Wld. Hlth. Org.* **71**(3–4):385–8.

Sexton, D. J., Burgdorfer, W., Thomas, L., Norment, B. R. (1976) Rocky Mountain spotted fever in Mississippi: survey for spotted fever antibodies in dogs and for spotted fever group rickettsiae in dog ticks. *Am. J. Epidemiol.* **103**(2):192–7.

Sexton, D. J., Rollin, P. E., Breitschwerdt, E. B. *et al.* (1997) Life-threatening Cache Valley virus infection. *N. Engl. J. Med.* **336**(8):547–50.

Shadick, N. A., Liang, M. H., Phillips, C. B., Fossel, K., Kuntz, K. M. (2001) The cost-effectiveness of vaccination against Lyme disease. *Arch. Intern. Med.* **161**:554–61.

Shadomy, S. V., Waring, S. C., Martin-Filho, O. A., Oliveira, R. C., Chappell, C. L. (2004) Combined use of enzyme-linked immunosorbent assay and flow cytometry to detect antibodies to *Trypanosoma cruzi* in domestic canines in Texas. *Clin. Diagn. Lab. Immunol.* **11**(2):313–19.

Shah, M. K. (1999) Human pulmonary dirofilariasis: review of the literature. *South. Med. J.* **92**(3):276–9.

Shanson, D. C., Gazzard, B. G., Midgley, J. *et al.* (1983) *Streptobacillus moniliformis* isolated from blood in four cases of Haverhill fever. *Lancet* **2**(8341):92–4.

Shaw, S. E., Kenny, M. J., Tasker, S., Birtles, R. J. (2004) Pathogen carriage by the cat flea *Ctenocephalides felis* (Bouche) in the United Kingdom. *Vet. Microbiol.* **102**(3–4):183–8.

Shcherbakova, S. A., Vyshemirskii, O. I., Liapin, M. W. (1997) A circulation study of California serogroup and Batai viruses (fam. Bunyaviridae, genus *Bunyavirus*) on the territory of Saratov Province. *Med. Parazit.* (*Mosk*) **1**:51–2.

Sherman, R. A. (2000) Wound myiasis in urban and suburban United States. *Arch. Intern. Med.* **160**(13):2004–14.

Shpynov, S., Parola, P., Rudakov, N. *et al.* (2001) Detection and identification of spotted fever group rickettsiae in *Dermacentor* ticks from Russia and Central Kazakhstan. *Eur. J. Clin. Microbiol. Infect. Dis.* **20**(12):903–5.

Shuvalova, E. P., Antonov, M. M. (1982) Aspects of clinical symptoms and therapy of imported malaria. *Klin. Med.* (*Mosk*). **60**(1):30–4.

Sideris, V., Karagouni, E., Papadopoulos, G., Garifallou, A., Dotsika, E. (1996) Canine visceral leishmaniasis in the great Athens region, Greece. *Parasite* **3**(2):125–30.

Sideris, V., Papadopoulos, G., Dotsika, E., Karagouni, E. (1999) Asymptomatic canine leishmaniasis in Greater Athens area, Greece. *Eur. J. Epidemiol.* **15**(3):271–6.

Siebinga, J. T., Jongejan, F. (2000) Tick-borne fever (*Ehrlichia phagocytophila* infection) on a dairy farm in Friesland. *Tijdschr Diergeneeskd* **125**(3):74–80.

Silverman, M., Law, B., Carson, J. (1991) A case of insect borne tularemia above the tree line. *Arctic Med. Res.* **50**(suppl.): 377–9.

Simser, J. A., Palmer, A. T., Fingerle, V. *et al.* (2002) *Rickettsia monacensis* sp. nov., a spotted fever group Rickettsia, from ticks (*Ixodes ricinus*) collected in a European city park. *Appl. Environ. Microbiol.* **68**(9):4559–66.

Sinski, E., Rijpkema, S. G. (1997) Prevalence of *Borrelia burgdorferi* infection in *Ixodes ricinus* ticks at urban and suburban forest habitats. *Przeg. Epidemiol.* **51**(4):431–5.

Sixl, W., Batikova, M., Stunzner, D. *et al.* (1973) Haemagglutination-inhibiting antibodies against arboviruses in animal sera, collected in some regions in Austria. II. *Zentralbl. Bakteriol. [Orig A]* **224**(3): 303–8.

Sixl, W., Kock, M., Withalm, H., Stunzner, D. (1989) Serological investigations of the hedgehog (*Erinaceus europaeus*) in Styria. 2. Report. *Geogr. Med. Suppl.* **2**:105–8.

Sixl, W., Stunzner, D., Withalm, H. (1976) Tests for anthropozoonoses in sera of farmers in eastern-Austria. I. Communication. *Zentralbl. Bakteriol. [Orig A]* **234**(2):265–70.

Skarphedinsson, S., Jensen, P. M., Kristiansen, K. (2005) Survey of tickborne infections in Denmark. *Emerg. Infect. Dis.* **11**(7):1055–61.

Skarphedinsson, S., Sogaard, P., Pedersen, C. (2001) Seroprevalence of human granulocytic ehrlichiosis in high-risk groups in Denmark. *Scand. J. Infect. Dis.* **33**(3):206–10.

Skotarczak, B., Cichocka, A. (2001) The occurrence DNA of *Babesia microti* in ticks *Ixodes ricinus* in the forest areas of Szczecin. *Folia Biol.* (Krakow) **49**(3–4): 247–50.

Smith, R., O'Connell, S., Palmer, S. (2000) Lyme disease surveillance in England and Wales, 1986–1998. *Emerg. Infect. Dis.* **6**(4):404–7.

Smithburn, K. C., Hughes, T. P., Burke, A. W., Paul, J. H. (1940) A neurotropic virus isolated from the blood of a native of Uganda. *Am. J. Trop. Med.* **20**(4):471–92.

Snow, K. (1987) Control of mosquito nuisance in Britain. *J. Am. Mosq. Control Assoc.* **3**(2):271–5.

Sokolova, M. I., Snow, K. (2002) Malaria vectors in European Russia. *Eur. Mosq. Bull.* **12**:1–5.

Solano-Gallego, L., Morell, P., Arboix, M., Alberola, J., Ferrer, L. (2001) Prevalence of *Leishmania infantum* infection in dogs living in an area of canine leishmaniasis endemicity using PCR on several tissues and serology. *J. Clin. Microbiol.* **39**(2):560–3.

Sorvillo, F. J., Gondo, B., Emmons, R. *et al.* (1993) A surburban focus of endemic typhus in Los Angeles County: association with seropositive domestic cats and opossums. *Am. J. Trop. Med. Hyg.* **48**(2):269–73.

Spach, D. H., Kanter, A. S., Dougherty, M. J. *et al.* (1995) *Bartonella* (*Rochalimaea*) *quintana* bacteremia in inner-city patients with chronic alcoholism. *N. Engl. J. Med.* **332**(7):424–8.

Spielman, A. (1994) The emergence of Lyme disease and human babesiosis in a changing environment. *Ann. N. Y. Acad. Sci.* **740**:146–56.

Spielman, A., Levine, J. F., Wilson, M. L. (1984) Vectorial capacity of North American *Ixodes* ticks. *Yale J. Biol. Med.* **57**(4):507–13.

Sprenger, D., Wuithiranyagool, T. (1986) The discovery and distribution of *Aedes albopictus* in Harris County, Texas. *J. Am. Mosq. Control Assoc.* **2**(2):217–19.

Spyridaki, I., Psaroulaki, A., Loukaides, F. *et al.* (2002) Isolation of *Coxiella burnetii* by a centrifugation shell-vial assay from ticks collected in Cyprus: detection by nested polymerase chain reaction (PCR) and by PCR-restriction fragment length polymorphism analyses. *Am. J. Trop. Med. Hyg.* **66**(1):86–90.

Stafford, K. C. III, Bladen, V. C., Magnarelli, L. A. (1995) Ticks (Acari: Ixodidae) infesting wild birds (Aves) and white-footed mice in Lyme, CT. *J. Med. Entomol.* **32**(4):453–66.

Stanczak, J., Racewicz, M., Kubica-Biernat, B. *et al.* (1999) Prevalence of *Borrelia burgdorferi* sensu lato in *Ixodes ricinus* ticks (Acari, Ixodidae) in different Polish woodlands. *Ann. Agric. Environ. Med.* **6**(2):127–32.

Steere, A. C., Broderick, T. F., Malawista, S. E. (1978) Erythema chronicum migrans and Lyme arthritis: epidemiologic evidence for a tick vector. *Am. J. Epidemiol.* **108**(4):312–21.

Steere, A. C., Coburn, J., Glickstein, L. (2004) The emergence of Lyme disease. *J. Clin. Invest.* **113**(8):1093–101.

Steere, A. C., Grodzicki, R. L., Kornblatt, A. N. *et al.* (1983) The spirochetal etiology of Lyme disease. *N. Engl. J. Med.* **308**(13):733–40.

Steere, A. C., Hardin, J. A., Malawista, S. E. (1978a) Lyme arthritis: a new clinical entity. *Hosp. Pract.* **13**(4):143–58.

Steere, A. C., Malawista, S. E., Snydman, D. R. *et al.* (1977) Lyme arthritis: an epidemic of oligoarticular arthritis in children and adults in three Connecticut communities. *Arthritis Rheum.* **20**(1):7–17.

Steiert, J., Gilfoy, F. (2002) Infection rates of *Amblyomma americanum* and *Dermacentor variabilis* by *Ehrlichia chaffeensis* and *Ehrlichia ewingii* in southwest Missouri. *Vector-Borne Zoon. Dis.* **2**(2):53–60.

Stein, A., Purgus, R., Olmer, M., Raoult, D. (1999) Brill–Zinsser disease in France. *Lancet* **353**(9168):1936.

Steketee, R. W., Eckman, M. R., Burgess, E. C. *et al.* (1985) Babesiosis in Wisconsin. A new focus of disease transmission. *J. Am. Med. Assoc.* **253**(18):2675–8.

Stelmach, I., Jerzynska, J., Stelmach, W. *et al.* (2002) Cockroach allergy and exposure to cockroach allergen in Polish children with asthma. *Allergy* **57**(8):701–5.

Stephenson, J. R., Barrett, A. D., Lee, J. M., Wilton-Smith, P. D. (1984) Antigenic variation among members of the tick-borne encephalitis complex. *J. Gen. Virol.* **65**(1):81–9.

Stich, A., Zwicker, M., Steffen, T., Kohler, B., Fleischer, K. (2003) Old age as risk factor for complications of malaria in non-immune travellers. *Dtsch. Med. Wochenschr.* **128**(7):309–14.

Stojcevic, D., Zivicnjak, T., Marinculic, A. *et al.* (2004) The epidemiological investigation of *Trichinella* infection in brown rats (*Rattus norvegicus*) and domestic pigs in Croatia suggests that rats are not a reservoir at the farm level. *J. Parasitol.* **90**(3):666–70.

Stoppacher, R., Adams, S. P. (2003) Malaria deaths in the United States: case report and review of deaths, 1979–1998. *J. Forensic Sci.* **48**(2):404–8.

Stratigos, J., Tosca, A., Nicolis, G., Papavasiliou, S., Capetanakis, J. (1980) Epidemiology of cutaneous leishmaniasis in Greece. *Int. J. Dermatol.* **19**(2):86–8.

Strle, F. (1999) Lyme borreliosis in Slovenia. *Zentralbl. Bakteriol.* **289**(5–7):643–52.

Strle, F., Stantic-Pavlinic, M. (1996) Lyme disease in Europe. *N. Engl. J. Med.* **334**(12):803.

Stronegger, W. J., Leodolter, K., Rasky, E., Freidl, W. (1998) Representative early summer meningoencephalitis vaccination rates of school children in Styria. *Wien. Klin. Wochenschr.* **110**(12):434–40.

Strong, R. P. (1944) Plague. In *Stitt's Diagnosis, Prevention and Treatment of Tropical Diseases*, 7th Edition, pp. 651–710. Philadelphia: Blackiston Co.

Stuen, S., Van De Pol, I., Bergstrom, K., Schouls, L. M. (2002) Identification of *Anaplasma phagocytophila* (formerly *Ehrlichia phagocytophila*) variants in blood from sheep in Norway. *J. Clin. Microbiol.* **40**(9):3192–7.

Stunzner, D., Hubalek, Z., Halouzka, J. *et al.* (1998) Prevalence of *Borrelia burgdorferi* s.I. in *Ixodes ricinus* ticks from Styria (Austria) and species identification by PCR–RFLP analysis. *Zentralbl. Bakteriol. Mikrobiol. Hyg. A* **288**(4):471–8.

Sudia, W. D., Newhouse, V. F., Beadle, I. D. *et al.* (1975) Epidemic Venezuelan equine encephalitis in North America in 1971: vector studies. *Am. J. Epidemiol.* **101**(1):17–35.

Sumption, K. J., Wright, D. J. M., Cutler, S. J., Dale, B. A. S. (1995) Human ehrlichiosis in the UK. *Lancet* **346**(8988):1487–8.

Suss, J., Schrader, C., Falk, U., Wohanka, N. (2004) Tick-borne encephalitis (TBE) in Germany – epidemiological data, development of risk areas and virus prevalence in field-collected ticks and in ticks removed from humans. *Int. J. Med. Microbiol.* **293**(suppl. 37):69–79.

Tablante, N. L., Lane, V. M. (1989). Wild mice as potential reservoirs of *Salmonella dublin* in a closed dairy herd. *Can. Vet. J.* **30**:590–2.

Taplin, D., Meinking, T. L. (1997) Treatment of HIV-related scabies with emphasis on the efficacy of ivermectin. *Semin. Cutan. Med. Surg.* **16**(3):235–40.

Tarasevich, I. V., Moteyunas, L. I., Rehacek, J. *et al.* (1981) Occurrence of rickettsioses in the Lithuanian Soviet Socialist Republic. *Folia Parasitol.* (Praha) **28**(2):169–77.

Tarasevich, I., Rydkina, E., Raoult, D. (1998) Outbreak of epidemic typhus in Russia. *Lancet* **352**(9134):1151.

Tea, A., Alexiou-Daniel, S., Arvanitdou, M., Diza, E., Antoniadis, A. (2003) Occurrence of *Bartonella henselae* and *Bartonella quintana* in a healthy Greek population. *Am. J. Trop. Med. Hyg.* **68**(5):554–6.

Telford, S. R. 3rd, Korenberg, E. I., Goethert, H. K. *et al.* (2002) Detection of natural foci of babesiosis and granulocytic ehrlichiosis in Russia. *Zh. Mikrobiol., Epidemiol. Immunobiol.* **6**:21–25.

Telford, S. R., III, Armstrong, P. M., Katavolos, P. *et al.* (1997) A new tick-borne encephalitis-like virus infecting New England deer ticks, *Ixodes dammini*. *Emerg. Infect. Dis.* **3**(2):165–70.

Tena, D., Perez Simon, M., Gimeno, C. *et al.* (1998) Human infection with *Hymenolepis diminuta*: case report from Spain. *J. Clin. Microbiol.* **36**(8):2375–6.

Tesh, R. B. (1989) The epidemiology of *Phlebotomus* (sandfly) fever. *Isr. J. Med. Sci.* **25**:214–17.

Tesh, R. B., Saidi, S., Gajdamovic, S. J., Rodhain, F., Vesenjak-Hirzan, J. (1976) Serological studies on the epidemiology of sandfly fever in the Old World. *Bull. Wld. Hlth. Org.* **54**(6):663–74.

Thang, H. D., Elsas, R. M., Veenstra, J. (2002) Airport malaria: report of a case and a brief review of the literature. *Neth. J. Med.* **60**(11):441–3.

Thompkins, L. S. (1997) Of cats, humans, and Bartonella (editorial). *N. Engl. J. Med.* **337**(26):1915–17.

Tikhomirov, E. (1999) Epidemiology and distribution of plague. In *Plague Manual*, pp. 11–42.WHO/CDS/CSR/EDC/99.2. Geneva: WHO.

Toivanen, A., Laine, M., Luukkainen, R., Oksi, J., Vainionpää, R. (2002) Prevalence of antibodies against a Sindbis-related (Pogosta) virus, a potential cause of chronic arthritis. *Arthritis Res.* **4**(suppl. 1):6.

Tomich, P. Q., Barnes, A. M., Devick, W. S., Higa, H. H., Haas, G. E. (1984) Evidence for the extinction of plague in Hawaii. *Am. J. Epidemiol.* **119**(2):261–73.

Tomov, A., Tsvetkova, E., Kryakov, I., Ivanova. K. (1994) First isolation of the agent of Lyme disease, *Borrelia burgdorferi* in Bulgaria. *Infectology* **31**(2):29–31.

Torres, M. I. (1997) Impact of an outbreak of dengue fever: a case study from rural Puerto Rico. *Human Org.* **56**(1):19–27.

Tovey, E. R. (1992) Allergen exposure and control. *Exp. Appl. Acarol.* **16**(1–2):181–202.

Traub, R. L., Wisseman, C. L. Jr., Farhang Azad, A. (1978) The ecology of murine typhus – a critical review. *Trop. Dis. Bull.* **75**(4):237–317.

Tselentis, Y., Gikas, A., Chaniotis, B. (1994) Kala-azar in Athens basin. *Lancet* **343**(8913):1635.

Tselentis, Y., Psaroulaki, A., Maniatis, J., Spyridaki, I., Babalis, T. (1996) Genotypic identification of murine typhus rickettsia in rats and their fleas in an endemic area of Greece by the polymerase chain reaction and restriction fragment length polymorphism. *Am. J. Trop. Med. Hyg.* **54**(4):413–17.

Turcinov, D., Kuzman, I., Puljiz, I. (2002) The Brill–Zinsser disease still occurs in Croatia: retrospective analysis of 25 hospitalized patients. *Lijec Vjesn* **124**(10):293–6.

Turell, M. J., Bailey, C. L., Beaman, J. R. (1988) Vector competence of a Houston, Texas strain of *Aedes albopictus* for a Rift Valley Fever virus. *J. Am. Mosq. Control Assoc.* **4**(1):94–6.

Turell, M. J., Beaman, J. R. (1992) Experimental transmission of Venezuelan equine encephalomyelitis virus by a strain of *Aedes albopictus* (Diptera: Culicidae) from New Orleans, Louisiana. *J. Med. Entomol.* **29**(5):802–5.

Turell, M. J., Dohm, D. J., Sardelis, M. R. *et al.* (2005) An update on the potential of north American mosquitoes (Diptera: Culicidae) to transmit West Nile virus. *J. Med. Entomol.* **42**(1):57–62.

Turner, S., Lines, S., Chen, Y., Hussey, L., Agius, R. (2005) Work-related infectious disease reported to the Occupational Disease Intelligence Network and The Health and Occupation Reporting network in the UK (2000–2003). *Occup. Med. (Lond).* **55**(4):275–81.

Tzamouranis, N., Schnur, L. F., Garifallou, A., Pateraki, E., Serie, C. (1984) Leishmaniasis in Greece I. Isolation and identification of the parasite causing human and canine visceral leishmaniasis. *Ann. Trop. Med. Parasitol.* **78**(4):363–8.

Ulrich, R., Meisel, H., Schutt, M. *et al.* (2004) Prevalence of hantavirus infections in Germany. *Bundesgesundheitsblatt Gesundheitsforschung Gesundheitsschutz.* **47**(7):661–70.

Ungureanu, A., Popovici, V., Catanas, F. *et al.* (1990) Isolation of Bhanja virus in Romania. *Arch. Roumaines. Pathol. Exper. Microbiol.* **49**(2):139–45.

Usinger, R. L. (1944) Entomological phases of recent dengue epidemic in Honolulu. *Publ. Hlth. Rep.* **59**(13):423–30.

Utz, J. T., Apperson, C. S., MacCormack, J. N. *et al.* (2003) Economic and social impacts of La Crosse encephalitis in western North Carolina. *Am. J. Trop. Med. Hyg.* **69**(5):509–18.

Valassina, M., Cusi, M. G., Valensin, P. E. (2003a) A Mediterranean arbovirus: the Toscana virus. *J. Neurovirol.* **9**(6):577–83.

Valassina, M., Meacci, F., Valensin, P. E., Cusi, M. G. (2000) Detection of neurotropic viruses circulating in Tuscany: the incisive role of Toscana virus. *J. Med. Virol.* **60**(1):86–90.

Valassina, M., Valentini, M., Pugliese, A., Valensin, P. E., Cusi, M. G. (2003b) Serological survey of Toscana virus infections in a high-risk population in Italy. *Clin. Diagn. Lab. Immunol.* **10**(3):483–4.

Valle, S. L. (1988) Sexually transmitted diseases and the use of condoms in a cohort of homosexual men followed since 1983 in Finland. *Scand. J. Infect. Dis.* **20**(2):153–61.

Van den Ende, J., Lynen, L., Elsen, P. *et al.* (1998) A cluster of airport malaria in Belgium in 1995. *Acta Clin. Belge* **53**(4):259–63.

Van den Ende, J., Morales, I., Van Den Abbeele, K. *et al.* (2000) Changing epidemiological and clinical aspects of imported malaria in Belgium. *J. Travel Med.* **8**(1):19–25.

Van Herck, K., Zuckerman, J., Castelli, F. *et al.* (2003) Travelers' knowledge, attitudes, and practices on prevention of infectious diseases: results from a pilot study. *J. Travel Med.* **10**(2):75–8.

Van Hest, N. A., Smit, F., Verhave, J. P. (2002) Underreporting of malaria incidence in The Netherlands: results from a capture-recapture study. *Epidemiol. Infect.* **129**(2):371–3.

Van Vliet, J. A., Samsom, M., van Steenbergen, J. E. (1998) Causes of spread and return of scabies in health care institutes; literature analysis of 44 epidemics. *Ned. Tijdschr. Geneeskd.* **142**(7):354–7.

Vapalahti, K., Mustonen, J., Lundkvist, A. *et al.* (2003) Hantavirus infections in Europe. *Lancet Infect. Dis.* **3**(10):653–61.

Vazeille-Falcoz, M., Adhami, J., Mousson, L., Rodhain, F. (1999) *Aedes albopictus* from Albania: a potential vector of dengue viruses. *J. Am. Mosq. Control Assoc.* **15**(4):475–8.

Velo, E., Bino, S., Kuli-Lito, G. *et al.* (2003) Recrudescence of visceral leishmaniasis in Albania: retrospective analysis of cases during 1997 to 2001 and results of an entomological survey carried out during 2001 in some districts. *Trans. R. Soc. Trop. Med. Hyg.* **97**: 288–9.

Vervloet, D., Pradal, M., Porri, F., Charpin, D. (1991) The epidemiology of allergy to house dust mites. *Rev. Mal. Resp.* **8**(1):59–65.

Vesenjak-Hirjan, J., Punda-Polic, V., Calisher, C. H. *et al.* (1989) Natural foci of infective agents on Mljet. 1. Arbovirus infections. In *Island of Mljet – Ecology and Health Conditions* (eds B. Kesic & J. Vesenjak-Hirjan), pp. 83–115. Zagreb: Yugoslav Acad. Arts Sci.

Vesenjak-Hirjan, J., Punda-Polic, V., Dobe, M. (1991) Geographical distribution of arboviruses in Yugoslavia. *J. Hyg. Epidemiol. Microbiol. Immunol.* **35**(2):129–40.

Viel, J. F., Giraudoux, P., Abrial, V., Bresson-Hadni, S. (1999) Water vole (*Arvicola terrestris* Scherman) density as risk factor for human alveolar echinococcosis. *Am. J. Trop. Med. Hyg.* **61**(4):559–65.

Vieyres, C., Allal, J., Coisne, D. *et al.* (1987) European aspects of Lyme disease. 8 cases. *Presse Med.* **16**(2):59–62.

Villari, P., Spielman, A., Komar, N., McDowell, M., Timperi, R. J. (1995) The economic burden imposed by a residual case of eastern encephalitis. *Am. J. Trop. Med. Hyg.* **52**(1):8–13.

Vinetz, J. M., Glass, G. E., Flexner, C. E., Mueller, P., Kaslow, D. C. (1996) Sporadic urban leptospirosis. *Ann. Intern. Med.* **125**(10):794–8.

Vinograd, I. A., Beletskaia, G. V., Chumachenko, S. S., Ardamatskaia, T. B., Rogochii, E. G. (1982) Isolation of West Nile virus in the Southern Ukraine. *Vopr. Virusol.* **27**(5):55–7.

Von Allmen, S. D., Lopez-Correa, R. H., Woodall, J. P. *et al.* 1979. Epidemic dengue fever in Puerto Rico, 1977: a cost analysis. *Am. J. Trop. Med. Hyg.* **28**(6):1040–4.

Voorhorst, R., Spieksma-Boezeman, M. I., Spieksma, F.Th. M. (1964). Is a mite (*Dermatophagoides* sp.) the producer of the house-dust allergen? *Allergy Asthma* **10**:329–34.

Vorobjeva, M. S., Vorontsova, T. V., Arumova, E. A., Paschshepkina, M. N., Tarasevich, L. A. (2001) Morbidity and epidemiology of tick-borne encephalitis in the Russian Federation. *EpiNorth 2* **2**:23–4.

Voss, A. (2000) Tick-borne encephalitis in Hungary. *ISW TBE Reports*.

Vutchev, D. (2001) Tertian malaria outbreak three decades after its eradication. *Jpn. J. Infect. Dis.* **54**:79–80.

Waddell, D. (1990) Yellow fever in Europe in the early 19th century – Cadiz 1819. *Rept. Proc. Scott. Soc. Hist. Med.* **1990–92**:20–34.

Wagner, B. (1999) Borreliose und FSME: Gefahr durch Zeckenstiche! *Der Hausarzt* **8**(99):34.

Walker, D. H., Dumler, J. S. (1996) Emergence of the ehrlichioses as human health problems. *Emerg. Infect. Dis.* **2**(1):18–29.

Walker, D. H., Feng, H. M., Saada, J. I. *et al.* (1995) Comparative antigenic analysis of spotted fever group rickettsiae from Israel, and other closely related organisms. *Am. J. Trop. Med. Hyg.* **52**(6):569–76.

Walker, D. H., Fishbein, D. B. (1991) Epidemiology of rickettsial diseases. *Eur. J. Epidemiol.* **7**(3):237–45.

Walker, E. D., Grayson, M. A., Edman, J. D. (1993) Isolation of Jamestown Canyon and snowshoe hare viruses (California serogroup) from *Aedes* mosquitoes in western Massachusetts. *J. Am. Mosq. Control Assoc.* **9**(2):131–4.

Walker, G. J. A., Johnstone, P. W. (2000) Interventions for treating scabies. *Arch. Dermatol.* **136**(3):387–9.

Weber, D. J., Walker, D. H. (1991) Rocky Mountain spotted fever. *Infect Dis. Clin North Am.* **5**(1):19–35.

Webster, J. P. (1994) Prevalence and transmission of *Toxoplasma gondii* in wild brown rats, *Rattus norvegicus*. *Parasitology* **108**(4):407–11.

Webster, J. P., Lloyd, G., Macdonald, D. W. (1995) Q fever (*Coxiella burnetii*) reservoir in wild brown rat (*Rattus norvegicus*) populations in the UK. *Parasitology* **110**: 31–5.

Webster, J. P., Lloyd, G., Macdonald, D. W. (1995) Q fever (*Coxiella burnetti*) in wild brown rats (*Rattus norvegicus*) populations in the UK. *Parasitology* **110**(1):31–5.

Webster, P., Frandsen, F. (1994) Prevalence of antibodies to *Borrelia burgdorferi* in Danish deer. *APMIS* **102**(4):287–90.

Wedincamp, J. Jr., Foil, L. D. (2002) Vertical transmission of *Rickettsia felis* in the cat flea (*Ctenocephalides felis* Bouche). *J. Vector Ecol.* **27**(1):96–101.

Weissenbock, H., Hubalek, Z., Halouzka, J. *et al.* (2003) Screening for West Nile virus infections of susceptible animal species in Austria. *Epidemiol. Infect.* **131**(2):1023–7.

Weissenbock, H., Kolodziejek, J., Fragner, K. *et al.* (2003) Usutu virus activity in Austria, 2001–2002. *Microbes Infect.* **5**(12):1132–6.

Weissenbock, H., Kolodziejek, J., Url, A., Lussy, H., Rebel-Bauder, B., Nowotny, N. (2002) Emergence of Usutu virus, an African mosquito-borne *Flavivirus* of the Japanese encephalitis virus Group, central Europe. *Emerg. Infect. Dis.* **8**(7):652–6.

Wells, J. D. (1991) *Chrysomya megacephala* (Diptera:Calliphoridae) has reached the continental United States: review of its biology, pest status, and spread around the world. *J. Med. Entomol.* **28**(3):471–3.

West, D. P. (2004) Head lice treatment costs and the impact on managed care. *Am. J. Manag. Care.* **10**(suppl. 9):S277–82.

Wetsteyn, J. C. F. M., Kager, P. A., van Gool, T. (1997) The changing pattern of imported malaria in the Academic Medical centre, Amsterdam. *J. Travel Med.* **4**(4):171–5.

Wharton, G. W. (1976). House dust mites. *J. Med. Entomol.* **12**(6):577–621.

WHO (1967) *WHO/FAO Expert Committee on Zoonoses*. WHO Tech. Report Ser. no. 378.

WHO (1988) Dust mite allergens and asthma: a worldwide problem. *Bull. Wld. Hlth. Org.* **66**(6):769–80.

WHO (1990) Lyme disease – 1988. *Weekly Epidemiol. Rec.* **65**(1):3–4.

WHO (1996) West Nile fever in Romania 2nd report. *WHO EURO Commun. Dis. Report* **13**:10–11.

WHO (1998) Crimean–Congo haemorrhagic fever. *WHO Fact Sheet* 208.

WHO (2000) *The Leishmaniases and Leishmania/HIV co-infections.* Fact Sheet No. **116**, 4 pp. Geneva: WHO.

WHO (2001) Imported human outbreak of trichinellosis, Italy. *Weekly Epidemiol Rec.* **76**(13):97–8.

WHO (2002b) CCHF in Kosovo again. *Health Action in Kosovo* No. 55 (Document). Geneva: WHO

Wichmann, O., Jelinek, T. (2003) Surveillance of imported dengue infections in Europe. *Eurosurveillance Weekly* **7**:32.

Wiechmann, I., Grupe, G. (2004) Detection of *Yersinia pestis* DNA in two early medieval skeletal finds from Aschheim (Upper Bavaria, 6th century A. D.). *Am. J. Phys. Anthropol.* **126**(1):48–55.

Wiley, J. S., Fritz, R. F. (1948) Tentative report on expanded murine typhus fever control operations in southern states. *Am. J. Trop. Med.* **28**(4):589–97.

Williams, D., Rolles, C. J., White, J. E. (1986) Lyme disease in a Hampshire child-medical curiosity or the beginning of an epidemic? *Br. Med. J.* **292**:1560–1.

Williams, J. P., Chitre, M., Sharland, M. (2002) Increasing *Plasmodium falciparum* malaria in southwest London: a 25 year observational study. *Arch. Dis. Child.* **86**(6):428–30.

Williams, M. C., Simpson, D. I., Haddow, A. J., Knight, E. M. (1964) The isolation of West Nile virus from man and of Usutu virus from the bird-biting mosquito *Mansonia aurites* (Theobald) in the Entebbe area of Uganda. *Ann. Trop. Med. Parasitol.* **58**(3):367–74.

Wilske, B., Steinhuber, R., Bergmeister, H. *et al.* (1987) Lyme borreliosis in South Germany. Epidemiologic data on the incidence of cases and on the epidemiology of ticks (*Ixodes ricinus*) carrying *Borrelia burgdorferi*. *Dtsch. Med. Wochenschr.* **112**(45):1730–6.

Wilson, J. H., Rubin, H. L., Lane, T. J., Gibbs, E. P. (1986) A survey of eastern equine encephalomyelitis in Florida horses: prevalence, economic impact, and management practices, 1982–1983. *Prev. Vet. Med.* **4**(3):261–71.

Woody, N. C., Woody, H. B. (1955) American trypanosomiasis (Chagas' disease). First indigenous case in the United States. *J. Am. Med. Assoc.* **159**:676–7.

Work, T. H. (1964) Serological evidence of arbovirus infection in the Seminole Indians of southern Florida. *Science* **145**(3629):270–2.

Yabsley, M. J., Varela, A. S., Tate, C. M. *et al.* (2002) *Ehrlichia ewingii* infection in white-tailed deer (*Odocoileus virginianus*). *Emerg. Infect. Dis.* **8**(7):668–71.

Yanagihara, R. (1990) Hantavirus infection in the United States: epizootiology and epidemiology. *Rev. Infect. Dis.* **12**(3):449–57.

Yanagihara, R., Daum, C. A., Lee, P. W. *et al.* (1987) Serological survey of Prospect Hill virus infection in indigenous wild rodents in the USA. *Trans. Roy. Soc. Trop. Med. Hyg.* **81**(1):42–5.

Zahler, M., Gothe, R., Rinder, H. (1996) *Dermacentor* ticks in France and Germany. Molecular biological differences in species, ecology and epidemiological implications. *Tierarztliche Praxis* **24**(3):209–11.

Zeman, P., Bene, C. (2004) A tick-borne encephalitis ceiling in Central Europe has moved upwards during the last 30 years: possible impact of global warming? *Int. J. Med. Microbiol.* **293**(suppl. 37):48–54.

Zeman, P., Pazdiora, P., Cinatl, J. (2002) HGE antibodies in sera of patients with TBE in the Czech Republic. *Int. J. Med. Microbiol.* **291**(suppl. 33): 190–3.

Zeman, P., Vitkova, V., Markvart, K. (1990) Joint occurrence of tick-borne encephalitis and Lyme borreliosis in the Central Bohemian Region of Czechoslovakia. *Cske. Epidemiol. Mikrobiol. Immunol.* **39**(2):95–105.

Zhioua, E., Postic, D., Rodhain, F., Perez-Eid, C. (1996) Infection of *Ixodes ricinus* (Acari: Ixodidae) by *Borrelia burgdorferi* in Ile de France. *J. Med. Entomol.* **33**(4):694–7.

Zientara, S. (2000) Le virus West Nile, un arbovirus reemergent en Europe. *Equ'Idee* **39**:25–6.

Zimmermann, H., Koch, D. (2005) Epidemiology of tick-borne encephalitis (TBE) in Switzerland 1984 to 2004. *Ther. Umsch.* **62**:719–25.

Zivanovic, B., Ler, Z., Cekanac, R. (1991) Risk of infection with *Borrelia burgdorferi* in an area endemic for Lyme disease. *Vojnosanit Pregl.* **48**(5):400–4.

Zohrabian, A., Meltzer, M. I., Ratard, R. *et al.* (2004) West Nile virus economic impact, Louisiana, 2002. *Emerg. Infect. Dis.* **10**(10):1736–44.

Zore, A., Trilar, T., Rusic-Sabijic, E., Avbhc-Supanc (2003) *Borrelia burgdoerferi* sensu lato isolated from engorged *Ixodes ricinus* ticks infesting birds in Slovenia, In VII International Potsdam Symposium on Tick-borne Diseases (IPS-VII).

Zukowski, K. (1989) Scabies in Poland 1966–1986. *Wiad. Parazytol.* **35**(2):151–9.

Zumpt, F. (1965) *Myiasis in Man and Animals in the Old World. A Textbook for Physicians, Veterinarians and Zoologists.* London: Butterworths.

Index

Note: page numbers in *italics* refer to figures

Learning Resources
Centre